Springer Series on Signals and Communication Technology

## Signals and Communication Technology

**Wireless Communications: 2007 CNIT Thyrrenian Symposium**
S. Pupolin
ISBN 978-0-387-73824-6

**Adaptive Nonlinear Systems Identification: The Volterra and Wiener Model Approaches**
T. Ogunfunmi
ISBN 978-0-387-26328-1

**Wireless Networks Security**
Y. Xiao, X. Shen, and D.Z. DU (Eds.)
ISBN 978-0-387-28040-0

**Satellite Communications and Navigation Systems**
E. Del Re and M. Ruggieri
ISBN: 0-387-47522-2

**Wireless Ad Hoc and Sensor Networks**
A Cross-Layer Design Perspective
R. Jurdak
ISBN 0-387-39022-7

**Cryptographic Algorithms on Reconfigurable Hardware**
F. Rodriguez-Henriquez, N.A. Saqib, A. Díaz Pérez, and C.K. Koc
ISBN 0-387-33956-6

**Multimedia Database Retrieval**
A Human-Centered Approach
P. Muneesawang and L. Guan
ISBN 0-387-25627-X

**Broadband Fixed Wireless Access**
A System Perspective
M. Engels and F. Petre
ISBN 0-387-33956-6

**Distributed Cooperative Laboratories**
Networking, Instrumentation, and Measurements
F. Davoli, S. Palazzo and S. Zappatore (Eds.)
ISBN 0-387-29811-8

**The Variational Bayes Method in Signal Processing**
V. Šmídl and A. Quinn
ISBN 3-540-28819-8

**Topics in Acoustic Echo and Noise Control**
Selected Methods for the Cancellation of Acoustical Echoes, the Reduction of Background Noise, and Speech Processing
E. Hänsler and G. Schmidt (Eds.)
ISBN 3-540-33212-x

**EM Modeling of Antennas and RF Components for Wireless Communication Systems**
F. Gustrau, D. Manteuffel
ISBN 3-540-28614-4

**Interactive Video Methods and Applications**
R. I Hammoud (Ed.)
ISBN 3-540-33214-6

**ContinuousTime Signals**
Y. Shmaliy
ISBN 1-4020-4817-3

**Voice and Speech Quality Perception**
Assessment and Evaluation
U. Jekosch
ISBN 3-540-24095-0

**Advanced ManMachine Interaction**
Fundamentals and Implementation
K.-F. Kraiss
ISBN 3-540-30618-8

**Orthogonal Frequency Division Multiplexing for Wireless Communications**
Y. (Geoffrey) Li and G.L. Stüber (Eds.)
ISBN 0-387-29095-8

**Circuits and Systems Based on Delta Modulation**
Linear, Nonlinear and Mixed Mode Processing
D.G. Zrilic ISBN 3-540-23751-8

**Functional Structures in Networks**
AMLn—A Language for Model Driven Development of Telecom Systems
T. Muth ISBN 3-540-22545-5

**RadioWave Propagation for Telecommunication Applications**
H. Sizun ISBN 3-540-40758-8

**Electronic Noise and Interfering Signals**
Principles and Applications
G. Vasilescu ISBN 3-540-40741-3

**DVB**
The Family of International Standards for Digital Video Broadcasting, 2nd ed.
U. Reimers ISBN 3-540-43545-X

**Digital Interactive TV and Metadata**
Future Broadcast Multimedia
A. Lugmayr, S. Niiranen, and S. Kalli
ISBN 3-387-20843-7

**Adaptive Antenna Arrays**
Trends and Applications
S. Chandran (Ed.) ISBN 3-540-20199-8

continued on page 347

# Wireless Communications 2007 CNIT Thyrrenian Symposium

**Edited by:**

**Silvano Pupolin**

Edited by:

Silvano Pupolin
University of Padova
Padova, Italy

ISBN 978-1-4419-4476-4 e-ISBN 978-0-387-73825-3

Printed on acid-free paper.



9 8 7 6 5 4 3 2 1

springer.com

**The workshop is organized by:**

consorzio nazionale
interuniversitario
per le telecomunicazioni

**and is technically co-sponsored by:**

# Preface

The 18th Tyrrhenian Workshop on digital communications is devoted to wireless communications. In the last decade, wireless communications research boosted launching new standards and proposing new techniques for the access technology. We moved from the UTRA standard capable to transmit 0.5 bit/s/Hz to WLAN which is promising 2.7 bit/s/Hz. Now wireless communication systems are facing a flourishing of new proposal moving from multiple antennas at transmitter and receiver side (MIMO systems), to new powerful Forward Error Correction Codes, to adaptive radio resource management algorithms. The new challenge, however, is the move towards multimedia communications and IP technology. This move implies efforts in several new aspects. First of all an open network, as IP is, imposes the necessity of a secure network, to guarantee the privacy of the ongoing communications, avoid the use of the networks by unauthorized customers, avoid the misuses and the charge to third parties of the cost of the connection. Also, quality of service (QoS) of the communications is becoming a must in IP networks which are carrying services which need a guaranteed QoS as telephony, real time services, etc. To get this new target some form of access control to the network must be setup. Recently, new form of communication networks has appeared to collect data for several applications (sensor networks, ad hoc networks, etc.) and they need a connection with a backbone network which could be a wireless one with a larger range than the sensor or ad hoc networks. These new networks are helpful for monitoring applications, and for actuation of some measure to permit a regular use of the available resources. An example could be the use of a road lane in one or opposite direction in different hours of the day as traffic condition requires.

This workshop is trying to put together all these new aspects of wireless communication systems.

It is organized in five sessions entitled: “4G wireless systems”, “ad hoc and cellular networks”, “security and applications in wireless networks”, “QoS and efficiency in multimedia heterogeneous wireless networks”, and “wireless sensor networks”.

The papers that will be presented represent an up-to-date critical analysis of the state of the art in each of the five areas and they will represent a reference for future development.

As final remarks, we express our gratitude to the session organizers, H. Ogawa, NiCT, Japan; M. Zorzi, University of Padova and CNIT, Italy; A. Prasad, DoCOMO Eurolabs, Germany; A. Jamalipur, University of Sydney, Australia; Shu Kato, NiCT, Japan, which have been in charge of selecting the papers for the workshops.

Also, thanks are to S. Basagni for the publicity action and to T. Erseghe for the hard job of collecting all the papers and checking all the final materials for the preparation of this book.

Shingo Ohmori, NiCT, Japan
Silvano Pupolin, University of Padova and CNIT, Italy
Workshop Co-Chairs

# Contents

**Session 1 - 4G Wireless Systems**

**1 Spatial Detection and Multistage Decoding for LST-MLC MIMO Systems**
*Maurizio Magarini and Arnaldo Spalvieri* .......................... 3

**2 Iterative (Turbo) Joint Rate and Data Detection in Coded CDMA Networks**
*Stefano Buzzi and Stefania Sardellitti* .............................. 19

**3 Hybrid ARQ Based on Rateless Coding for UTRAN LTE Wireless Systems**
*Lorenzo Favalli, Matteo Lanati, and Pietro Savazzi* .................. 29

**4 On the Performance of Transmit Antenna Selection with OSTBC in Ricean MIMO Channels**
*Saeed Kaviani, Chintha Tellambura, and Witold A. Krzymień* ......... 39

**5 A Packet Detection Algorithm for the UWB Standard ECMA 368**
*Tomaso Erseghe, Nicola Laurenti, Valentina Rizzi, and Roberto Corvaja* 51

**6 Low-Rate Predictive Feedback for the OFDM MIMO Broadcast Channel**
*Nevio Benvenuto, Ermanna Conte, Stefano Tomasin, and Matteo Trivellato* ........................................ 65

**Session 2 - Ad-Hoc and Cellular Networks**

**7 Interferer Nulling Based on Neighborhood Communication Patterns**
*Robert Vilzmann, Jörg Widmer, Imad Aad, and Christian Hartmann* ... 81

**8 On the Beneficial Effects of Cooperative Wireless Peer-to-Peer Networking**
*L. Militano, F.H.P. Fitzek, A. Iera, and A. Molinaro* . . . . . . . . . . . . . . . . . 97

**9 Relay Quality Awareness in Mesh Networks Routing**
*Claudio Casetti, Carla Fabiana Chiasserini, and Marco Fiore* . . . . . . . . . . 111

**10 Fundamental Bound on the Capacity of Ad Hoc Networks with Conventional Hop-by-Hop Routing**
*Anthony Acampora, Michael Tan, and Louisa Ip* . . . . . . . . . . . . . . . . . . . . . 127

**11 A Stochastic Non-Cooperative Game for Energy Efficiency in Wireless Data Networks**
*Stefano Buzzi, H. Vincent Poor, and Daniela Saturnino* . . . . . . . . . . . . . . . 135

---

**Session 3 - Security and Applications in Wireless Networks**

---

**12 Security Overheads for Signaling in Beyond-3G Networks**
*Dario S. Tonesi, Alessandro Tortelli, and Luca Salgarelli* . . . . . . . . . . . . . . 153

**13 Mobility and Key Management in SAE/LTE**
*Anand R. Prasad, Julien Laganier, Alf Zugenmaier, Mortaza S. Bargh, Bob Hulsebosch, Henk Eertink, Geert Heijenk, and Jeroen Idserda* . . . . . . 165

**14 Enhanced Operation Modes in IEEE 802.16 and Integration with Optical MANs**
*Isabella Cerutti, Luca Valcarenghi, Dania Marabissi, Filippo Meucci, Laura Pierucci, Luca Simone Ronga, Piero Castoldi, and Enrico Del Re* . . . . . . . . . . . . . . . . . . . . . . . . . . . . . . . . . . . . . . . . . . 179

**15 WPAN Applications and System Performance**
*Yoshinori Nishiguchi, Ryuhei Funada, Yozo Shoji, Hiroshi Harada, and Shuzo Kato* . . . . . . . . . . . . . . . . . . . . . . . . . . . . . . . . . . . . . . . . . . . . . 195

**16 Improving User Relocatability, Practicality, and Deployment in the Web Stream Customizer System**
*Jesse Steinberg and Joseph Pasquale* . . . . . . . . . . . . . . . . . . . . . . . . . . . . . 201

**17 Cross-Layer Error Recovery Optimization in WiFi Networks**
*Dzmitry Kliazovich, Nadhir Ben Halima, and Fabrizio Granelli* . . . . . . . . 213

---

**Session 4 - Qos and Efficiency in Multimedia Heterogeneous Wireless Networks**

---

**18 Technology-Independent Service Access Point for QoS Interworking**
*Mario Marchese, Maurizio Mongelli, Vincenzo Gesmundo, and Annamaria Raviola* . . . . . . . . . . . . . . . . . . . . . . . . . . . . . . . . . . . . . 225

**19 A Rate-Controlled VoIP System Based on Wireless Mesh Network Infrastructure: Design Issues and Performance Analysis**
*Francesco Licandro, Carla Panarello, and Giovanni Schembra* ......... 235

**20 Toward the QoS Support in 4G Wireless Systems**
*A.L. Ruscelli and G. Cecchetti* ......... 245

**21 A Scheduling Algorithm for Providing QoS Guarantees in 802.11e WLANs**
*G. Cecchetti and A.L. Ruscelli* ......... 253

**22 Mobility Management QoS Measures Evaluation for Next Generation Mobile Data Networks**
*Kumudu S. Munasinghe and Abbas Jamalipour* ......... 265

**23 Wireless Resource Allocation Considering Value of Frequency for Multi-Band Mobile Communication Systems**
*Hidenori Takanashi, Rihito Saito, Dorsaf Azzabi, Yoshikuni Onozato, and Yoshitaka Hara* ......... 281

Session 5 - Wireless Sensor Networks

**24 Wireless Sensor Networks and SNMP: Data Publication Over and IP Network**
*L. Berruti, L. Denegri, and S. Zappatore* ......... 295

**25 Self-Localization of Wireless Sensor Nodes By Means of Autonomous Mobile Robots**
*Andrea Zanella, Emanuele Menegatti, and Luca Lazzaretto* ......... 309

**26 An Experimental Study of Aggregator Nodes Positioning in Wireless Sensor Networks**
*Laura Galluccio, Alessandro Leonardi, Giacomo Morabito, and Sergio Palazzo* ......... 321

**27 SignetLab$^2$: A Modular Management Architecture for Wireless Sensor Networks**
*Riccardo Crepaldi, Albert F. Harris III, Andrea Zanella, and Michele Zorzi* ......... 331

# Session 1

## 4G Wireless Systems

1

# Spatial Detection and Multistage Decoding for LST-MLC MIMO Systems

Maurizio Magarini and Arnaldo Spalvieri

Dipartimento di Elettronica e Informazione, Politecnico di Milano
P.zza L. da Vinci, 32, 20133 Milano, Italy
{magarini,spalvier}@elet.polimi.it

**Summary.** Layered space–time (LST) coding schemes based on multilevel coding (MLC) represent a good approach to achieve high bandwidth and power efficiency in wireless transmission over multiple-input multiple-output (MIMO) channels. The combination of spatial detection algorithms and multistage decoding (MSD) is required at the receiver to perform soft detection and decoding. Since the complexity of the soft detection and decoding process may be impractical for many systems, we are interested in developing low complexity schemes providing a good tradeoff between performance and complexity. In this paper we compare the performance of two different LST-MLC architectures where MSD at the receiver is combined with different suboptimal spatial detection techniques.

**Key words:** MIMO systems, Layered space–time coding, Multilevel coding, Multistage decoding

## 1 Introduction

The ever increasing demand for high bandwidth and/or power efficiency in wireless communications leads to the introduction of architectures based on multiple antenna elements both at the transmitter and at the receiver [1]. Layered space–time (LST) coding schemes based on multilevel coding (MLC) is a suitable approach to achieve these expected efficiency benefits [2–5]. MLC, introduced in [6] together with the concept of multistage decoding (MSD), represents the optimum capacity achieving approach when the separation of coding and modulation is considered in single-input single-output systems [7]. In [2] it is shown that MLC also constitutes the optimum coded modulation scheme for transmission over multiple-input multiple-output (MIMO) channels when multiple-antenna signaling is regarded as multidimensional modulation.

The combination of MLC and LST can be realized in the same way as conventional block or convolutional coding is introduced in an LST transmission scheme [1]. Among such architectures, the horizontal LST (HLST)

approach has the advantage of being easily incorporated into existing systems and rendering the implementation of the decoding process at the receiver less complex. Depending on the position of the multilevel encoder in the transmitter chain, there are two alternative approaches to implement an HLST-MLC architecture. With the first, proposed in [2] and called here separate HLST-MLC (S-HLST-MLC), the information sequence is demultiplexed in $n_T$ substreams which are separately encoded by $n_T$ multilevel encoders. Although a multidimensional mapping might be applied, for a practical implementation blocks of bits at the output of each multilevel encoder are mapped according to their significance to one constituent PSK/QAM symbol. In contrast to [2], where code rates of the component encoders are chosen according to the constellation-constrained capacity at each level, we consider the case where identical multilevel encoders are used on the separate transmit antenna branches. In the second approach, called joint HLST-MLC (J-HLST-MLC), the information sequence is encoded by the same multilevel encoder used on the separate branches of S-HLST-MLC scheme. Coded bits at the same level are then demultiplexed in $n_T$ substreams and mapped, according to their significance, to the constituent PSK/QAM symbols transmitted over the $n_T$ transmit antenna elements.

The signal at the receiving antenna elements consists of a spatial superposition of the transmitted multilevel encoded symbols scaled by the fading coefficients and corrupted by additive white Gaussian noise. Spatial detection techniques combined with MSD are required at the receiver to perform soft detection and decoding. Since the complexity of the soft detection and decoding process may be impractical for many systems, we are interested in developing low complexity receiving schemes providing a good tradeoff between performance and complexity. In particular, we will compare the performance of the two HLST-MLC schemes where the reduction of the complexity of the soft detection-decoding process is achieved through the use of different suboptimal low-complexity spatial detection algorithms.

The spatial detection stage is responsible for generating the soft information to be passed to the MSD. The optimum spatial detector, which is the maximum likelihood detector (MLD), has a complexity proportional to $M^{n_T}$, where $M$ denotes the number of points of the PSK/QAM constellation. The complexity of the MLD can be prohibitively large when the number of transmitting antennas and constellation points is high. The receiver architecture proposed in [10] for an uncoded LST system, also known as vertical Bell Layered Space–Time (V-BLAST) detector, is a practical nonlinear detection technique that allows the detection of the $n_T$ substreams while keeping the complexity low. In such a scheme symbols are detected sequentially according to the well known ordered successive interference suppression and cancellation process (OSIC). Despite its detection simplicity, the main drawback of the V-BLAST approach is that the diversity order in the early stages is lower than in the next ones. This contributes to enhancing the performance gap between V-BLAST and MLD (in the latter the diversity order is equal to $n_T$).

Several sub-optimal detection strategies can be devised to reduce the performance gap. In particular, we are interested in the performance obtained when the detection is done by using the V-BLAST coset detector (V-BLAST-CD) of [8]. The V-BLAST-CD is obtained by extending the principle of reduced state sequence estimation [9], based on mapping by set partitioning (MSP), to perform the detection in LST transmission systems using non-binary constellations. In [8] it is shown that the V-BLAST-CD greatly outperforms the conventional V-BLAST detector at the cost of a slight increase of complexity. In particular, from low-to-intermediate signal-to-noise ratio (SNR) the performance of the V-BLAST-CD is the same as that of MLD, while at high SNR the V-BLAST-CD still provides a significant performance gain over the V-BLAST detector.

The paper is organized as follows. The model of the MIMO system we focus on is given in Sect. 2. Section 3 introduces the two HLST-MLC schemes we have considered throughout the paper. In Sect. 4 we illustrate the two LST-MSD receivers implementing detection and decoding of the information bits transmitted by using the two HLST-MLC schemes. A description of the suboptimal low-complexity spatial detection algorithms and the associated soft metric computations we have used in the LST-MSD receivers is given in Sect. 5. Experimental results are then shown in Sect. 6 and conclusions are drawn in Sect. 7.

## 2 System Model

We consider the equivalent discrete-time complex-baseband representation of a flat fading MIMO channel with $n_T$ transmitting antennas and $n_R \geq n_T$ receiving antennas. The received signal vector at $k$-th time instant is

$$\mathbf{r}_k = \mathbf{H}_k \mathbf{x}_k + \mathbf{n}_k, \tag{1}$$

where $\mathbf{x}_k$ is the $n_T \times 1$ vector of transmitted complex symbols drawn from a square $M$-QAM constellation, $\mathbf{n}_k$ is the $n_R \times 1$ noise vector whose entries are temporally and spatially i.i.d. complex Gaussian random variables (RVs) with zero mean and variance $\sigma_n^2$ and $\mathbf{H}_k$ is the $n_R \times n_T$ channel matrix whose elements are spatially i.i.d. RVs having uniform-distributed phase and Rayleigh-distributed magnitude with average power equal to 1. $\mathbf{H}_k$ is independent of both $\mathbf{x}_k$ and $\mathbf{n}_k$ and it is assumed perfectly known to the receiver. We assume a block-fading channel, where $\mathbf{H}_k$ assumes a constant value over a coded symbol frame and then changes to a new value. The average radiated power from each antenna is fixed to $1/n_T$. Thus, the total average radiated power is fixed to 1 and it turns out to be independent of the total number of transmitting antennas. The average SNR per transmitted symbol at the receiver is defined as $\mathrm{SNR} = n_R/(n_T \sigma_n^2)$.

# 3 HLST-MLC Transmission Schemes

A description of the two HLST-MLC transmission schemes we focus on throughout the paper is given in Sects. 3.1 and 3.2.

## 3.1 S-HLST-MLC

In the S-HLST-MLC scheme the input information sequence is demultiplexed into $n_T$ substreams which are separately multilevel encoded, modulated by an $M$-QAM modulator and transmitted in parallel from $n_T$ antennas at the same time and frequency. Each substream in turn is demultiplexed in $m = 1/2 \cdot \log_2 M$ parallel sequences which are distributed to the component encoders according to their relative rate. The modulator uses mapping by set partitioning, where a binary partition is considered at each partitioning level for each dimension of the QAM constellation. Let $\{b_{\Re;m}, \ldots, b_{\Re;1}\}$ and $\{b_{\Im;m}, \ldots, b_{\Im;1}\}$ be, respectively, the binary labels for the real part, $x_{\Re}$, and the imaginary part, $x_{\Im}$, of the QAM constellation symbols $x$. Bits are listed from the most (MSB, first entry with index $m$) to the least significant (LSB, last entry with index 1). The block diagram of the S-HLST-MLC scheme is shown in Fig. 1. We assume that MLC on the separate branches uses identical component codes at the same partitioning level.

## 3.2 J-HLST-MLC

In contrast to S-HLST-MLC, in the J-HLST-MLC transmission scheme the input information sequence is encoded by using only one multilevel encoder.

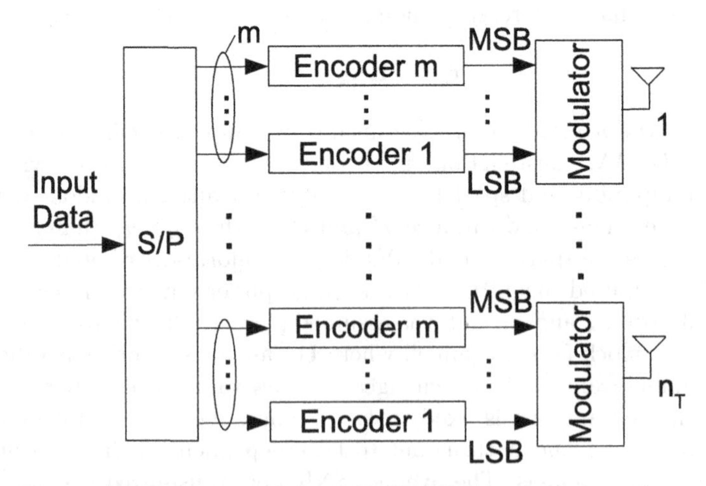


**Fig. 1.** Block diagram of the S-HLST-MLC transmission scheme

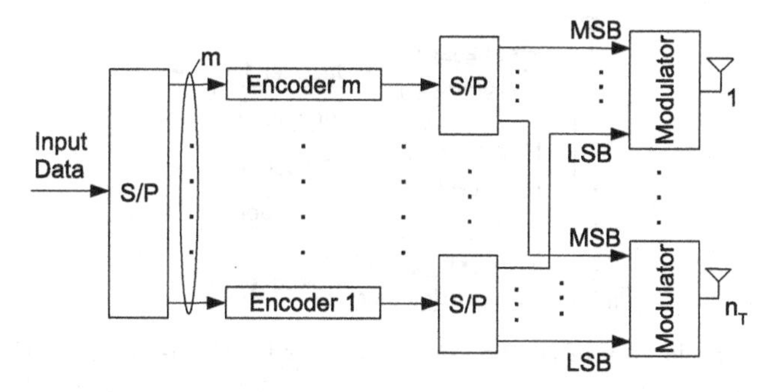


**Fig. 2.** Block diagram of the J-HLST-MLC transmission scheme

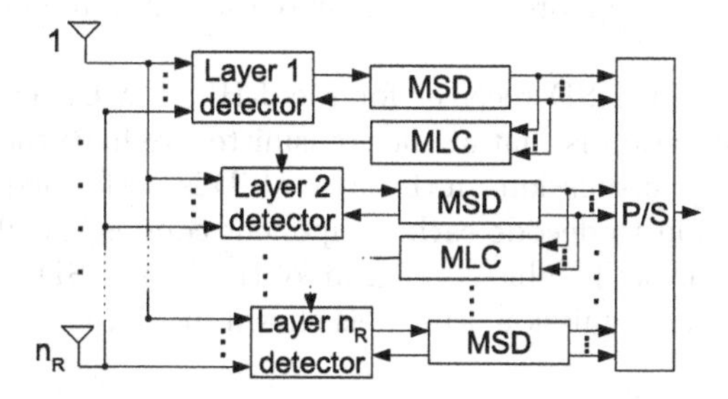


**Fig. 3.** Block diagram of the LST-MSD receiver for the S-HLST-MLC scheme

Then, the sequence of coded bits at the output of each component encoder is demultiplexed into $n_T$ substreams, each of which is sent to a separate transmitting branch. The demultiplexed coded bits received at each transmitting branch are sorted according to their significance from the MSB to the LSB both on the real and imaginary axis, modulated by an $M$-QAM modulator and finally transmitted from $n_T$ antennas at the same time and frequency. The mapping is the same as that used in each separate branch of the S-HLST-MLC scheme. The block diagram of the J-HLST-MLC scheme is shown in Fig. 2.

## 4 Design of the LST-MSD Receivers

The general form of the LST-MSD receiver used for the S-LST-MLC transmission scheme is depicted in Fig. 3. As the figure shows, the detector is first employed to provide soft metrics to the MSD on layer 1. The decoded data of layer 1 is then multilevel re-encoded and used to cancel the contribution of layer 1 in the received signal vector. The resulting received signal vector is used for the detection and decoding of data from layer 2. The process continues until all layers are decoded. It is worth noting that the cancellation step just described coincides with that of the V-BLAST detector for an uncoded transmission. However, in this case the cancellation step of the hard symbol

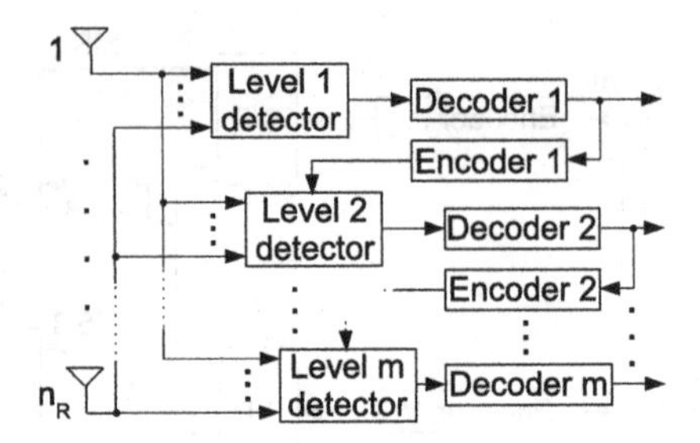


**Fig. 4.** Block diagram of the LST-MSD receiver for the J-HLST-MLC scheme

estimate can exploit the more reliable symbol estimate provided by the MSD at previous layer.

As far as the LST-MSD receiver for the J-HLST-MLC is concerned, the fundamental observation is that at the transmitter we have the concatenation of MLC with LST transmission. In this case MSD is performed by considering that soft information to decode each component code is calculated using the spatial detection block. The block diagram of the LST-MSD receiver used for the J-HLST-MLC transmission scheme is shown in Fig. 4.

## 5 Soft Metric Computation

A description of the soft metric computation for the different suboptimal low complexity detectors, considered in the spatial detection stage of the two LST-MSD receivers, is given in the following subsections.

### 5.1 The Linear Minimum Mean Square Error (LMMSE) Detector

The goal of the LMMSE interface is that of performing a separation of the transmitted substreams by minimizing the variance of the error between the transmitted vector $\mathbf{x}_k$ and the receiver's estimate

$$\hat{\mathbf{x}}_k = \mathbf{G}\mathbf{r}_k, \tag{2}$$

where $\mathbf{G}$ is the $n_T \times n_R$ linear weighting matrix. The error vector is defined as

$$\mathbf{e}_k = \mathbf{x}_k - \hat{\mathbf{x}}_k = \mathbf{x}_k - \mathbf{G}\mathbf{r}_k. \tag{3}$$

By minimizing the mean square error defined as

$$\mathrm{MSE} = E\{\|\mathbf{e}_k\|^2\} = E\{\|\mathbf{x}_k - \mathbf{G}\mathbf{r}_k\|^2\},$$

one obtains the following MMSE weight matrix [1]

$$\mathbf{G}_{\mathrm{MMSE}} = (\mathbf{H}^\dagger\mathbf{H} + n_T\sigma_n^2\mathbf{I}_{n_T})^{-1}\mathbf{H}^\dagger, \tag{4}$$

where $\mathbf{H}$ is the constant channel, $\mathbf{I}_{n_T}$ is an $n_T \times n_T$ identity matrix and $\dagger$ denotes conjugate transposition. The covariance matrix of the error vector in (3) is

$$\mathbf{P} = E\{\mathbf{e}_k \, \mathbf{e}_k^\dagger\} = (\mathbf{H}^\dagger \mathbf{H} + n_T \sigma_n^2 \mathbf{I}_{n_T})^{-1}. \tag{5}$$

The soft decision statistic for the symbol sent from antenna $i$ at time $k$ is given by

$$y_{k;i} = \mathbf{G}_{\mathrm{MMSE},i} \mathbf{r}_k, \tag{6}$$

where $\mathbf{G}_{\mathrm{MMSE},i}$ is the $i$-th row of matrix $\mathbf{G}_{\mathrm{MMSE}}$.

### Separate LST-MSD Receiver

Considering the case of the LST-MSD receiver for the S-HLST-MLC scheme, the spatial detector first provides soft metrics to the MSD on layer 1. At each decoding stage on layer 1 the detector examines the soft decision statistic $y_{k;1}$ and compares its real and imaginary parts $y_{k;\Re;1}$, $y_{k;\Im;1}$ with the two subsets of the real and imaginary parts of the constellation symbols at that stage, outputting soft values to indicate their relative closeness. The decoder then uses the detector output to decode that component code, thereby selecting one of the two subsets on the real and imaginary axes. The information about the two selected subsets is sent to the detector on layer 1 that uses it to compute the new soft metrics to be passed to the next stage component decoder. After the decoding of the last component code is completed, the decoded bits of layer 1 are multilevel re-encoded and used to cancel the contribution of the decided symbol $\hat{x}_k^{(1)}$ on layer 1 in the received signal vector $\mathbf{r}_k$. Thus, treating $\hat{x}_k^{(1)}$ as a known quantity, we obtain the following reduced order problem

$$\mathbf{r}_k^{(2)} = \mathbf{r}_k^{(1)} - \mathbf{h}_1 \hat{x}_{k;1} = \mathbf{H}^{(2)} \mathbf{x}_k + \mathbf{n}_k, \tag{7}$$

where $\mathbf{r}_k^{(1)} = \mathbf{r}_k$, $\mathbf{h}_1$ denotes the first column of $\mathbf{H}$ and $\mathbf{H}^{(2)}$ is the deflated version of $\mathbf{H}$ obtained by setting to zero the $n_R$ entries of the first column. This modified received signal, denoted by $\mathbf{r}_k^{(2)}$, has a lower level of interference and this will increase the probability of correct detection at the successive detection step. The received signal vector $\mathbf{r}_k^{(2)}$ is used for detection and MSD of data from layer 2. The process continues until all layers are decoded.

The optimum soft metrics to feed from the spatial detector to the decoder on layer $i$ for coded bits $b_{k;\Re;p}^{(i)}$ and $b_{k;\Im;p}^{(i)}$, $p = 1, \ldots, m$, are given by the log likelihood ratios (LLRs) of the a posteriori probabilities computed as [5]

$$\begin{aligned}
\lambda_{k;\Re;p}^{(i)} &= \log \frac{\Pr\left(b_{k;\Re;p}^{(i)} = 0 | y_{k;\Re;i}, \hat{b}_{k;\Re;p-1}^{(i)}, \cdots, \hat{b}_{k;\Re;1}^{(i)}\right)}{\Pr\left(b_{k;\Re;p}^{(i)} = 1 | y_{k;\Re;i}, \hat{b}_{k;\Re;p-1}^{(i)}, \cdots, \hat{b}_{k;\Re;1}^{(i)}\right)} \\
&= \log \left( \frac{\sum_{x_{\Re;p}:b_{\Re;p}=0} e^{-\frac{1}{2\sigma_n^2}(y_{k;\Re;i} - x_{\Re;p})^2}}{\sum_{x_{\Re;p}:b_{\Re;p}=1} e^{-\frac{1}{2\sigma_n^2}(y_{k;\Re;i} - x_{\Re;p})^2}} \right)
\end{aligned} \tag{8}$$

and

$$\lambda_{k;\Im;p}^{(i)} = \log \frac{\Pr\left(b_{k;\Im;p}^{(i)} = 0 | y_{k;\Im;i}, \hat{b}_{k;\Im;p-1}^{(i)}, \cdots, \hat{b}_{k;\Im;1}^{(i)}\right)}{\Pr\left(b_{k;\Im;p}^{(i)} = 1 | y_{k;\Im;i}, \hat{b}_{k;\Im;p-1}^{(i)}, \cdots, \hat{b}_{k;\Im;1}^{(i)}\right)}$$
$$= \log \left( \frac{\sum_{x_{\Im;p}:b_{\Im;p}=0} e^{-\frac{1}{2\sigma_n^2}(y_{k;\Im;i} - x_{\Im;p})^2}}{\sum_{x_{\Im}:b_{\Im;p}=1} e^{-\frac{1}{2\sigma_n^2}(y_{k;\Im;i} - x_{\Im;p})^2}} \right), \quad (9)$$

where $\hat{b}_{k;\Re;l}^{(i)}$ ($\hat{b}_{k;\Im;l}^{(i)}$) is the bit decision at level $l$ for the real (imaginary) part of the transmitted symbol $x_{k;\Re}^{(i)}$ ($x_{k;\Im}^{(i)}$) and $x_{\Re;p}$ ($x_{\Im;p}$) denotes the real (imaginary) part of constellation symbols in the subset obtained from the partitioning at level $p$. The summation at the numerator of (8) and (9) is performed on constellation points whose binary label is 0 while that at the denominator considers symbols having 1 as binary label.

Close approximations of the above LLRs can be obtained considering the constellation points in each subset at the nearest distance from the received signal $y_{k;i}$

$$\lambda_{k;\Re;p}^{(i)} \approx \frac{1}{2\sigma_n^2}\left(y_{k;\Re;i} - x_{\Re;p}(b_{\Re;p} = 0)\right)^2 - \frac{1}{2\sigma_n^2}\left(y_{k;\Re;i} - x_{\Re;p}(b_{\Re;p} = 1)\right)^2 \quad (10)$$

and

$$\lambda_{k;\Im;p}^{(i)} \approx \frac{1}{2\sigma_n^2}\left(y_{k;\Im;i} - x_{\Im;p}(b_{\Im;p} = 0)\right)^2 - \frac{1}{2\sigma_n^2}\left(y_{k;\Im;i} - x_{\Im;p}(b_{\Im;p} = 1)\right)^2, \quad (11)$$

where $x_{\Re;p}(b_{\Re;p} = b)$ ($x_{\Im;p}(b_{\Im;p} = b)$) is the real (imaginary) part of the constellation symbols in the subset at partitioning level $p$ whose binary label is $b$, with $b = 0, 1$.

### Joint LST-MSD Receiver

When the LMMSE detector is used in the LSD-MSD receiver for the J-HLST-MLC scheme, at each decoding stage the spatial detector computes the LLRs of all bits associated to the same partitioning level on the real and imaginary axis of the $n_T$ $M$-QAM transmitted symbols. The computation of the LLRs is done by using (10) and (11), the difference being that now the decision statistic $y_i$ does not benefit from the cancellation as in the ST-MSD used for the S-HLST-MLC transmission scheme.

Note that, in contrast to the separate LST-MSD, with the joint LST-MSD receiver the coefficients of $\mathbf{G}_{\text{MMSE}}$ are only computed once, as $\mathbf{H}$ remains unchanged. This will reduce system performance, saving some computational cost.

### 5.2 The V-BLAST Detector

In the original implementation of V-BLAST, the received vector elements are linearly weighted to null the interference from the yet undetected symbols [10]. This approach leads to the so-called zero-forcing (ZF) V-BLAST. It is worth noting that ZF V-BLAST disregards the effects of noise and removes only the interference components. A better performance is obtained if one minimizes the effects of total disturbance, that is, interference plus noise. This approach leads to MMSE V-BLAST [1].

According to the OSIC process implemented by the MMSE V-BLAST detector, an ordering has to be introduced to establish which signal has to be detected first. Clearly, in this case, the best symbol estimate would be the one that corresponds to the smallest MSE, that is, the one for which the $i$-th element of the main diagonal of the error covariance matrix $\mathbf{P}$ in (5) is the lowest. This leads to an ordering criterion where the symbol with the lowest MSE is detected first.

#### Separate LST-MSD Receiver

When the MMSE V-BLAST detector with ordering and successive interference cancellation is used as a spatial detection block in the LST-MSD receiver for the S-HLST-MLC scheme, we are exactly in the situation seen in Sect. 5.2 for the LMMSE detector. The only difference with the LMMSE case is that now an ordering has been introduced. The LLRs are computed as reported in (10) and (11).

#### Joint LST-MSD Receiver

When the ordered MMSE V-BLAST detector is used as a spatial detector in the LSD-MSD receiver for the J-HLST-MLC scheme, the soft decision statistic $y_i$ at the different detection steps is computed as reported in (6). In contrast to the separate LSD-MSD receiver, the hard symbol estimate $\hat{x}_{k;i}$ subtracted from the received signal vector in (7) is now obtained from hard decision of the ordered MMSE V-BLAST detector in the previous detection step. Also in this case the soft metrics for the MSD detector are computed by using (10) and (11).

### 5.3 The V-BLAST-CD

The principle behind MSP is a geometric partitioning of the signal constellation in subsets of diminishing size, in such a way that minimum Euclidean distance within the subsets increases down the partition chain. A square $M$-QAM constellation can be seen as a finite set of points carved out from the two-dimensional integer lattice $\mathbb{Z}^2$. Similarly, the signal transmitted from the $n_T$ antennas can be seen as belonging to a multidimensional constellation carved out from $\mathbb{Z}^{2n_T}$. The partitioning of the constellation in subsets corresponds to the partitioning of the lattice into a sublattice and its cosets. In the V-BLAST-CD a partition $\mathbb{Z}/q\mathbb{Z}$ is considered in each dimension of the square $M$-QAM constellation for each transmitted substream, where $q$ is an integer power of two such that $2 \leq q \leq \sqrt{M}$. The partitioning studied here divides the multidimensional constellation into $q^{2n_T}$ subsets, each containing $M^{n_T}/q^{2n_T}$ vectors. We consider the application of the ordered MMSE V-BLAST detector to each multidimensional subconstellation. Hence our detector includes $q^{2n_T}$ MMSE V-BLAST detectors. In each step of the MMSE V-BLAST detection process, a scalar decision is taken on a symbol belonging to the subset associated to the transmitted substream considered at that step for the given multidimensional subconstellation. At the end of the MMSE V-BLAST detection process, a decision is taken on a vector by collecting the scalar decisions taken at each step by the MMSE V-BLAST detector. Having applied the $q^{2n_T}$ MMSE V-BLAST detectors, a list of $q^{2n_T}$ candidate vectors is generated by considering the set of V-BLAST decisions. The list of candidate vectors is used to compute the soft information for the component error correction decoders. For the implementation of the V-BLAST-CD in the present work we consider the case $q = 2$.

### Separate LST-MSD Receiver

Consider the soft metric calculation when the MMSE V-BLAST-CD is used as spatial detector in the LST-MSD receiver for the S-HLST-MLC scheme. The optimum soft inputs to the binary error decoder corresponding to bits $b_{k;\Re;p}^{(i)}$ and $b_{k;\Im;p}^{(i)}$, $p = 1, \ldots, m$, on layer $i$ can be written, respectively, as

$$\begin{aligned}
\lambda_{k;\Re;p}^{(i)} &= \log \frac{\Pr\left(b_{k;\Re;p}^{(i)} = 0 | \mathbf{r}_k^{(i)}, \mathbf{H}, \hat{b}_{k;\Re;p-1}^{(i)}, \cdots, \hat{b}_{k;\Re;1}^{(i)}\right)}{\Pr\left(b_{k;\Re;p}^{(i)} = 1 | \mathbf{r}_k^{(i)}, \mathbf{H}, \hat{b}_{k;\Re;p-1}^{(i)}, \cdots, \hat{b}_{k;\Re;1}^{(i)}\right)} \\
&= \log \left( \frac{\sum_{\hat{\mathbf{x}}_{k;r}^{(i)} \in \mathcal{X}_{k;r}^{(i)} : b_{\Re;p}^{(i)} = 0} e^{-\frac{1}{2\sigma_n^2} \left\| \mathbf{r}_k^{(i)} - \mathbf{H}\hat{\mathbf{x}}_{k;r}^{(i)} \right\|^2}}{\sum_{\hat{\mathbf{x}}_{k;r}^{(i)} \in \mathcal{X}_{k;r}^{(i)} : b_{\Re;p}^{(i)} = 1} e^{-\frac{1}{2\sigma_n^2} \left\| \mathbf{r}_k^{(i)} - \mathbf{H}\hat{\mathbf{x}}_{k;r}^{(i)} \right\|^2}} \right)
\end{aligned} \tag{12}$$

and

$$\begin{aligned}
\lambda_{k;\Im;p}^{(i)} &= \log \frac{\Pr\left(b_{k;\Im;p}^{(i)} = 0 | \mathbf{r}_k^{(i)}, \mathbf{H}, \hat{b}_{k;\Im;p-1}^{(i)}, \cdots, \hat{b}_{k;\Im;1}^{(i)}\right)}{\Pr\left(b_{k;\Im;p}^{(i)} = 1 | \mathbf{r}_k^{(i)}, \mathbf{H}, \hat{b}_{k;\Im;p-1}^{(i)}, \cdots, \hat{b}_{k;\Im;1}^{(i)}\right)} \\
&= \log \left( \frac{\sum_{\hat{\mathbf{x}}_{k;r}^{(i)} \in \mathcal{X}_{k;r}^{(i)} : b_{\Im;p}^{(i)} = 0} e^{-\frac{1}{2\sigma_n^2} \left\| \mathbf{r}_k^{(i)} - \mathbf{H}\hat{\mathbf{x}}_{k;r}^{(i)} \right\|^2}}{\sum_{\hat{\mathbf{x}}_{k;r}^{(i)} \in \mathcal{X}_{k;r}^{(i)} : b_{\Im;p}^{(i)} = 1} e^{-\frac{1}{2\sigma_n^2} \left\| \mathbf{r}_k^{(i)} - \mathbf{H}\hat{\mathbf{x}}_{k;r}^{(i)} \right\|^2}} \right),
\end{aligned} \tag{13}$$

where $\hat{\mathbf{x}}_{k;r}^{(i)}$ is a vector taken from the list $\mathcal{X}_{k;r}^{(i)}$ whose elements are the $2^{2(n_T-i+1)}$ candidate vectors obtained by applying the MMSE V-BLAST-CD at the $i$-th detection and MSD step. The summation at the numerator of (12) and (13) is performed, respectively, on vectors $\hat{\mathbf{x}}_{k;r}^{(i)}$ whose real and imaginary part of the $i$-th symbol has binary label 0 while for the summation at the denominator we considering vectors $\hat{\mathbf{x}}_{k;r}^{(i)}$ whose real and imaginary part of the $i$-th symbol has binary label 1.

Close approximations of the above LLRs can be obtained by considering the vectors in each subset $\hat{\mathbf{x}}_{k;r}^{(i)}$ at the nearest distance from the received signal $\mathbf{r}_k^{(i)}$

$$\lambda_{k;\Re;p}^{(i)} \approx \frac{1}{2\sigma_n^2}\left\|\mathbf{r}_k^{(i)} - \mathbf{H}\hat{\mathbf{x}}_{k;r}^{(i)}(b_{\Re;p}^{(i)} = 0)\right\|^2 - \frac{1}{2\sigma_n^2}\left\|\mathbf{r}_k^{(i)} - \mathbf{H}\hat{\mathbf{x}}_{k;r}^{(i)}(b_{\Re;p}^{(i)} = 1)\right\|^2 \tag{14}$$

and

$$\lambda_{k;\Im;p}^{(i)} \approx \frac{1}{2\sigma_n^2}\left\|\mathbf{r}_k^{(i)} - \mathbf{H}\hat{\mathbf{x}}_{k;r}^{(i)}(b_{\Im;p}^{(i)} = 0)\right\|^2 - \frac{1}{2\sigma_n^2}\left\|\mathbf{r}_k^{(i)} - \mathbf{H}\hat{\mathbf{x}}_{k;r}^{(i)}(b_{\Im;p}^{(i)} = 1)\right\|^2 . \tag{15}$$

After the decoding of the last component code at layer $i$ is completed, the decoded bits are multilevel re-encoded and used to cancel the contribution of the decided symbol $\hat{x}_k^{(i)}$ on layer $i$ in the received signal vector $\mathbf{r}_k^{(i)}$. The resulting received signal vector $\mathbf{r}_k^{(i+1)}$ is used by the V-BLAST-CD to generate the new list of candidate vectors $\mathcal{X}_{k;r}^{(i+1)}$. All the $2^{2(n_T-i+1)}$ candidate vectors contained in $\mathcal{X}_{k;r}^{(i+1)}$ have in common symbols $x_k^{(l)} = \hat{x}_k^{(l)}$, $l = 1, \ldots, i$, that is, symbols obtained from the application of MSD in the previous steps. The process continues until all layers are decoded.

### Joint LST-MSD Receiver

As for the previous two cases of LMMSE and MMSE V-VLAST, in the LSD-MSD receiver for the J-HLST-MLC scheme we consider the implementation of the conventional ordered LMMSE V-BLAST-CD. Also for this case the soft metrics for the MSD detector are computed by using (14) and (15) where the received signal vector $\mathbf{r}_k$ and the list of $2^{2n_T}$ candidate vectors $\mathbf{x}_{k;r}$ are used in place of $\mathbf{r}_k^{(i)}$ and the $\mathbf{x}_{k;r}^{(i)}$.

## 6 Simulation Results

Consider a $2 \times 2$ MIMO system using 16 and 64-QAM modulations. As a component encoder on the first level we consider a tailbiting version of the (133, 171) convolutional code with rate 1/2. At the higher levels we assume an uncoded transmission. The length of the coded frame is equal to 64 space–time symbols.

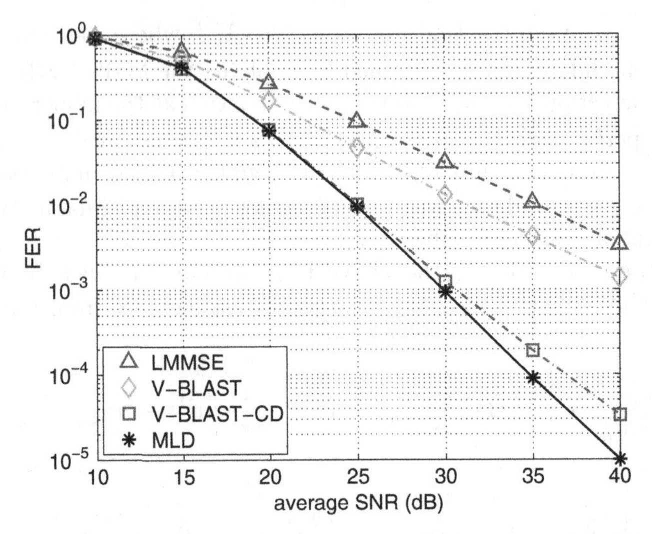


**Fig. 5.** Performance comparison of LMMSE, V-BLAST and V-BLAST-CD for a $2 \times 2$ J-HLST-MLC with 16-QAM. The first level is encoded with a tailbiting version of the (133,171) convolutional code with rate 1/2 and the second level is uncoded

Figure 5 reports the frame error rate (FER) versus SNR obtained when the spatial detection schemes described in Sect. 4 are used in the LST-MSD receiver for the J-HLST-MLC scheme with 16-QAM. As a reference we have also reported the performance obtained with MLD. The FER is measured by transmitting a sequence of $10^5$ frames. We observe that the performance of the V-BLAST-CD algorithm is close to that of the MLD for a FER greater than $10^{-3}$. The performance of the S-HLST-MLC scheme is shown in Fig. 6. Note that the performance of S-HLST-MLC is virtually the same as that of J-HLST-MLC with V-BLAST and V-BLAST-CD detection, while for LMMSE detection the S-HLST-MLC outperforms the J-HLST-MLC.

The performance obtained for the J-HLST-MLC and the S-HLST-MLC schemes using 64-QAM modulation are reported in Figs. 7 and 8, respectively. Due to the high complexity of the MLD, in this case we have reported as a reference the performance obtained by using the sphere detector (SD) of [11]. From the results presented in the figures we observe a behavior similar to that of the 16-QAM case.

From the presented results, we conclude that from low-to-intermediate SNR the V-BLAST-CD provides a performance close to the optimum both for the J-HLST-MLC and the S-HLST-MLC. Therefore, both the J-HLST-MLC and the S-HLST-MLC schemes can be adopted when this detector is used in the receiver. In case of LMMSE detection the best performance is obtained for the S-HLST-MLC architecture.

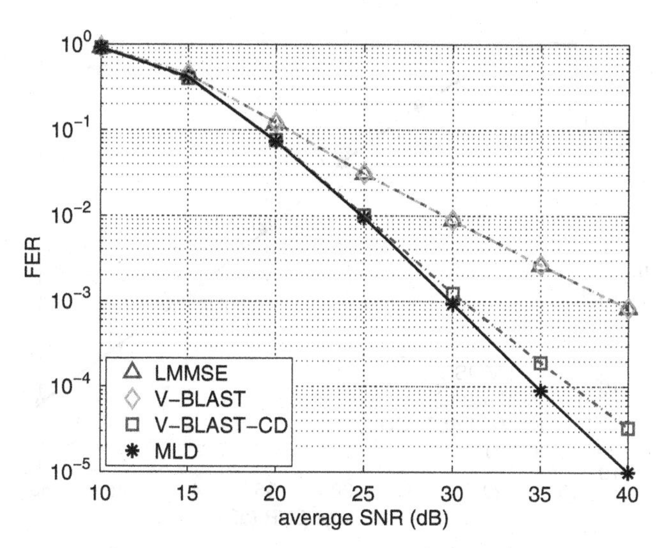


**Fig. 6.** Performance comparison of LMMSE, V-BLAST and V-BLAST-CD for a $2 \times 2$ S-HLST-MLC with 16-QAM. The first level is encoded with a tailbiting version of the (133,171) convolutional code with rate 1/2 and the second level is uncoded

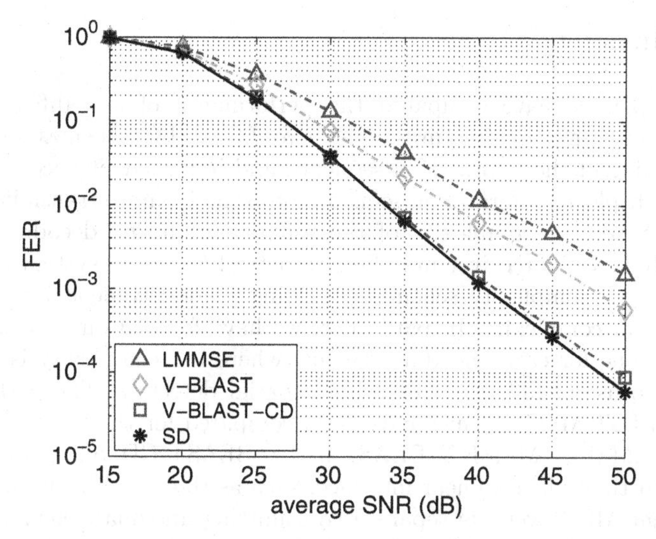


**Fig. 7.** Performance comparison of LMMSE, V-BLAST and V-BLAST-CD for a $2 \times 2$ J-HLST-MLC with 64-QAM. The first level is encoded with a tailbiting version of the (133,171) convolutional code with rate 1/2 while the second and third levels are uncoded

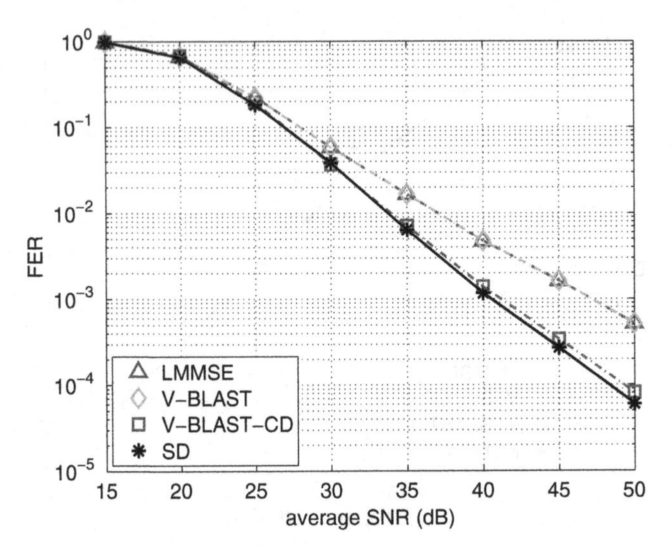


**Fig. 8.** Performance comparison of LMMSE, V-BLAST and V-BLAST-CD for a $2 \times 2$ S-HLST-MLC with 64-QAM. The first level is encoded with a tailbiting version of the (133,171) convolutional code with rate 1/2 while the second and third levels are uncoded

# 7 Conclusions

In this paper we have compared the performance of two different LST-MLC transmitting architectures. The two LST-MLC schemes implement joint encoding of the information sequence and independent encoding of the same demultiplexed information sequence onto the separate transmit antenna branches. Since the complexity of the soft detection and decoding process implemented by the receiver may be impractical for many systems, we have considered low complexity schemes providing a good tradeoff between performance and complexity. In particular, we have focused on two LST-MSD structures, one for each encoding scheme, where low complexity is achieved through the use of different suboptimal spatial detectors. The performance of the two LST-MLC schemes have been compared for the following spatial detectors: LMMSE, MMSE V-BLAST and V-BLAST-CD. Simulation results have shown that when a linear interface, such as the LMMSE, is considered, independent MLC over the separate transmitting antenna elements outperforms joint encoding. On the other hand, when we use a nonlinear spatial detector separate and joint encoding LST-MLC schemes provide similar performance.

## References

1. Vucetic B, Yuan J (2003) Space-time coding. John Wiley & Sons Ltd, West Sussex
2. Lampe LHJ, Schober R, Fischer RFH (2004) Multilevel coding for multiple-antenna transmission. IEEE Trans Wireless Commun 3: 203–208
3. Lamarca M, Lou H (2005) Spectrally efficient MIMO transmission schemes based on multilevel codes. In Proc. 6th IEEE Workshop on Signal Processing Advances in Wireless Communications: 1063–1067
4. Rezk D, Wang XF (2006) Improved layered space time architecture through unequal power allocation and multistage decoding. In Proc. IEEE Canadian Conference on Electrical and Computer Engineering: 932–935
5. Martin PA, Rankin DM, Taylor DP (2006) Multi-dimensional space–time multilevel codes. IEEE Trans Wireless Commun 5: 3287–3295
6. Imai H, Hirakawa S (1977) A new multilevel coding method using error correcting codes. IEEE Trans Inform Theory 23: 371–377
7. Wachsmann U, Fischer RFH, Huber JB (1999) Multilevel codes: theoretical concepts and practical design rules. IEEE Trans Inform Theory 44: 927–946
8. Magarini M, Spalvieri A (2006) Coset detection in MIMO systems based on spatial multiplexing. IEEE Commun Lett 10: 390–392
9. Eyuboglu MV, Qureshi SUH (1998) Reduced-state sequence estimation with set partitioning and decision feedback. IEEE Trans Commun 36: 13–20
10. Wolniansky PW, Foschini GJ, Golden GD, Valenzuela RA (1998) V-BLAST: an architecture for realizing very high data rates over the rich-scattering wireless channel. In Proc. IEEE International Symposium on Signal, Systems and Electronics: 295–300
11. Damen O, Chkeif A, Belfiore JC (2000) Lattice code decoder for space–time codes. IEEE Commun Lett 4: 161–163

2

# Iterative (Turbo) Joint Rate and Data Detection in Coded CDMA Networks

Stefano Buzzi and Stefania Sardellitti

Università degli Studi di Cassino
03043 Cassino (FR), Italy
buzzi@unicas.it, s.sardellitti@unicas.it

**Summary.** A convolutionally coded CDMA data network wherein each user may transmit at one out of a set of possible data rates is considered in this chapter. The problem of joint detection of the data-rate of each user and of the corresponding information symbols is considered. In particular, it is shown here that the so-called turbo principle can be used also for joint rate and data detection, and indeed an iterative (turbo) procedure is derived wherein the single-user decoders, and the data-rate detectors for each user exchange information in order to achieve lower and lower detection error probabilities. Numerical results show that the proposed approach is effective and achieves satisfactory performance.

## 1 Introduction

In this chapter the problem of detecting the data-rate and the information symbols transmitted by each user in a convolutionally coded variable-rate direct-sequence code division multiple access system (DS/CDMA) is considered. Variable-rate CDMA systems wherein each user transmits at one out of a set of available data-rates are nowadays of widespread interest, since they are able to support transmission of heterogeneous kinds of data , such as streaming video, voice, bulk data, etc., with different data-rates. These systems are part of current 3G wireless networks, and, also, they can support adaptive rate strategies that allow to achieve a larger system throughput [1, 2] than a fixed data-rate CDMA systems. Obviously, in a variable-rate CDMA system the receiver generally needs knowledge of the actual data-rate of each user in order to perform data detection, and this information is usually conveyed on a separate multiplexed channel. Alternatively, it may be extracted in some way from the incoming signal, an approach that has been only recently addressed in the literature.

In particular, existing studies in this area have mainly addressed the problem of frame rate detection with regard to a specific second-generation (2G) or 3G cellular standard [3, 4]. A key remark, however, is that none of these studies explicitly takes into account the presence of the multiple access interference (MAI). As a further example, the study [5] considers an iterative data detection algorithm for a single user system. More recently, the paper [6] has addressed the problem of blind rate-detection only (i.e. with no joint data detection) for variable data-rate CDMA systems, and the problem of *joint* detection of the data-rate and of the information symbols in such systems has been considered by the authors in [7, 8]. In particular, in [7] optimal (i.e. based on the maximum likelihood strategy) non-blind detectors have been proposed; in [8], instead, the focus has been on simpler receiver structures, that are amenable to blind implementations and with a computational complexity that is much lower than that of the receivers in [7].

In this chapter the case in which the users' information symbols are convolutionally coded is considered, and an iterative (turbo) procedure [9, 10] is devised in order to obtain reliable estimates of both the data rate and the information symbols for the user of interest. The contribution of this chapter is thus to show that the so-called "turbo principle" can be used also in multiuser joint data-rate and symbol detection with a very attractive performance.

The rest of this chapter is organized as follows. Next section contains the considered system model, while the turbo detection procedure is described in Sect. 3. Finally, numerical results are illustrated in Sect. 4, and conclusions are draws in Sect. 5.

## 2 System Model

We consider a synchronous convolutionally coded DS/CDMA network with $K$ active users employing BPSK modulation (extension to larger cardinality signaling constellations is trivial); we also assume that the propagation channel introduces frequency-flat fading, that each user may transmit, in each data-frame, at one out of $S$ available data rates $R_1 < R_2 < \ldots < R_S$, and that the data rates $R_2, \ldots, R_S$ are all integer multiples of the lowest data-rate $R_1$, i.e. we have

$$R_i = m_i R_1, \qquad \forall i = 2, \ldots, S \tag{1}$$

where $m_i$ are positive integers. In each frame slot, each user, based on the kind of data to be transmitted, and, possibly, on the propagation channel state, selects one out of the $S$ available data-rates for its transmission. The multirate access scheme that is here considered is the variable spreading length (VSL) [11], i.e. all the signature waveforms have the same chip interval $T_c$ and different spreading lengths. In particular, the signatures corresponding to the rate $R_i$ have a spreading length equal to $1/(R_i T_c) = T_{b,i}/T_c$, with $T_{b,i}$ denoting the symbol interval for the users transmitting at rate $R_i$. Based on the above assumptions, the complex envelope of the received signal in a given frame slot is written as

$$r(t) = \sum_{k=0}^{K-1} \sum_{p_k=0}^{\frac{Pr_k}{R_1}-1} A_k \alpha_k(\lfloor pR_1/r_k \rfloor) c_k(p_k) s_k^{r_k}(t - \frac{p_k}{r_k}) + n(t) \, . \tag{2}$$

In this equation, $r_k$ is a random variate taking on values in the set $\{R_1, \dots, R_S\}$, and denoting the data-rate of the $k$th active user, $P$ is the number of data-bits transmitted in each frame by the users at rate $R_1$ (thus implying that $Pr_k/R_1$ is the number of bits transmitted by the users at rate $r_k$), $c_k(\cdot) \in \{+1, -1\}$ denotes the coded bit stream transmitted by the $k$th user, $\alpha_k(\cdot)$ is a complex gain accounting for the channel propagation effects,[1] $\lfloor \cdot \rfloor$ denotes the usual floor function, while the waveform $n(t)$ represents the ambient noise, that we assume to be a complex zero-mean Gaussian white random process with power spectral density $2\mathcal{N}_0$. Moreover, $s_k^{r_k}(t)$ and $A_k$ are respectively the rate-$r_k$ signature waveform and the amplitude of the $k$th user transmitted signal. Denoting by $N_{r_k} = 1/(r_k T_c)$ the processing gain for the $k$th user (recall that in the VSL access strategy the processing gain for each user is tied to its data-rate), by $\{\beta_k^{r_k}(n)\}_{n=0}^{N_{r_k}-1}$ the spreading code (at rate $r_k$) of the $k$th user, and by $\phi_{T_c}(\cdot)$ a square-root raised cosine bandlimited waveform, we have

$$s_k^{r_k}(t) = \sum_{n=0}^{N_{r_k}-1} \beta_k^{r_k}(n) \phi_{T_c}(t - nT_c) \, ,$$

assuming that $s_k^{r_k}(t)$ are unit-energy waveform. At the receiver, the signal $r(t)$ is passed through a filter matched to the waveform $\phi_{T_c}(t)$ and sampled at rate $1/T_c$. The resulting samples can be stacked in a $PN_{R_1}$-dimensional vector, which we denote by $\boldsymbol{r}$, and which is expressed as

$$\boldsymbol{r} = \sum_{k=0}^{K-1} \sum_{p_k=0}^{Pr_k/R_1-1} A_k \alpha_k(\lfloor pR_1/r_k \rfloor) c_k(p_k) \boldsymbol{s}_k^{r_k}(p_k) + \boldsymbol{n} \, . \tag{3}$$

In the above equation, the vector $\boldsymbol{s}_k^{r_k}(p_k)$ is the discrete-time version of the waveform $s_k^{r_k}(t - p_k/r_k)$ (notice that this vector, although being $PN_{R_1}$-dimensional, has only $N_{r_k}$ non-zero entries), while the vector $\boldsymbol{n}$ is the discrete-time version of the ambient noise, and is a white complex zero-mean Gaussian random vector with covariance matrix $E[\boldsymbol{n}\boldsymbol{n}^H] = 2\mathcal{N}_0 \boldsymbol{I}_{PN_{R_1}}$ (with $(\cdot)^H$ denoting conjugate transpose and $\boldsymbol{I}_n$ the identity matrix of order $n$). Now the $PN_{R_1}$-dimensional vector $\boldsymbol{r}$ can be split in $P$ distinct $N_{R_1}$-dimensional vectors, $\boldsymbol{r}(0), \dots, \boldsymbol{r}(P-1)$, i.e. $\boldsymbol{r} = [\boldsymbol{r}^T(0), \dots, \boldsymbol{r}^T(P-1)]^T$ with

$$\boldsymbol{r}(p) = \sum_{k=0}^{K-1} A_k \alpha_k(p) \boldsymbol{S}_k{}^{r_k} \boldsymbol{c}_k{}^{r_k}(p) + \boldsymbol{n}(p) \qquad p = 0, \dots, P-1, \tag{4}$$

[1] Note that we are considering a block fading channel model, i.e. the channel gain is constant over intervals of length not smaller than $T_{b,1}$.

wherein the $N_{R_1} \times (r_k/R_1)$-dimensional matrix ${\boldsymbol{S}_k}^{r_k}$ is expressed as

$$\boldsymbol{S}_k^{r_k} = \boldsymbol{I}_{r_k/R_1} \otimes [\beta_k^{r_k}(0), \ldots, \beta_k^{r_k}(N_{r_k}-1)]^T \tag{5}$$

with $\otimes$ denoting the Kronecker product and

$$\boldsymbol{c}_k^{r_k}(p) = \left[c_k\left(p\frac{r_k}{R_1}\right), \ldots c_k\left((p+1)\frac{r_k}{R_1}-1\right)\right]^T . \tag{6}$$

## 3 Detection Algorithm

A block scheme of the proposed detection algorithm for the $k$th user is reported in Fig. 1. The vector $\boldsymbol{r}(p)$ is pruned out of the multiple access interference (MAI)[2] and then is fed to a bank of linear minimum mean square error (MMSE) receivers, each one keyed to one of the available data rates; the output of each MMSE filter is fed to a soft-input-soft-output (SISO) decoder, implementing the well-known BCJR algorithm [12], that gives at the output soft estimates of the coded information symbols. These estimates are then used through the discrete expectation-maximization algorithm [5] in order to take a decision on the most probable data-rate for the $k$th user. These steps are repeated for each user, so that at the end of each iteration we have the most probable data-rate and the soft estimates of the corresponding coded symbols; these quantities are used at the following iteration in order to build estimates of the multiple access interference, say $\hat{\boldsymbol{m}}_0(\cdot), \ldots, \hat{\boldsymbol{m}}_{K-1}(\cdot)$, for each

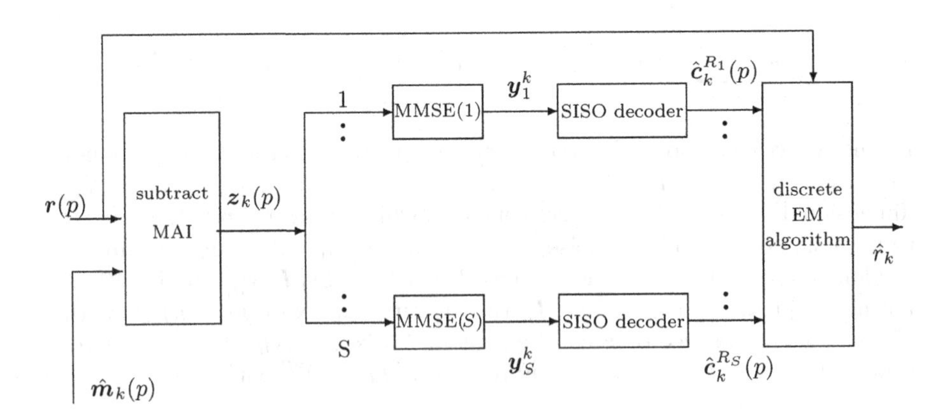


**Fig. 1.** Block scheme of the $k$th user receiver

[2] More details on how the estimate $\hat{\boldsymbol{m}}_k(p)$ of the MAI for the $k$th user is formed will be given in the sequel of the paper.

user. Note that, obviously, at the beginning of the algorithm no MAI cancellation is carried out (indeed, no MAI estimate is yet available), and the vector $\boldsymbol{r}(p)$ is directly fed to the bank of MMSE filters.

A deeper description of the algorithm is given in the following. We begin with considering the first iteration and, then, the subsequent iterations are described.

### 3.1 The Bank of MMSE Filters at the First Iteration

According to the detection strategy proposed in [8], the data vector $\boldsymbol{r}(p)$ is fed to a bank of MMSE filters, each keyed to one of the available data rates. Denoting by $\boldsymbol{D}_n^k(p)$ the MMSE filter for the $k$th user and keyed to the $n$th data rate (with $n = 1, \ldots, S$), and denoting by $\boldsymbol{y}_n^k(p)$ the output of this MMSE filter,[3] we have

$$\boldsymbol{y}_n^k(p) = \boldsymbol{D}_n^{k\,H}(p)\boldsymbol{r}(p) \ , \quad n = 1, \ldots, S \ , \tag{7}$$

with $(\cdot)^H$ denoting conjugate transpose and

$$\boldsymbol{D}_n^k(p) = A_k \alpha_k(p) \boldsymbol{R}_{k,n}^{-1} \boldsymbol{S}_k^{R_n} \ . \tag{8}$$

In the above equation, $\boldsymbol{R}_{k,n}$ is the data covariance matrix under the hypothesis that the $k$th user is transmitting at rate $R_n$, i.e. we have

$$\boldsymbol{R}_{k,n} = A_k^2 \sigma_{\alpha_k}^2 \boldsymbol{S}_k^{R_n} \boldsymbol{S}_k^{R_n\,H} + \boldsymbol{M}_k \tag{9}$$

with

$$\boldsymbol{M}_k = \frac{1}{S} \sum_{j \neq k} \sum_{\ell=1}^{S} A_j^2 \sigma_{\alpha_j}^2 \boldsymbol{S}_j^{R_\ell} \boldsymbol{S}_j^{R_\ell\,H} + 2\mathcal{N}_0 \boldsymbol{I}_{N_{R_1}} \ . \tag{10}$$

the covariance matrix of the $k$th user's MAI plus thermal noise. In the above equation, $\sigma_{\alpha_j}^2$ is the mean square value of the channel coefficient $\alpha_j(p)$. Note that in (10) the covariance matrix of the MAI has been averaged with respect to the data rates of the interfering users too; indeed, at the first iteration no prior information on the data rates of the interfering users is available, and thus averaging with respect to the available data rates seems to be the only reasonable option.

### 3.2 The SISO Decoders at the First Iteration

The output of each MMSE filter is then fed to a bank of SISO decoders implementing the BCJR algorithm [12]. For the sake of brevity, we do not dwell on the description of this algorithm, that can be found in many digital

[3] Note that for $n = 1$ $\boldsymbol{y}_n^k(p)$ is a scalar quantity and not a vector; likewise, $\boldsymbol{D}_n^k(p)$ for $n = 1$ is a vector and not a matrix.

communications textbooks. What suffices to say here is that at the output of each SISO decoder we have soft estimates for the coded information symbols for each rate hypothesis. Otherwise stated, at the output of the SISO decoder on the $k$th user's receiver' $n$th finger, we have the soft estimates $\hat{c}_k^{R_n}(0), \hat{c}_k^{R_n}(1), \ldots$ of the information symbols for the hypothesis that the user $k$ is transmitting at the rate $R_n$.

### 3.3 The Rate Estimation at the First Iteration

The outputs of the SISO decoders are then fed to a discrete EM algorithm that takes a decision on the most likely data-rate for the $k$th user according to the maximum likelihood rule

$$\hat{r}_k = \arg\max_{r_k} E_{c_k^{r_k}(\cdot)}\left[\log p(\boldsymbol{r}(p)|\boldsymbol{c}_k^{r_k}(\cdot), r_k)\right], \tag{11}$$

with $E_{c_k^{r_k}(\cdot)}[\cdot]$ denoting statistical expectation with respect to the coded information symbols, and $p(\boldsymbol{r}(p)|\cdot)$ denoting the conditional likelihood of the data vector. Since the data vector $\boldsymbol{r}(p)$ is conditionally Gaussian, upon letting $\boldsymbol{Q}_{k,n} = \boldsymbol{S}_k^{R_n\,H}\boldsymbol{M}_k^{-1}\boldsymbol{S}_k^{R_n}$, it can be shown that maximization (11) is equivalent to

$$\begin{aligned}\hat{r}_k = \arg\max\nolimits_{r_k} \Bigg[ & -\boldsymbol{r}^H(p)\boldsymbol{M}_k^{-1}(p)\boldsymbol{r}(p) \\ & +2\Re\left(\boldsymbol{r}^H(p)\boldsymbol{M}_k^{-1}(p)A_k\alpha_k(p)\boldsymbol{S}_k^{R_n}\hat{c}_k^{R_n}(p)\right) - A_k^2|\alpha_k(p)|^2 \\ & \times\left(\sum_{i,j=1,i\neq j}^{m_n} \hat{c}_k^{R_n}(i)\hat{c}_k^{R_n\,*}(j)\boldsymbol{Q}_{k,n}(i,j) + \mathrm{trace}(\boldsymbol{Q}_{k,n})\right)\Bigg] .\end{aligned} \tag{12}$$

### 3.4 The Following Iterations

At the end of the first iteration we have, for each user, a partial estimate $\hat{r}_k$ of its rate and a soft estimate of the corresponding coded symbols, i.e. $\hat{c}_k^{\hat{r}_k}(0), \hat{c}_k^{\hat{r}_k}(1), \ldots$. At the $k$th user receiver, MAI is cancelled according to the relation

$$\boldsymbol{z}_k(p) = \boldsymbol{r}(p) - \sum_{j\neq k} A_j \boldsymbol{S}_j^{\hat{r}_j}\boldsymbol{c}_j^{\hat{r}_j}(p)\alpha_j(p) , \tag{13}$$

i.e. we have

$$\begin{aligned}\boldsymbol{z}_k(p) = & A_k\alpha_k(p)\boldsymbol{S}_k^{r_k}\boldsymbol{c}_k^{r_k} \\ & + \sum_{j\neq k} A_j\alpha_j(p)\left(\boldsymbol{S}_j^{r_j}\boldsymbol{c}_j^{r_j} - \boldsymbol{S}_j^{\hat{r}_j}\boldsymbol{c}_j^{\hat{r}_j}\right) + \boldsymbol{n}(p) .\end{aligned} \tag{14}$$

The output of the MMSE filter is now written as

$$\boldsymbol{y}_n^k(p) = \boldsymbol{D}_n^{k\,H}(p)\boldsymbol{z}_k(p) , \quad n = 1, \ldots, S \tag{15}$$

with

$$D_n^k(p) = A_k \alpha_k(p) G_{k,n}^{-1} S_k^{R_n} \ . \tag{16}$$

In the above equation, $G_{k,n}$ is the covariance matrix of $z_k(p)$ under the hypothesis that the $k$th user is transmitting at rate $R_n$, and given the available estimates of the data rates and coded symbols for the interfering users, i.e.

$$\begin{aligned} G_{k,n} &= A_k^2 \sigma_{\alpha_k}^2 S_k^{R_n} E[c_k^{R_n}(p) c_k^{R_n\,H}(p)] S_k^{R_n\,H} + 2\mathcal{N}_0 I_{N_{R_1}} \\ &+ \sum_{j \neq k} A_j^2 \sigma_{\alpha_j}^2 S_j^{\hat{r}_j} (I_{\hat{r}_j/R_1} - \text{diag}(c_j^{\hat{r}_j}(p) \odot c_j^{\hat{r}_j\,*}(p)) S_j^{\hat{r}_j\,H} \ . \end{aligned} \tag{17}$$

In the above equation, $\odot$ denotes componentwise product, while the statistical expectation $E[c_k^{R_n}(p) c_k^{R_n\,H}(p)]$ is such that its $(i,j)$th entry is equal to 1 for $i = j$ and is equal to $\hat{c}_k^{R_n}(i) \hat{c}_k^{R_n\,*}(j)$ for $i \neq j$.

# 4 Numerical Results

In this section we show some numerical results on the performance of the proposed detection algorithm. In Figs. 2 and 3 the probability of erroneous rate detection (PERD) and of bit error probability conditioned on correct rate

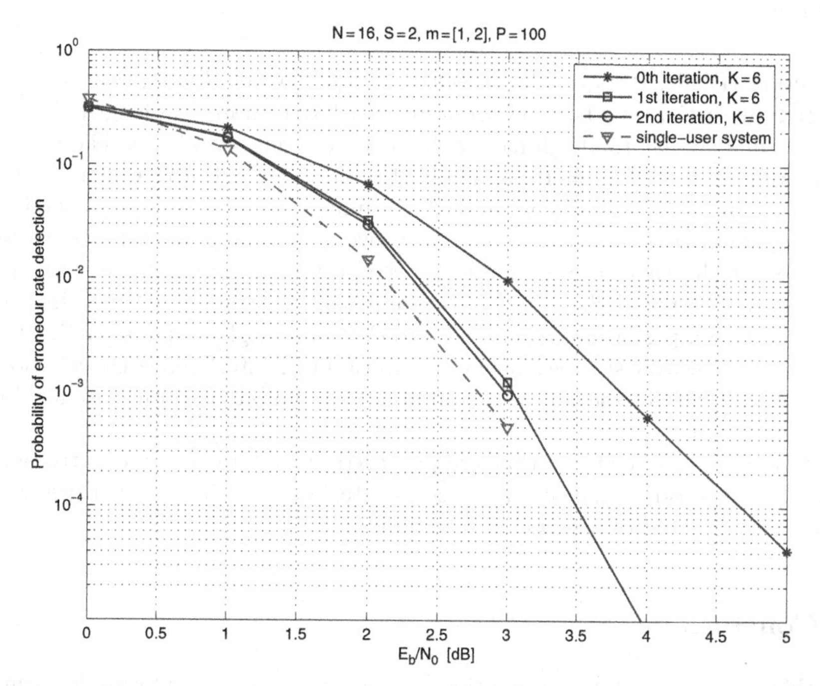


**Fig. 2.** Probability of erroneous rate detection vs. the received energy contrast

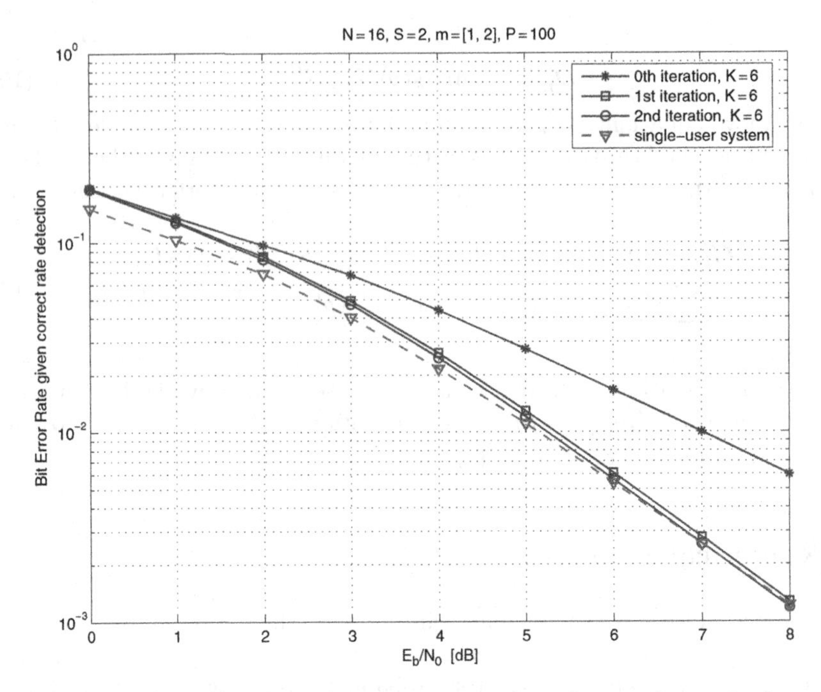


**Fig. 3.** Probability of bit error given correct rate detection vs. the received energy contrast

detection (BER|CRD) vs. the received energy contrast $E_b/\mathcal{N}_0$ has been reported. In particular a DS/CDMA system with data frames of length $P=100$ on a frequency-flat Rayleigh fading channel has been considered. The largest processing gain is $N_{R_1} = 16$, and $S = 2$ data rates with a rate ratio vector given by $\boldsymbol{m} = [1, 2]$ has been used. The number of users is $K = 6$, while a convolutional code of rate 1/2 has been used. For comparison purposes we also report the curve corresponding to the performance of a single-user system. It can be noted that the turbo rate detection strategy achieves in only one iteration a performance very close to that of a single user system, i.e. the proposed detection strategy achieves almost ideal interference cancellation. From Fig. 3, it is also seen that the proposed system bit error rate also attains the single user bound.

Overall it can be stated that the proposed procedure is very effective and is able to get rid of the multiple access interference achieving performance levels close to that of a single user system.

## 5 Conclusions

In this chapter an iterative multiuser joint data-rate and information symbol detection algorithm for coded DS/CDMA networks has been proposed.

The proposed receiver structure consists, for each user, of a bank of MMSE receivers, each one matched to one of the available data-rates, followed by a bank of SISO decoders, and, finally, by a discrete EM algorithm that chooses, for each user, the most probable data-rate at that iteration. The estimated data rates and coded information symbols provided at the end of each iteration are used in the following one in order to cancel multiple access interference and to improve the detection reliability. Numerical results have shown that the proposed strategy is very effective, thus proving that the turbo principle can be successfully exploited in order to detect the data-rate and the information symbols in a coded CDMA network.

## References


1. A. Goldsmith and S. Chua, "Adaptive coded modulation for fading channels," *IEEE Trans. Commun.*, Vol. 46, pp. 595–602, May 1998
2. S. A. Jafar and A. Goldsmith, "Adaptive multirate CDMA for uplink throughput maximization," *IEEE Trans. Wireless Commun.*, Vol. 2, pp. 218–228, March 2003
3. G. Yang and S. Kellel, "Optimal frame rate detection for CDMA mobile stations with variable-rate data transmission," *Proc. of 1999 Fifth Asia-Pacific Conf. Commun. and Fourth Optoelectron. and Commun. Conf.*, Bejing, China, October 1999
4. I. Held and A. Chen, "Reduced complexity blind rate detection for second and third generation cellular CDMA systems," *Proc. 1999 IEEE Global Commun. Conf. (GLOBECOM '99)*, Rio de Janeiro (Brasil), December 1999
5. H. Wymeersch and M. Moeneclaey, "ML-based blind symbol rate detection for multi-rate receivers," *Proc. of 2005 IEEE Int. Conf. Commun. (ICC 2005)*, Vol. 3, pp. 2092–2096, May 2005
6. S. Buzzi and A. De Maio, "Code-aided blind detection of the transmission rate for multirate direct-sequence CDMA systems," *IEEE Trans. Signal Process.*, Vol. 53, pp. 2182–2192, June 2005
7. S. Buzzi and S. Sardellitti, "Joint rate and data detection in variable-rate CDMA systems," *Proc. of IEEE Radio and Wireless Symp. (RWS '06)*, San Diego (CA), USA, January 2006
8. S. Buzzi and S. Sardellitti, "Blind MMSE-based receivers for rate and data detection in variable-rate CDMA systems," *Proc. of the 14th European Signal Process. Conf.*, Florence, Italy, September 2006
9. X. Wang and H. V. Poor, "Iterative (Turbo) soft interference cancellation and decoding for coded CDMA," *IEEE Trans. Commun.*, Vol. 47, pp. 1046–1061, July 1999
10. N. Noels et al., "Turbo synchronization: an EM algorithm interpretation," *Proc. of 2003 IEEE Int. Conf. Commun. (ICC 2003)*, Vol. 4, pp. 2933–2937, May 2003
11. S. Buzzi, M. Lops and A.M. Tulino, "Blind adaptive multiuser detection for asynchronous dual-rate DS/CDMA systems," *IEEE J. Selected Areas Commun.*, Vol. 19, pp. 233–244, February 2001
12. L. R. Bahl, J. Cocke, F. Jelinek and J. Raviv, "Optimal decoding of linear codes for minimizing symbol error rate," *IEEE Trans. Inform. Theory*, Vol. 20, pp. 284–287, March 1974

# 3

# Hybrid ARQ Based on Rateless Coding for UTRAN LTE Wireless Systems

Lorenzo Favalli, Matteo Lanati, and Pietro Savazzi

Dipartimento di Elettronica, Università degli Studi di Pavia, Italy
lorenzo.favalli@unipv.it, matteo.lanati@unipv.it,pietro.savazzi@unipv.it

**Summary.** In this work an incremental redundancy hybrid automatic repeat request (H-ARQ) scheme, based on rateless coding, is introduced for future extensions of third generation mobile systems. The proposed solution is tested by simulating an UMTS terrestrial radio access network (UTRAN) long term evolution (LTE) cell environment and compared to the standardized hybrid ARQ scheme. Performance results are obtained in terms of average throughput and cumulative density functions (CDF) of the average delays. The main contributions of this work are a novel cell traffic simulation model for rateless coded transmissions and the application of such methods to UTRAN LTE hybrid ARQ systems.

## 1 Introduction

During the last decade we assisted to a quick evolution in radio access interfaces, so nowadays cellular systems can offer to potential customers not only voice services but also data-oriented applications and multimedia contents. Advanced 3G and 4G wireless networks, e.g. WiMax or UMTS LTE, offer more resources employing technologies such as OFDM and HARQ. Growing interest in OFDM is due to its efficiency and granularity in spectrum management and to its robustness against frequency selective fading. HARQ is obtained mixing ARQ mechanism with forward error correction (FEC) protection coding in order to avoid waste of resources during retransmissions. As a matter of fact, protecting user information with a FEC code allows to minimize the number of retransmissions while an ARQ strategy permits to limit the redundancy introduced with respect to a FEC-only solution. Unfortunately, only a discrete set of coding rates can be used, so the efficiency can be seen as a staircase function of the channel quality. From this point of view, efficiency can be further increased replacing FEC with fountain codes [1]. In this type of codes, a source having to transmit a certain amount of data, divided in $N$ packets, sends these packets as a continuous stream that flows through the channel, from which the analogy with a water fountain. The receiver needs to collect at least $N$ encoded packet, not necessarily in the same order as they are produced, to correctly decode the source. A fountain code is rateless because the number of encoded packets is not fixed and the coding rate of the transmission depends on the number of packets collected.

In this article we focus on a generic fountain code, adapted to work in an UMTS LTE single cell environment. Since we are interested in downlink traffic, the digital fountain source is a Node B. Packets from upper layers are split in PDUs (packet data units), each one transmitted in one of the eight Stop and Wait (SaW) processes making up the HARQ scheme. In case of failure, the Node B keeps on transmitting redundant information of the PDU till a positive acknowledgement is received. This message tells the sender to switch to another PDU and also carries the feedback information. Note that a rateless code scheme can also work ignoring channel state information.

## 2 System Overview

### 2.1 UTRAN LTE Description

UTRAN LTE is the new version of radio access interface proposed by Third Generation Partnership Project (3GPP) to achieve flexibility in service provisioning and cost reduction in network deployment and planning. Main novelty is the introduction of OFDM, together with a reduced time transmission interval (TTI) of 0.5 ms. In each TTI, six or seven OFDM symbols are transmitted, depending on the cyclic-prefix, 4.7 or 16.7 μs respectively. The choice depends on the cell coverage and it has also to take into account the nature of the service (unicast or broadcast) [2].

Sub-carrier spacing is fixed equal to 15 kHz, consequently, varying the number of sub-carriers used for transmission, a spectrum of 1.25, 2.5, 5, 10, 15 or 20 MHz can be allocated. Such a flexibility is very useful, especially if the new architecture has to share some portions of the spectrum with the previous system. Available sub-carriers are subdivided in groups of 25 elements: each set represents a resource unit (RU), assigned to a single user for the entire duration of a TTI.

UTRAN LTE also aims to reduce the round trip time between mobile users and Radio Network Controller (RNC) to a value not greater than 10 ms. For this purpose 3GPP lowered TTI and adopted the HARQ scheme already proposed for High Speed Downlink Packet Access (HSDPA) [3].

### 2.2 Hybrid ARQ

As already explained, HARQ consists in simultaneous employment of retransmission scheme and error protection coding, merging redundant information on a transport block and multiple transmissions, at the cost of greater memory requirements. There are two HARQ strategies: chase combining (CC) and incremental redundancy (IR). In CC the sender, in case of negative acknowledgement, retransmits the transport blocks, as in a traditional ARQ scheme. The receiver does not discard erroneous blocks, but it combines them with the replicated ones, exploiting time diversity. In IR, each retransmission supplies additional redundant information with respect to previous attempt, obtained, for example, varying the puncturing scheme. A retransmitted block can be self decodable or not (if it contains systematic bits), however it contributes to increase the coding rate.

HARQ management is located in Node B, in a new entity called Medium Access Control – High Speed (MAC-HS), while the RNC still provides PDUs retransmission at radio link control (RLC) level, as we can see in Fig. 1.

In case of error on the radio channel, there is no need to wait for the RNC to send again a PDU, but it is possible to locally recover the failure, minimizing delays. The number of MAC-HS entities in a Node B equals the number of users and each entity has to keep track of eight SaW processes associated to that particular user. To simplify management, a SaW protocol has been chosen to regulate data retransmissions of a user, but, to avoid waste of resources, multiple processes run simultaneously. It is necessary to specify in a control channel which user is going to receive information carried by the transport channel in successive TTIs, and which one of its SaW process is involved.

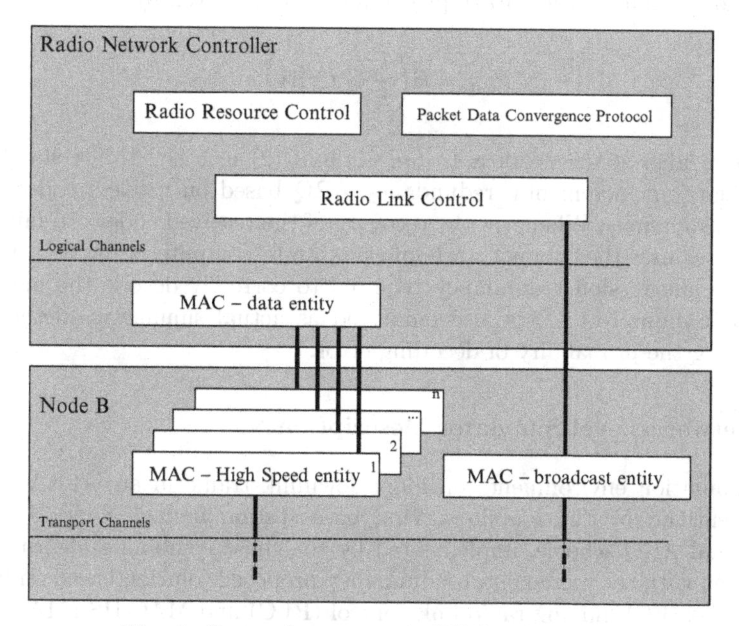


**Fig. 1.** Protocol structure in RNC and Node B

# 3 Problem Formulation

## 3.1 Rateless Coding Modelling

Recent works show that rateless codes may be used as an alternative of Hybrid ARQ schemes [4], especially for communication systems with delay constraints. Fountain [5] and Raptor codes [6] are the representatives of such techniques available in literature, and are considered rateless, in the sense that the number of encoded symbols that can be generated from the source message is potentially limitless. Furthermore, in presence of good channel conditions, a little redundancy may be enough to recover the transmitted message. Considering the transmissions of coded messages over a block fading channel, the realized capacity of a sequence $\mathbf{h} = (h_1, h_2, \ldots, h_n)$ of channel state information, may be computed as [4]

$$C(\mathbf{h}) = \frac{1}{n} \sum_{i=1}^{n} \log_2(1 + \gamma |h_i|^2) \;, \tag{1}$$

where $\gamma$ is the transmitted SINR, including the fading multiplicative noise, while $C(\mathbf{h})$ may be viewed as the transmission rate supported by $\mathbf{h}$. When using a codeword of length $n$ to transmit $k$ symbols of information, the theoretical lower limit of the outage probability may be given by [4]

$$P_{\text{out}} = P\left(\frac{k}{n} > C(\mathbf{h})\right) \,. \tag{2}$$

The main idea of this work is to use (1) and (2) in a cell traffic simulator, to evaluate an incremental redundancy ARQ based on rateless coded message transmissions. Whenever the recovery of the received codeword fails, the transmitter uses the channel state information to compute, using (1), the additional transmission redundancy required to correctly decode the message. The lower bound of (2) is also employed as actual simulation method for computing the probability of decoding error.

## 3.2 Network-Level Simulator Description

The simulation environment considers downlink traffic in an UTRAN LTE cell, identified by the mobile serving base station named Node B [7]. A traditional ARQ scheme, implemented by the chase combining algorithm, is compared with the incremental redundancy proposed solution based on Fountain Codes (FC), mixing radio link control (RLC) and MAC-HS (MAC-High Speed) ARQ mechanisms. The simulator consists of four basic blocks as shown in Fig. 2.

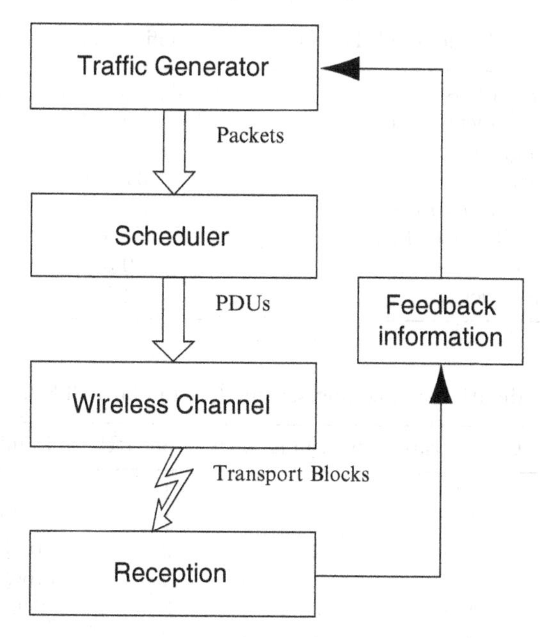


**Fig. 2.** Cell traffic simulator scheme

First of all, a traffic generator produces packets according to some common profiles such as CBR, web traffic or speech activity. Packets are then passed to the second block, the scheduler, which has to select the users to be served adopting two possible scheduling procedures: round robin or proportional fair. Another task accomplished by this block is the segmentation of packets in PDUs, before transmitting them over the radio channel. The PDU size is quite different if a FC approach is used instead of a traditional HARQ one.

In this case, all PDUs have a fixed constant size adapted to fit a resource unit (RU, corresponding to 25 sub-carriers) for a given code rate and modulation order: for example, if a code rate 2/3 and QPSK are used, the PDU size is 200 bits. Depending on the channel state information (CSI), PDUs are eventually grouped to form a transport block, then protected by turbocoding.

On the contrary, using HARQ based on a FC, the size of the PDU is decided every TTI depending on the rate estimated for a successful transmission at the current CSI. If the rate requested is lower than one third, a more robust modulation is chosen, if available.

The third block refers to the transmission of PDUs over a block fading wireless channel, where fading attenuation values are updated every TTI, taking into account a correlation coefficient. This model follows the specifications [8, 9], reported in Table 1. Finally, the last block has to decide whether the received codewords are correct or erroneous and it has also to provide the feedback information. In the traditional HARQ scheme simulation, the decision process needs to know the probability of error, analytically evaluated

**Table 1.** Cell environment specifications

| | |
|---|---|
| Average path loss | $128.1 + 37.6 * \log_{10}(d)$; $d$ in km |
| Slow fading, mean value | 0 dB |
| Slow fading, variance | 5 dB |
| Mobile speed | $3\,\mathrm{km\,h}^{-1}$ |
| Number of sub-carriers per RU | 25 |
| Number of RUs per TTI | 48 |
| TTI duration | 0.5 ms |
| Total TX Power | 25 W |

**Table 2.** Modulation and coding scheme for standard LTE hybrid ARQ

| $E_b/N_0$ | Coding rate | Res. unit | PDU per RU | Modulation |
|---|---|---|---|---|
| −0.52 | 1/3 | 2 | 1 | QPSK |
| 3.15 | 2/3 | 1 | 1 | QPSK |
| 4.7 | 1/3 | 2 | 2 | 16QAM |
| 6.7 | 2/3 | 1 | 2 | 16QAM |
| 7.35 | 1/3 | 2 | 3 | 64QAM |
| 8.65 | 2/3 | 1 | 3 | 64QAM |

considering the modulation used and a corrective factor due to error protection coding. The SNR represents the CSI and it is sent back to the base station, in order to determine from a finite and discrete set of modulation and coding schemes (see Table 2) the one suitable for the next transmission. An eventual retransmission involves the entire transport block.

In the HARQ scheme simulation based on FC, the error probability is not needed, instead the realized capacity of the wireless medium is calculated by using (1). If the rate used to transmit a PDU is lower than the capacity, it is assumed that the applied redundancy was sufficient to ensure a correct reception. In the other case, there is no need to entirely retransmit the PDU, it is sufficient to send a few redundancy bits to reduce the total codeword rate below the capacity. Remaining bit positions in the TTI frame can be used for new data, protected accordingly to current CSI. The return channel information is the computed realized capacity, so new PDUs may be sent with the right redundancy, based on the last CSI estimation, to avoid errors.

# 4 Simulation Results

The two HARQ approaches have been simulated in the same channel conditions reported in Table 1. The input traffic considered represents a very worst case with 150 users transmitting at a constant rate of $350\,\mathrm{Kb\,s}^{-1}$.

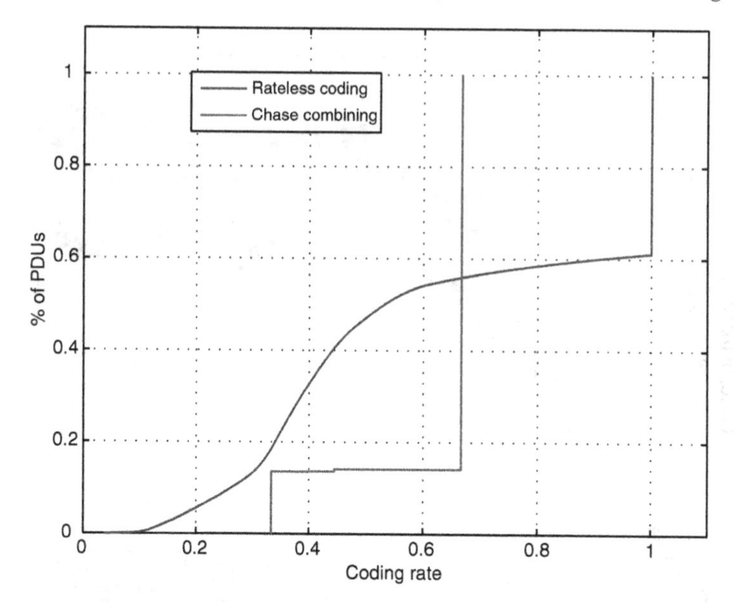


**Fig. 3.** CDF of coding rate

In order to better explain the differences between the two HARQ algorithms, the number of transmitted PDUs vs. the coding rate is plotted in Fig. 3. As it can be seen, traditional HARQ can only exploit two values, namely 1/3 and 2/3, while FC coding rate varies with continuity from 0.1 to 1.

The total goodput of the base station is higher using the new approach as reported in Fig. 4. Using a fixed scheme, a rate of 2/3 has to be used at least, while, in the proposed one, only the really needed redundancy is sent. If the CSI at the transmitter is good enough, data can be sent with low protection or even unprotected. Another consequence of a more efficient use of resources is the reduction of inter-arrival time, as it can be seen by comparing results of Fig. 5, showing cumulative density distribution of time intervals between the beginning of a PDU transmission and its correct reception.

## 5 Conclusions

Simulation results show that the novel proposed HARQ scheme may be applied to UTRAN LTE systems, obtaining a substantial gain with respect to a more traditional approach. Furthermore, the idea of using channel capacity computation as a simplified model for network aspect simulations, seems to be promising. Ongoing works are devoted to try using network coding theoretic analyses in order to compute asymptotic performance limits of the proposed technique.

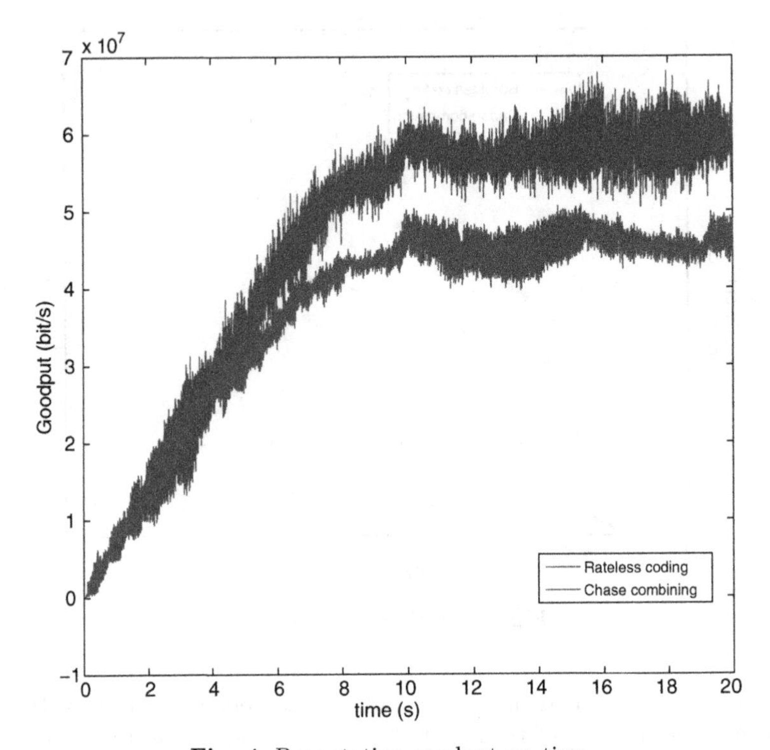


**Fig. 4.** Base station goodput vs. time

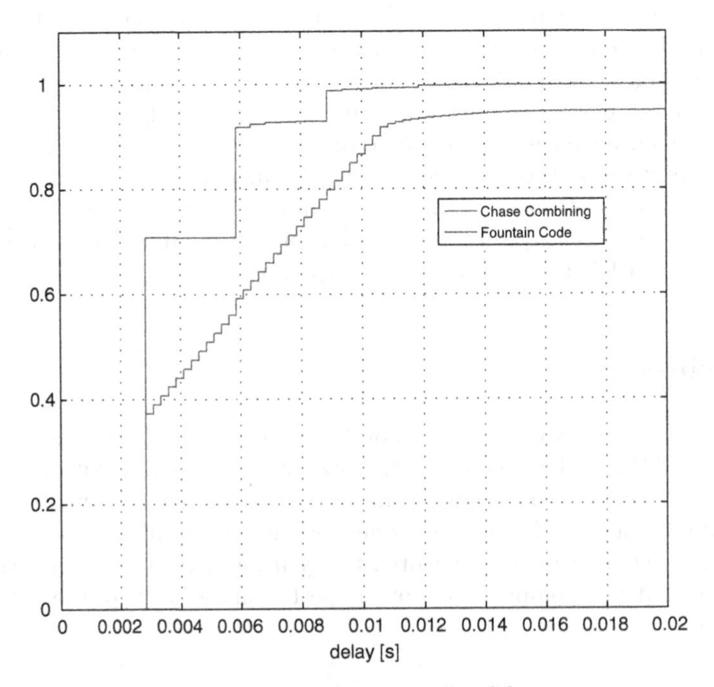


**Fig. 5.** CDF of transmission delays

## References

1. Luby M (2002) LT Codes. Proc. of 43rd IEEE Symposium on Foundations of Computer Science (FOCS), November 16–19: 271–280
2. Ekström H, Furuksär A, Karlsson J, Meyer M, Parkvall S, Torsner J, Wahlqvist M (2006) Technical Solutions for the 3G Long-Term Evolution. IEEE Commun. Mag., Vol. 44, Issue 3, March: 38–45
3. 3GPP TR 25.308, v. 5.2.0, "High Speed Downlink Packet Access (HSDPA) Overall description," release 5. http://www.3gpp.org
4. Castura J (2006) Rateless coding over fading channels. IEEE Commun. Letters, Vol. 10, No. 1, Jan.: 46–48
5. MacKay DJC (2005) Fountain codes. IEEE Proc. Commun., Vol. 152, No.6, December: 1062–1068
6. Shokrollahi A (2006) Raptor codes. IEEE Trans. Information Theory, Vol. 52, No.6, June: 2552–2567
7. 3GPP TR 25.814, v. 7.1.0, "Physical layer aspects for evolved universal terrestrial radio access (UTRA)," release 7. http://www.3gpp.org
8. Recommendation ITU-R M.1225, Guidlines for evaluation of radio transmission technologies FOR IMT-2000 (Question ITU-R 39/8)
9. Jeon WS, Jeong DJ, Kim B (2004) Packet Scheduler for Mobile Internet Services Using High Speed Downlink Packet Access. IEEE Trans. on Wireless Commun., Vol. 3, No. 5, September: 1789–1801

4

# On the Performance of Transmit Antenna Selection with OSTBC in Ricean MIMO Channels

Saeed Kaviani[1], Chintha Tellambura[1], and Witold A. Krzymień[1,2]

[1] Department of Electrical and Computer Engineering, University of Alberta Edmonton, Alberta, Canada T6G 2V4 {saeedk,chintha,wak}@ece.ualberta.ca
[2] Also with TRLabs, Edmonton, Alberta, Canada

**Summary.** In this chapter, the performance of orthogonal space–time block codes (OSTBCs) with transmit antenna selection (TAS) in a Ricean fading channel is analyzed. Out of the total of $L_t$ transmit antennas, the receiver selects $N$ antennas that maximize the received signal-to-noise ratio (SNR). A low rate feedback channel from the receiver to the transmitter is available to convey the indices of the selected transmit antennas. Bit error rate (BER) of Gray coded $M$-ary modulations is derived. Asymptotic performance analysis (high SNR) of the scheme is also carried out. We also present BER approximations and derive the diversity order and coding gain of the system. It is shown that the full diversity order of OSTBC is preserved in a Ricean fading environment when TAS is employed.

**Key words:** Antenna selection, diversity, multiple-input multiple-output systems, Ricean channel, space–time codes, bit error rate performance

## 1 Introduction

The performance of wireless communication systems operating in fading environments can be improved substantially by using multiple transmit and receive antennas [1–4]. However, the increase in the number of antennas results in high hardware complexity, due to the requirement for multiple radio frequency (RF) chains (consisting of power amplifiers, mixers, and analog/digital converters) associated with the multiple transmit–receive antenna pairs. Thus, antenna selection, where transmission and/or reception is performed by selecting a subset of antennas, is a promising approach to achieve the goal of providing multiple-antenna performance benefits while significantly reducing system complexity.

It is conventional wisdom that antenna selection at the receive side reduces complexity. More interestingly, at the transmit side, suitable antenna selection not only reduces complexity, but also improves the performance of multiple-input multiple-output (MIMO) systems [5] at the cost of a minimal amount of feedback. Moreover, full diversity space–time codes can be used together with antenna selection.

To improve the reliability of transmission, several transmit diversity techniques have been proposed, such as orthogonal space–time block codes (OSTBCs) [2, 6, 7]. OSTBCs are particularly attractive since they not only extract the maximum spatial diversity order, but also decouple the vector detection of the transmitted symbols into a set of scalar detection problems, thereby significantly reducing decoding complexity.

Transmit antenna selection (TAS) has recently sparked significant interest due to its relatively low complexity. The performance of TAS has been investigated by several authors. In [8] it is shown that full diversity can be achieved via antenna selection provided that there is no feedback error. This chapter also develops a criterion for optimal antenna subset selection at high signal-to-noise ratio (SNR). Symbol error rate (SER) performance of TAS combined with OSTBCs is analyzed in [9]. The asymptotic BER performance of single TAS and receive maximal-ratio combining (MRC) with generalized selection criterion is investigated in [10]. The exact BER performance of BPSK using the Alamouti code is derived in [11]. In [12], the scheme combining single transmit antenna selection and receiver maximal-ratio combining (the TAS/MRC scheme) is analyzed. The performance analysis of the Alamouti scheme with only two transmit antennas selection under uncorrelated Rayleigh fading channels is given in [13] for $M$-ary signals. In [14, 15], the closed-form BER expressions for independent/receive-correlated Rayleigh fading channels are given.

TAS performance over Ricean channels has so far not been available in the literature. We therefore present a theoretical performance analysis of TAS/OSTBC in a Ricean fading environment, and quantify the asymptotic performance at high SNR by deriving the diversity order and coding gain. We show that the diversity order with TAS is preserved for OSTBCs in Ricean fading channels with full spatial diversity.

*Notation*: The Frobenius norm of matrix $\mathbf{A}$ is denoted by $\|\mathbf{A}\|_{\mathrm{F}}$ and the Euclidean norm for vector $\mathbf{h}$ is $\|\mathbf{h}\| = (h_1^2 + \cdots + h_{L_\mathrm{r}}^2)^{1/2}$. A circularly symmetric complex Gaussian variable with mean $\mu$ and variance $\sigma^2$ is denoted by $z \sim \mathcal{CN}(\mu, \sigma^2)$. $\mathbb{C}$ represents the set of complex numbers. $\mathbb{E}_X[\cdot]$ is the expectation operator over a random variable $X$.

## 2 System Model

We consider a multiple-antenna wireless communication system with $L_\mathrm{t}$ transmit antennas and $L_\mathrm{r}$ receive antennas that operates over a Ricean flat fading channel. The channel is assumed constant over $T$ symbol periods. Full channel state information (CSI) is available at the receiver. $N$ $(N \leq L_\mathrm{t})$

transmit antennas out of $L_t$ are activated for the transmission of OSTBC signal matrices, while the remaining transmit antennas are inactive. The selection is performed using the receiver feedback. The indices of the selected transmit antennas are available through a low rate feedback channel from the receiver. In the presence of a line of sight (LOS) component between the transmitter and receiver, the MIMO channel, $\mathbf{H} \in \mathbb{C}^{L_r \times L_t}$, can be modeled by a sum of a zero mean complex Gaussian channel matrix and a fixed component matrix [16],

$$\mathbf{H} = \sqrt{\frac{K}{1+K}}\bar{\mathbf{H}} + \sqrt{\frac{1}{1+K}}\mathbf{H}_w \ , \tag{1}$$

where $\sqrt{K/(1+K)}\bar{\mathbf{H}} = \mathbb{E}[\mathbf{H}]$ is the fixed LOS component of the channel model and $\sqrt{1/(1+K)}\mathbf{H}_w$ is the fading component where $\mathbf{H}_w \in \mathbb{C}^{L_r \times L_t}$ is spatially white complex Gaussian channel matrix with entries $h_{ij} \sim \mathcal{CN}(0,1)$ being channel gains between the $i$th transmit and $j$th receive antennas. The entries of $\bar{\mathbf{H}} \in \mathbb{C}^{L_r \times L_t}$ are normalized to unit power (the magnitudes of all entries in $\bar{\mathbf{H}}$ are equal to one). $K$ is the Ricean factor of the MIMO channel. When $K = 0$, the channel becomes a Rayleigh fading MIMO channel. $\mathbf{H}_{\text{TAS}}$ consists of the channel gains for the $N$ selected transmit antennas and $L_r$ receive antennas. Thus, $\mathbf{H}_{\text{TAS}} \in \mathbb{C}^{L_r \times N}$ is a submatrix of the channel matrix $\mathbf{H}$.

Assume that $Q$ symbols $s_1, \ldots, s_Q$ with average energy equal to one, chosen from an $M$-ary signal constellations, are transmitted within the transmission matrix $\mathbf{X} \in \mathbb{C}^{N \times T}$ through $N$ selected antennas and over $T$ time slots. $\mathbf{X}$ is constructed as a form of OSTBCs introduced in [6, 7]. Hence, $Q$ different symbols are transmitted over $T$ time slots which means the symbol rate $R_s$ is defined as $R_s = Q/T$. The received signals are expressed as

$$\mathbf{Y} = \sqrt{\frac{E_s}{N}}\mathbf{H}_{\text{TAS}}\mathbf{X} + \mathbf{V} \ , \tag{2}$$

where $\mathbf{Y} \in \mathbb{C}^{L_r \times T}$ is the complex received signal matrix. $\mathbf{V} \in \mathbb{C}^{L_r \times T}$ is the additive noise matrix with independent and identically distributed $\mathcal{CN}(0, N_0)$ entries. The coefficient $\sqrt{E_s/N}$ ensures that the total transmitted energy per channel use is $E_s$ and it is independent of the number of transmit antennas.

When an OSTBC is used, the MIMO system is equivalent to $Q$ independent single-input single-output (SISO) systems defined as [6, 7]

$$r_q = \sqrt{\frac{E_s}{N}}\left(\frac{1}{R_s}\|\mathbf{H}_{\text{TAS}}\|_{\text{F}}^2\right) s_q + \nu_q, \ q = 1, \ldots, Q \ , \tag{3}$$

where $\nu_q \sim \mathcal{CN}\left(0, \frac{1}{R_s}\|\mathbf{H}_{\text{TAS}}\|_{\text{F}}^2 N_0\right)$ is the noise and $r_q$ is the received output. We conclude that the instantaneous received SNR per bit with an $M$-ary constellation is

$$\gamma_b(\rho) = \frac{E_s}{N_0} \frac{1}{R_s N \log_2 M} \|\mathbf{H}_{\text{TAS}}\|_{\text{F}}^2 = c\rho \|\mathbf{H}_{\text{TAS}}\|_{\text{F}}^2 , \quad (4)$$

where $\rho = \frac{E_s}{N_0}$ and $c = 1/(R_s N \log_2 M)$.

To maximize the instantaneous received SNR per bit in (4), transmit antennas must be selected to maximize $\|\mathbf{H}_{\text{TAS}}\|_{\text{F}}$. Suppose that $\mathbf{h}_j$ ($j = 1, 2, \ldots, L_{\text{t}}$) are columns of the channel matrix $\mathbf{H}$. The columns are sorted according to their norms $\|\mathbf{h}_{i_1}\| \geq \cdots \geq \|\mathbf{h}_{i_{L_{\text{t}}}}\|$, where $i_k \in \{1, 2, \ldots, L_{\text{t}}\}$, and $\mathbf{H}_{\text{TAS}}$ is formed as follows

$$\mathbf{H}_{\text{TAS}} = [\mathbf{h}_{i_1} \mathbf{h}_{i_2} \cdots \mathbf{h}_{i_N}]. \quad (5)$$

Therefore, the antenna selection criterion selects $N$ transmit antennas to maximize the instantaneous received SNR per bit and hence minimizes the error rate.

Let $\gamma_k = c\rho \|\mathbf{h}_k\|^2$, $k = 1, 2, \ldots, L_{\text{t}}$, be the scaled norms of the columns of $\mathbf{H}$. In an uncorrelated flat Ricean MIMO channel, $\gamma_k$s are sums of squared magnitudes of $2L_{\text{r}}$ statistically independent and Gaussian distributed random variables with non-zero mean and identical variance $c\rho/2(1+K)$. Thus, the received SNR corresponding to the $k$th transmit antenna is a non-central chi-squared distributed random variable with $2L_{\text{r}}$ degrees of freedom and the probability density function (pdf) [17]

$$f_{\gamma_k}(x) = \frac{1+K}{c\rho} \left\{ \frac{x(1+K)}{L_{\text{r}} c\rho K} \right\}^{\frac{L_{\text{r}}-1}{2}} \mathrm{e}^{-L_{\text{r}} K} \mathrm{e}^{-\frac{(1+K)x}{c\rho}} \times I_{L_{\text{r}}-1}\left( 2\sqrt{\frac{L_{\text{r}} K(1+K)x}{c\rho}} \right), \quad x \geq 0 \quad (6)$$

where $I_n(\cdot)$ is the modified Bessel function of the first kind represented by

$$I_n(x) = \left(\frac{x}{2}\right)^n \sum_{k=0}^{\infty} \frac{x^{2k}}{2^{2k} k! \Gamma(n+k+1)}. \quad (7)$$

The cumulative distribution function (cdf) of the non-central chi-square random variable with $2L_{\text{r}}$ degrees of freedom can be expressed as [17]

$$F_{\gamma_k}(x) = 1 - \mathcal{Q}_{L_{\text{r}}}\left( \sqrt{2L_{\text{r}} K}, \sqrt{\frac{2x(1+K)}{c\rho}} \right), \quad (8)$$

where $\mathcal{Q}$ is the generalized Marcum's function defined as [17]

$$\mathcal{Q}_{L_{\text{r}}}(\alpha, \beta) = \frac{1}{\alpha^{L_{\text{r}}-1}} \int_{\beta}^{\infty} y^{L_{\text{r}}} \mathrm{e}^{-\frac{y^2+\alpha^2}{2}} I_{L_{\text{r}}-1}(\alpha y) \mathrm{d}y. \quad (9)$$

From (4) and (5), the received SNR per bit can be written as

$$\gamma_b(\rho) = \sum_{k=1}^{N} \gamma_{(k)}(\rho), \tag{10}$$

where $\gamma_{(k)}(\rho) = c\rho\|\mathbf{h}_{i_k}\|^2$. The moment generating function (MGF) of $\gamma_b$ is given by [18]

$$\Phi_{\gamma_b}(s) = N\binom{L_t}{N}\int_0^{\infty} \mathrm{e}^{-sx} f_{\gamma_k}(x)[F_{\gamma_k}(x)]^{L_t-N}\left[\phi(s,x)\right]^{N-1} \mathrm{d}x, \tag{11}$$

where

$$\phi(s,x) = \int_x^{\infty} \mathrm{e}^{-st} f_{\gamma_k}(t)\mathrm{d}t. \tag{12}$$

Using the definition of generalized Marcum's $\mathcal{Q}$-function in (9) we can obtain a simplification of

$$\phi(s,x) = \frac{\exp\left\{-\frac{sL_r c\rho K}{1+K+sc\rho}\right\}}{\left(1+s\frac{c\rho}{1+K}\right)^{L_r}} \mathcal{Q}_{L_r}\left(\frac{\sqrt{2L_r K}}{1+s\frac{c\rho}{1+K}}, \sqrt{\frac{2x}{c\rho}(1+K+sc\rho)}\right). \tag{13}$$

Thus, the MGF can be obtained by substitution of $f_{\gamma_k}(x)$, $F_{\gamma_k}(x)$, and $\phi(s,x)$ into (11).

# 3 BER Analysis of $M$-ary Constellations

We first derive the BER for $M$-ary PAM with antenna selection and OSTBCs using Gray mapping. Then, using the BER results for $M$-PAM, we find the BER for $M$-QAM and an approximation for $M$-PSK.

## 3.1 Exact BER for $M$-ary PAM

In an additive white Gaussian noise (AWGN) channel, the exact BER of the $n$th bit in $M$-PAM with Gray mapping is given by [19]

$$P_M^{\mathrm{AWGN}}(n;\rho) = \frac{2}{M}\sum_{i=0}^{k_n} B_i \mathcal{Q}\left(D_i\sqrt{\gamma_b(\rho)}\right), \tag{14}$$

where

$$k_n = \left(1-\frac{1}{2^n}\right)M - 1, \tag{15a}$$

$$B_i = (-1)^{\left\lfloor \frac{i.2^{n-1}}{M}\right\rfloor}\left(2^{n-1} - \left\lfloor \frac{i.2^{n-1}}{M} + \frac{1}{2}\right\rfloor\right), \tag{15b}$$

$$D_i = (2i+1)\sqrt{\frac{6\log_2 M}{M^2-1}}. \tag{15c}$$

Thus, to obtain the average BER in a MIMO fading channel, we take the expectation with respect to the channel statistics

$$\begin{aligned} P_M(n;\rho) &= \frac{2}{M}\sum_{i=0}^{k_n} B_i \mathbb{E}_{\mathbf{H}_{\mathrm{TAS}}}\left[\mathcal{Q}\left(D_i\sqrt{\gamma_b(\rho)}\right)\right] \\ &= \frac{2}{M}\sum_{i=0}^{k_n} B_i \frac{1}{\pi}\int_0^{\pi/2} \Phi_{\gamma_b}\left(-\frac{D_i^2}{2\sin^2\theta}\right) \mathrm{d}\theta, \end{aligned} \tag{16}$$

where an equivalent form of the Gaussian $\mathcal{Q}(x)$ given in [20]

$$\mathcal{Q}(x) = \frac{1}{\pi}\int_0^{\pi/2} \exp\left(-\frac{x^2}{2\sin^2\theta}\right) \mathrm{d}\theta \tag{17}$$

has been used to obtain (16).

The exact average BER of TAS scheme with OSTBC in $M$-PAM is given by

$$P_M(\rho) = \frac{1}{\log_2 M}\sum_{n=1}^{\log_2 M} P_M(n;\rho). \tag{18}$$

### 3.2 Exact BER for $M$-ary QAM

Noting that rectangular or square QAM constellations can be composed of two independent PAM constellations [17, 19], we can easily find the average BER of an $M$-QAM constellation using the results of Sect. 3.1. Consider two independent PAM constellations: $I$-ary PAM for the in-phase component and $J$-ary PAM for the quadrature component, where $M = I \times J$. Then, using the BER result for $M$-PAM in (18), exact average BER of $M$-QAM is given by

$$P_M(\rho) = \frac{1}{\log_2(IJ)}\left(\sum_{n=1}^{\log_2 I} P_I(n;\rho) + \sum_{m=1}^{\log_2 J} P_J(m;\rho)\right). \tag{19}$$

### 3.3 Approximate BER for $M$-ary PSK

A tight approximation for the BER of the coherent $M$-ary PSK in AWGN channels is given by [21]

$$P_M^{\mathrm{AWGN}}(\rho) \simeq \frac{2}{\mathcal{M}}\sum_{i=1}^{\max(M/4,1)} \mathcal{Q}\left(\sqrt{2\gamma_b(\rho)}\sin\frac{(2i-1)\pi}{M}\right), \tag{20}$$

where $\mathcal{M} = \max(\log_2 M, 2)$. Hence, by using the expression for $\mathcal{Q}$-function in (17) and averaging over the TAS sub-channel, similar to the results in (16), the BER can be obtained as

$$P_M(\rho) \simeq \frac{2}{\mathcal{M}}\sum_{i=1}^{\max(M/4,1)} \int_0^{\pi/2} \Phi_{\gamma_b}\left(-\frac{\sin^2\frac{(2i-1)\pi}{M}}{\sin^2\theta}\right) \mathrm{d}\theta. \tag{21}$$

# 4 Asymptotic Performance Analysis

In this section, we derive the approximate BER expressions to evaluate the asymptotic performance of TAS with OSTBCs in Ricean fading.

We write a function $g(x)$ as $o(x^r)$ when $x \to x_0$ if

$$\lim_{x \to x_0} \frac{g(x)}{x^r} = 0. \tag{22}$$

For high SNR, if the BER can be written as

$$P_M(\rho) = (G_c \rho)^{-G_d} + o(\rho^{-G_d}), \quad \rho \to \infty, \tag{23}$$

we call $G_d$ the diversity order, and $G_c$ the coding gain of system. The diversity order determines the slope of the average BER curve vs. the average SNR $\rho$ at high SNR on a log–log scale, whereas the coding gain (in dB) determines the shift of the curve in SNR relative to a BER curve given by $\rho^{-G_d}$.

In [22], Wang and Giannakis developed a simple and general method to quantify the asymptotic performance of wireless transmission in fading channels. They showed that the asymptotic performance depends on the behavior of the pdf of the achievable SNR and thus on its MGF. If the MGF, $\Phi_{\gamma_b}(s)$, can be approximated by a single polynomial term for $s \to \infty$ as [22]

$$|\Phi_{\gamma_b}(s)| = b|s|^{-G_d} + o(s^{-G_d}) \tag{24}$$

then $G_d$ is the diversity order of the system at high SNRs and the coding gain can be derived from $b$ [22]. Thus, to obtain the coding gain and diversity order of our system, we need to approximate MGF as $s \to \infty$.

For small values of its argument, the modified Bessel function can be written as [20]

$$I_{L_r-1}(z) = \frac{(z/2)^{L_r-1}}{(L_r-1)!} + o(z^{L_r-1}), \quad z \to 0. \tag{25}$$

Hence, at $x \to 0$, the pdf of the received SNR corresponding to the $k$th transmit antenna (6) can be approximated as follows

$$f_{\gamma_k}(x) = \frac{\mathrm{e}^{-L_r K}(1+K)^{L_r}}{(L_r-1)!(c\rho)^{L_r}} x^{L_r-1} + o(x^{L_r-1}). \tag{26}$$

Hence, $\phi(s,x)$ in (11) becomes

$$\phi(s,x) \simeq \frac{\mathrm{e}^{-L_r K}\mathrm{e}^{-x\left(s+\frac{1+K}{c\rho}\right)}(1+K)^{L_r}}{\left(s+\frac{1+K}{c\rho}\right)^{L_r}(c\rho)^{L_r}} \sum_{k=0}^{L_r-1} \frac{x^k\left(s+\frac{1+K}{c\rho}\right)^k}{k!}. \tag{27}$$

Another form of the Marcum $\mathcal{Q}$-function proposed by [23] is the following

$$\mathcal{Q}_{L_\mathrm{r}}(\alpha,\beta) = \mathrm{e}^{-\frac{\alpha^2}{2}} \sum_{n=0}^{\infty} \frac{1}{n!}\left(\frac{\alpha^2}{2}\right)^n \mathrm{e}^{-\frac{\beta^2}{2}} \sum_{k=0}^{n+L_\mathrm{r}-1} \frac{1}{k!}\left(\frac{\beta^2}{2}\right)^k . \tag{28}$$

Thus, for $x \to 0$, the cdf (8) of the received SNR, $\gamma_k$, can be approximated as follows

$$\begin{aligned} F_{\gamma_k}(x) &= 1 - \mathrm{e}^{-L_\mathrm{r}K} \sum_{n=0}^{\infty} \frac{(KL_\mathrm{r})^n}{n!} \mathrm{e}^{-\frac{(1+K)x}{c\rho}} \sum_{k=0}^{n+L_\mathrm{r}-1} \frac{\left[\frac{x(1+K)}{c\rho}\right]^k}{k!} \\ &= \frac{\mathrm{e}^{-L_\mathrm{r}K}(1+K)^{L_\mathrm{r}}}{(c\rho)^{L_\mathrm{r}} L_\mathrm{r}!} x^{L_\mathrm{r}} + o(x^{L_\mathrm{r}}), \quad x \to \infty \end{aligned} \tag{29}$$

where we have used

$$\mathrm{e}^{-z} \sum_{k=0}^{n-1} \frac{z^k}{k!} = 1 - \frac{z^n}{n!} + o(z^n). \tag{30}$$

In order to obtain the approximation of the BER expressions and quantify diversity order and coding gain, we need to present the MGF in the form of (24). Therefore, we ignore the $o(x^{L_\mathrm{r}})$ term in the approximation above, and similarly to the results in [14, 15] we get

$$\begin{aligned} \Phi_{\gamma_b}(s) = \frac{N\binom{L_\mathrm{t}}{N} L_\mathrm{r} \mathrm{e}^{-L_\mathrm{t}L_\mathrm{r}K}(1+K)^{L_\mathrm{r}L_\mathrm{t}}}{(c\rho)^{L_\mathrm{r}L_\mathrm{t}}(L_\mathrm{r}!)^m N^{L_\mathrm{r}m}} \underbrace{\left[\sum_{j=0}^{(L_\mathrm{r}-1)(N-1)} \frac{a_j\,(L_\mathrm{r}m+j-1)!}{N^j}\right]}_{\Delta_N} \\ \times \frac{1}{(s+\frac{1+K}{c\rho})^{L_\mathrm{r}L_\mathrm{t}}} + o(s^{-L_\mathrm{r}L_\mathrm{t}}), \end{aligned} \tag{31}$$

where $m = L_\mathrm{t} - N + 1$ and $a_j$ is the coefficient of $x^j$, $j = 0, 1, \ldots, (L_\mathrm{r} - 1)(N - 1)$, in the expansion

$$\left[\sum_{k=0}^{L_\mathrm{r}-1} \frac{x^k}{k!}\right]^{N-1} . \tag{32}$$

$\Delta_N$ is a parameter depending on $N$, the number of transmit antennas selected. We substitute the approximate MGF into (16) to approximate the BER as

$$P_M(\rho) \simeq \frac{2N\binom{L_\mathrm{t}}{N}}{M\log_2 M} \cdot \frac{L_\mathrm{r}\mathrm{e}^{-L_\mathrm{t}L_\mathrm{r}K}\Delta_N}{(L_\mathrm{r}!)^m N^{L_\mathrm{r}m}} \sum_{n=1}^{\log_2 M} \sum_{i=0}^{k_n} B_i J_{L_\mathrm{r}L_\mathrm{t}}\left(\frac{D_i^2 c\rho}{2(1+K)}\right), \tag{33}$$

where [24, 25]

$$
\begin{aligned}
J_m(\nu) &= \frac{1}{\pi}\int_0^{\frac{\pi}{2}}\left(\frac{\sin^2\theta}{c+\sin^2\theta}\right)^m \mathrm{d}\theta \\
&= \frac{1}{2^m}\left[1-\sqrt{\frac{\nu}{1+\nu}}\right]^m \sum_{k=0}^{m-1}\frac{\binom{m-1+k}{k}}{2^k}\left[1+\sqrt{\frac{\nu}{1+\nu}}\right]^k . \qquad (34)
\end{aligned}
$$

For the purpose of finding the approximate MGF, we employ the following limit [24]

$$
\lim_{\nu\to\infty} J_m(\nu) = \binom{2m}{m}\frac{1}{2^{2m+1}}\frac{1}{\nu^m}. \qquad (35)
$$

Utilizing this limit, we find

$$
\begin{aligned}
\lim_{\rho\to\infty} P_M(\rho) &= \frac{N\binom{L_\mathrm{t}}{N}\binom{2L_\mathrm{r}L_\mathrm{t}}{L_\mathrm{r}L_\mathrm{t}}L_\mathrm{r}\mathrm{e}^{-L_\mathrm{r}L_\mathrm{t}K}\Delta_N}{M\log_2 M (L_\mathrm{r}!)^m N^{L_\mathrm{r}m}} \\
&\times \underbrace{\left[\sum_{n=1}^{\log_2 M}\sum_{i=0}^{k_n}\frac{B_i}{D_i^{2L_\mathrm{r}L_\mathrm{t}}}\right]}_{\Lambda_M}\frac{(1+K)^{L_\mathrm{r}L_\mathrm{t}}}{2^{L_\mathrm{r}L_\mathrm{t}}(c\rho)^{L_\mathrm{r}L_\mathrm{t}}} \qquad (36)
\end{aligned}
$$

where $\Lambda_M$ is a parameter depending on $M$, for $M$-ary modulation. Thus, comparing (36) to (23), the coding gain and diversity order of system can be obtained as

$$
\begin{aligned}
G_\mathrm{c} &= \left\{\frac{\binom{L_\mathrm{t}}{N}\binom{2L_\mathrm{r}L_\mathrm{t}}{L_\mathrm{r}L_\mathrm{t}}}{M\log_2 M}\cdot\frac{L_\mathrm{r}\mathrm{e}^{-L_\mathrm{r}L_\mathrm{t}K}\Delta_N\Lambda_M(1+K)^{L_\mathrm{r}L_\mathrm{t}}}{(L_\mathrm{r}!)^m N^{L_\mathrm{r}m-1}2^{L_\mathrm{r}L_\mathrm{t}}c^{L_\mathrm{r}L_\mathrm{t}}}\right\}^{-\frac{1}{G_\mathrm{d}}} \\
G_\mathrm{d} &= L_\mathrm{r}L_\mathrm{t}. \qquad (37)
\end{aligned}
$$

Clearly, if $N$ antennas corresponding to the largest received SNRs are selected, a full diversity order of $L_\mathrm{r}L_\mathrm{t}$ is obtained.

## 5 Numerical Results

In this section, numerical results are given for the selection of $N = 2$ transmit antennas. Alamouti scheme [2] is used over the selected antennas. In Figs. 1 and 2, the analytical approximations and the simulation results are compared; two transmit antennas out of $L_\mathrm{t} = 3$ and $L_\mathrm{t} = 4$ are selected. For the MIMO Ricean fading channels, Ricean $K$-factor $K = 1$ and $K = 2.5$ are considered and 16-QAM signal constellation is used. For the simulation examples and the numerical results, the fixed component matrix of the Ricean channel in (1) is given by $\bar{\mathbf{H}} = \mathbf{1}_{L_\mathrm{r}\times L_\mathrm{t}}$, where $\mathbf{1}_{L_\mathrm{r}\times L_\mathrm{t}}$ is a matrix of size $L_\mathrm{r}\times L_\mathrm{t}$ with all entries equal to one.

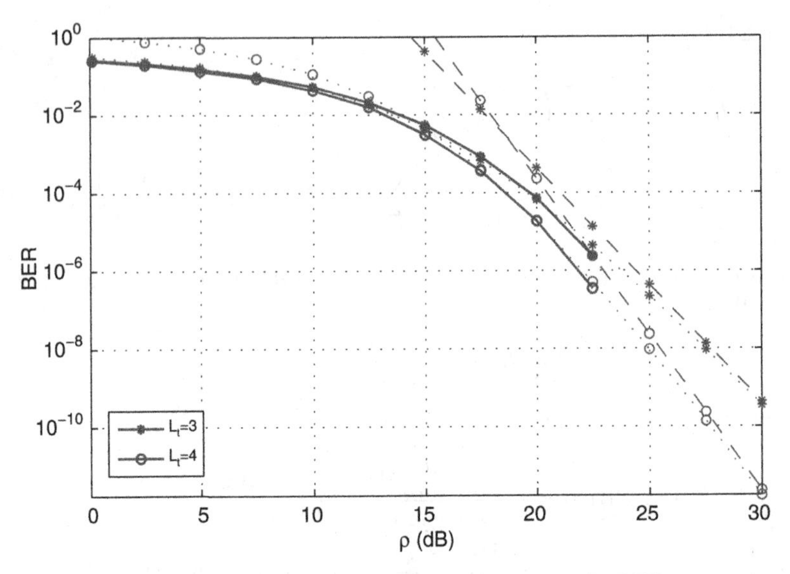


**Fig. 1.** Comparison of BER simulation results (*solid lines*), BER approximation given in (33), (*dashed lines*), and BER approximation given in (36), (*dotted lines*). Two antennas out of $L_t = 3, 4$ transmit antennas are selected and Alamouti scheme is used. The MIMO channel is Ricean flat fading with $K = 1$

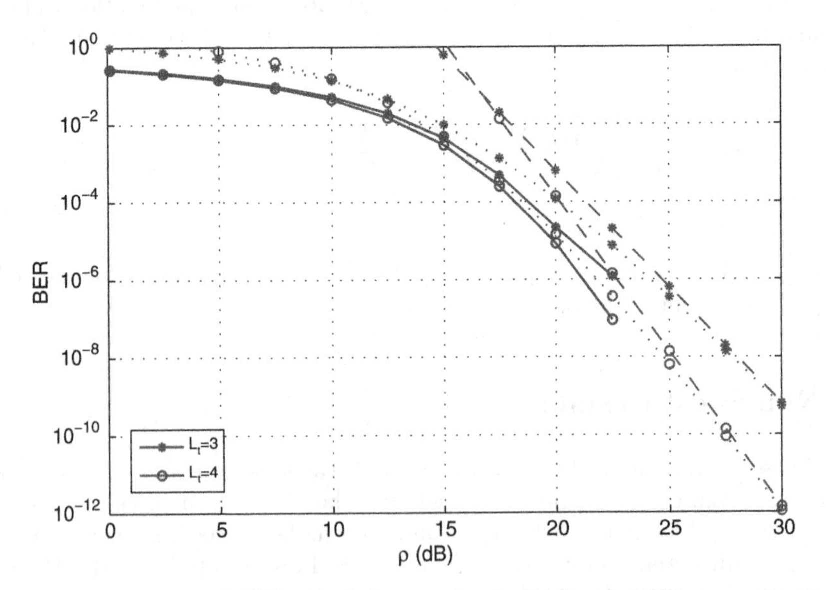


**Fig. 2.** Comparison of BER simulation results (*solid lines*), BER approximation given in (33), (*dashed lines*), and BER approximation given in (36), (*dotted lines*). Two antennas out of $L_t = 3, 4$ transmit antennas are selected and Alamouti scheme is used. The MIMO channel is Ricean flat fading with $K = 2.5$

The numerical results clearly show that the diversity order and coding gain derived analytically agree with the simulation results at high SNRs. Thus, full diversity order is preserved with TAS scheme using OSTBCs in a flat Ricean fading environment.

## 6 Conclusion

In this chapter we have investigated the performance of transmit antenna selection (TAS) with orthogonal space–time block coding (OSTBC) in Ricean flat fading MIMO channels. The exact BER for $M$-PAM and $M$-QAM, and an approximate BER for $M$-PSK have been derived. Our results are sufficiently general to handle an arbitrary number of antennas within Ricean fading environment. Moreover, we have derived the BER directly, and not via the symbol error probability. As expected, we have found that TAS with OSTBC achieves full diversity order asymptotically (i.e., $L_t$ not $N$), as if all the transmit antennas have been used.

## Acknowledgments

Funding for this work has been provided by the Rohit Sharma Professorship, TRLabs, the University of Alberta, and the Natural Sciences and Engineering Research Council (NSERC) of Canada.

## References

1. V. Tarokh, H. Jafarkhani, and A. Calderbank, "Space-time block coding for wireless communications: performance results," *IEEE J. Select. Areas Commun.*, vol. 17, pp. 451–460, Mar. 1999.
2. S. Alamouti, "A simple transmit diversity technique for wireless communications," *IEEE J. Select. Areas Commun.*, vol. 16, pp. 1451–1458, Oct. 1998.
3. G. J. Foschini and M. J. Gans, "On limits of wireless communications in a fading environment when using multiple antennas," *Wireless Personal Commun.*, vol. 6, no. 3, pp. 311–335, Mar. 1998.
4. E. Telatar, "Capacity of multi-antenna Gaussian channels," *Euro. Trans. Telecommun.*, vol. 10, no. 6, pp. 585–595, 1999.
5. S. Sanayei and A. Nosratinia, "Capacity of MIMO channels with antenna selection," *submitted to IEEE Trans. Inform. Theory*, Feb. 2005.
6. V. Tarokh, H. Jafarkhani, and A. Calderbank, "Space-time block codes from orthogonal designs," *IEEE Trans. Inform. Theory*, vol. 45, pp. 1456–1467, July 1999.
7. E. G. Larsson and P. Stoica, *Space-time Block Coding for Wireless Communications.* Cambridge University Press, 2003.
8. D. A. Gore and A. J. Paulraj, "MIMO antenna subset selection with space-time coding," *IEEE Trans. Signal Processing*, vol. 50, pp. 2580–2588, Oct. 2002.

9. D. J. Love, "On the probability of error of antenna-subset selection with space-time block codes," *IEEE Trans. Commun.*, vol. 53, pp. 1799–1803, Nov. 2005.
10. Z. Chen, "Asymptotic performance of transmit antenna selection with maximal-ratio combining for generalized selection criterion," *IEEE Commun. Lett.*, vol. 8, pp. 247–249, Apr. 2005.
11. Z. Chen, J. Yuan, B. Vucetic, and Z. Zhou, "Performance of Alamouti scheme with transmit antenna selection," *IEEE Int. Symposium on Personal, Indoor and Mobile Radio Commun. (PIMRC)*, vol. 2, pp. 1135–1141, Sept. 2004.
12. Z. Chen, J. Yuan, and B. Vucetic, "Analysis of transmit antenna selection/ maximal-ratio combining in Rayleigh fading channels," *IEEE Trans. Veh. Technol.*, vol. 54, no. 4, pp. 1312–1321, July 2005.
13. L. Yang and J. Qin, "Performance of Alamouti scheme with transmit antenna selection for M-ary signals," vol. 5, no. 12, pp. 3365–3369, Dec. 2006.
14. S. Kaviani and C. Tellambura, "Closed-form BER analysis for antenna selection using orthogonal space–time block codes," *IEEE Commun. Lett.*, vol. 10, pp. 704–706, Oct. 2006.
15. S. Kaviani and C. Tellambura, "Closed-form BER performance analysis for antenna selection using orthogonal space–time block codes," *Proc. IEEE Vehicular Technology Conf. (VTC)*, Sept. 2006.
16. A. Paulraj, R. Nabar, and D. Gore, *Introduction to Space-Time Wireless Communications.* Cambridge, United Kingdom: Cambridge University Press, May 2003.
17. J. G. Proakis, *Digital Communications*, 4th ed. New York: McGraw Hill, 2000.
18. A. Annamalai, G. K. Deora, and C. Tellambura, "Theoretical diversity improvement in GSC(N,L) receiver with nonidentical fading statistics," *IEEE Trans. Commun.*, vol. 53, pp. 1027–1035, June 2005.
19. K. Cho and D. Yoon, "On the general BER expression of one- and two-dimensional amplitude modulations," *IEEE Trans. Commun.*, vol. 50, pp. 1074–1080, July 2002.
20. M. S. Alouini and M. K. Simon, "An MGF-based performance analysis of generalized selection combining over Rayleigh fading channels," *IEEE Trans. Commun.*, vol. 48, pp. 401–415, Mar. 2000.
21. J. Lu, K. B. Letaief, J. C. Chuang, and M. L. Liou, "M-PSK and M-QAM BER computation using signal-space concepts," *IEEE Trans. Commun.*, vol. 47, pp. 181–184, Feb. 1999.
22. Z. Wang and G. Giannakis, "A simple and general parameterization quantifying performance in fading channels," *IEEE Trans. Commun.*, vol. 51, pp. 1389–1398, 2003.
23. G. M. Dillard, "Recursive computation of the generalized Q-function," *IEEE Trans. Aerosp. Electron. Syst.*, vol. 9, p. 614–615, July 1973.
24. M. K. Simon and M.-S. Alouini, *Digital Communication over Fading Channels: A unified approach to performance analysis*, 1st ed. New York: Wiley, 2000.
25. I. S. Gradshteyn and I. M. Ryzhik, *Table of Integrals, Series, and Products*, 5th ed. Academic Press, Inc., 1994.

# 5

# A Packet Detection Algorithm for the UWB Standard ECMA 368

Tomaso Erseghe, Nicola Laurenti, Valentina Rizzi, and Roberto Corvaja

Department of Information Engineering, University of Padua, Italy
erseghe,nil,corvaja@dei.unipd.it, rizzi.valentina@gmail.com

**Summary.** In this paper we present a packet detection technique developed in the context of a Italian national research project aiming at developing the "Enabling blocks for the integration in CMOS technology of a Multi-Band OFDM transceiver," adhering to the recent ECMA 368 standard. The paper deals with the digital section of the receiver, and specifically with the identification of a low complexity packet detection technique. The chosen approach is a correlation receiver followed by a threshold comparison. Its performance is tested in a realistic scenario taking into account channel dispersion and receiver inaccuracies.

## 1 Introduction

ECMA 368 [1] is a recently standardized Ultra wide band (UWB) communication system, issued in December 2005, covering the 3.1–10.6 GHz band. The modulation used is Multi-Band OFDM, which exploits an Orthogonal frequency division multiplexing (OFDM) over a $B = 528\,\text{MHz}$ bandwidth, plus a further Frequency hopping (FH) option for permitting simultaneous channel access by more than a single Wireless local area network (WLAN) or Wireless personal area network (WPAN). OFDM uses 128 sub-carriers and a 37 tap zero-prefix [2, 3] permitting effective inter-symbol-interference (ISI) rejection. Supported data rates range from 53.3 to $480\,\text{Mb}\,\text{s}^{-1}$ (at the physical layer), the latter being employed for very fast data download at a very short distance. Depending on the chosen data rate, a 4-Phase shift keying (PSK) or a 16-Quadrature amplitude modulation (QAM) constellation is used.

Today, ECMA 368 is supported by the WiMedia Alliance [4], and has been selected by the Bluetooth SIG and the USB Implementers Forum as the foundation of their high-speed wireless specifications for use in next generation consumer electronics, mobile and computer applications.

In this context, the Italian National project PRIN 2005-093524 *Enabling blocks for the integration in CMOS technology of a Multi-Band OFDM transceiver*, hereafter PRIN-UWB, aims at developing both the Radio frequency (RF) analog front-end, and the baseband digital receiver structure for ECMA 368.

In this paper, after a brief summary of the specific issues covered by the project in Sect. 2, we concentrate on digital algorithms for packet detection. Specifically, in Sect. 3 we discuss the packet detection approach that has been chosen, while closed form analytical results of its performance are given in Sect. 4, together with the identification of optimal settings for ECMA 368. Then, Sect. 5 reports the detection performance in realistic conditions, taking into account channel dispersion, but also other typical impairments, such as carrier and sampling frequency offsets, ADC quantization and phase noise effects.

## 2 The PRIN-UWB Project

The PRIN-UWB project involves researchers from four Italian universities (Pavia, Padova, Pisa and Bologna), over a two-year duration, with the purpose to study and design all the enabling blocks and subsystems of a high speed radio receiver for ECMA 368, working on all 14 bands, from 3.1 GHz to 10.6 GHz.

The cooperation between the four research units of the project allows to coordinate the system definition effort with the actual design and implementation of the critical building blocks. The various blocks to be designed are (see Fig. 1):

- Broad-band antenna
- RF front-end of the receiver (low noise amplifier and I/Q mixers)
- Fast hopping frequency synthesizer
- Base-band analog filter and variable gain amplifier (VGA)
- High speed 6 bit A/D converter
- Digital back-end receiver algorithms

The broad-band antenna has been characterized in terms of parameters of interest such as input impedance, radiation pattern and group delay, taking into account easiness of integration and fabrication. The chosen structure will

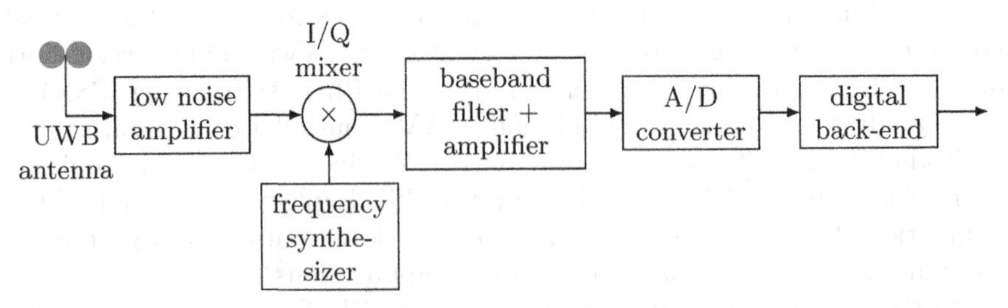


**Fig. 1.** Block diagram of the ECMA-368 receiver as designed within the PRIN-UWB project

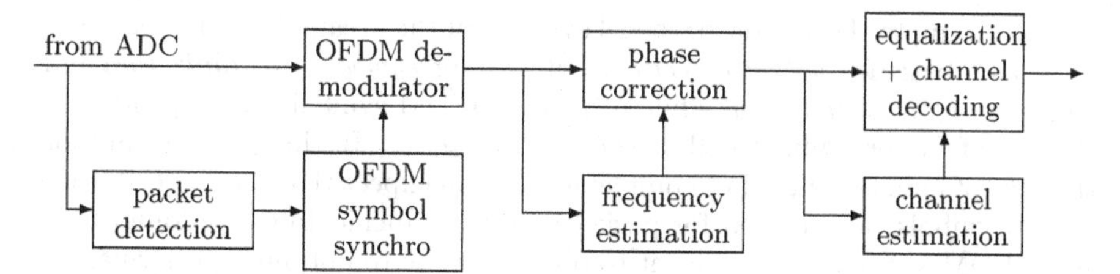


**Fig. 2.** Block diagram of the digital back-end as designed within the PRIN-UWB project

be numerically simulated in order to give design rules for novel antennas. The receiver front-end, the fast hopping synthesizer, the base-band VGA plus filter and the A/D are being designed, simulated, fabricated and characterized as single blocks.

The digital baseband algorithms have been studied with particular focus on low-complexity solutions. Indeed, although the literature on receiver algorithms for OFDM systems is very rich, the true challenge for Multiband systems is the unusually large bandwidth and transmission rate. Our attention is especially devoted to digital synchronization and tuning, channel estimation and equalization, and channel decoding, as illustrated in Fig. 2. Low complexity algorithms are designed, and the most critical parts will be implemented on FPGA.

## 3 Packet Detection Approach

### 3.1 Requirements of the Standard

In ECMA 368, packet detection can be accomplished by relying on the time domain portion of the packet preamble, which has a length of 24 OFDM symbols, and is built by repeating a *base time domain sequence* $\boldsymbol{s}_{\text{base}}$ of length 128, with flat spectrum occupancy, null DC, and good auto-correlation properties. Each OFDM symbol of the preamble is mapped onto a different frequency according to the frequency hopping patterns

$$\begin{aligned}
\boldsymbol{p}_1 &= \{f_1, f_2, f_3, f_1, f_2, f_3\}, \\
\boldsymbol{p}_2 &= \{f_1, f_3, f_2, f_1, f_3, f_2\}, \\
\boldsymbol{p}_3 &= \{f_1, f_1, f_2, f_2, f_3, f_3\}, \\
\boldsymbol{p}_4 &= \{f_1, f_1, f_3, f_3, f_2, f_2\}, \\
\boldsymbol{p}_5 &= \{f_1, f_1, f_1, f_1, f_1, f_1\}, \\
\boldsymbol{p}_6 &= \{f_2, f_2, f_2, f_2, f_2, f_2\}, \\
\boldsymbol{p}_7 &= \{f_3, f_3, f_3, f_3, f_3, f_3\},
\end{aligned}$$

all with periodicity of 6 symbols, and such that each frequency is re-used either in the next or every fourth symbol.

According to the standard, the 24 symbols of the preamble are supposed to serve as a basis for *packet detection*, but also for *symbol synchronization* and *frequency offset estimation*, while 6 further OFDM symbols are appended at the end of the preamble to allow *channel estimation.* In this context, and for the sake of maintaining a low complexity, we can expect that only 6 to 12 (out of 24) symbols can be used for packet detection. Incidentally, we found that two OFDM symbols are sufficient to obtain a close to optimum performance.

It is also relevant to us to recall that two operation modes are envisioned by the standard, namely: (1) *reception mode*, which is the situation where the receiver is synchronized to the ECMA 368 *superframe* structure, and the observation window is limited to 256 µs, i.e. $\simeq$ 820 OFDM symbols; (2) *Clear channel assessment (CCA) mode*, which is the situation where the receiver is searching a superframe to synchronize to, so that the observation window is an entire superframe, i.e. $\simeq 2 \cdot 10^5$ OFDM symbols. Each of such modes can in principle require different settings for the detection algorithm.

The minimum performance requested by the standard are expressed in terms of a Packet error rate (PER) of less than 8% at the minimum receiver sensitivity $\mathsf{P}_{\text{ref}} = -80.8$ dBm for the lowest bit rate, and with an ideal (non dispersive) channel.

### 3.2 Overview of the Literature and Chosen Approach

The literature on the subject of packet detection, often called *signal acquisition*, uses approaches based upon modifications of the correlation method, i.e. detection is unanimously based upon a correlating filter and a threshold comparison. These are obtained by application of a Generalized likelihood ratio test (GLRT) (e.g. see [5] and references therein) comparing the two hypotheses: $H_1$ the hypothesis of the signal being sent, and $H_0$ the hypothesis of the signal not being sent.

Application of the above concept can be found in many contexts, starting from the historical work of Polydoros and Weber [6]. For example, performance with adaptive thresholds is evaluated in [7, 8], use of verification procedures (multiple dwell search) is tested in [8]. More recently such ideas have been applied to UWB both in the context of impulse radio (IR) [9–12] and in Multiband-OFDM [13, 14] systems.

However, the scenarios that can be found in the literature are not applicable to the ECMA 368 context, since they refer to a situation where the signal to be detected has a periodic or cyclostationary structure, and the aim is that of reducing the mean acquisition time [6, 8, 15]. This scenario is typical of a continuous transmission, where acquisition time can be further reduced by use of clever search techniques alternative to the ordinary serial search method (e.g. see [16]).

In contrast, ECMA 368 uses a packet based transmission where only serial search can be employed, and where mean acquisition time is not an issue. Moreover, we need to activate the detection algorithm for every packet, hence we are rather interested in lowering the probability of Missed detection (MD). The chosen algorithm for ECMA 368 is a standard correlation method whose parameters are optimized in order to lower the probability of MD. This probability is defined in the present paper coherently with a packet based transmission.

### 3.3 Detection Criterion

The basic criterion we selected for packet detection is a standard GLRT criterion, where the two hypotheses of interest are

$$\begin{aligned} H_0 &: r_n = w_n \\ H_1 &: r_n = w_n + A\, s_{n-n_0} \end{aligned} \tag{1}$$

$H_1$ being the hypothesis of having sent the signal $\{s_n\}$, $n = 0, \ldots, M-1$ at time $n_0$, $H_0$ being the hypothesis of not having sent the signal, and $w_n$ being an Additive white Gaussian noise (AWGN) with variance $\sigma_w^2$.

In GLRT the output of the correlation between the received samples $r_n$ and the reference signal $s_n$ (i.e. the preamble), namely

$$c_m = \sum_n r_n\, s_{n-m}^* \,, \tag{2}$$

is compared to a threshold. Specifically, the criterion gives

$$\Lambda(n_0) = \frac{|c_{n_0}|^2}{\sigma_w^2\, E_s} \gtrless_{H_0}^{H_1} \lambda \tag{3}$$

with $\sigma_w^2 = 2N_0 B$ the variance of the Gaussian noise, $E_s$ the energy of $s_n$. Also, $s_n$ will be chosen to be the base sequence $\boldsymbol{s}_{\text{base}}$, so that a 128 tap filter is needed to perform correlation. Shorter filters are not taken into account because of the poor correlation performance of sub-sequences of $\boldsymbol{s}_{\text{base}}$. Longer filters can hardly be employed since, because of the FH pattern, they would require that parallel analog demodulation of $f_1, f_2, f_3$ is performed, which would exaggeratedly increase the receiver complexity.

When (3) is applied to a simplified AWGN context assuming perfect knowledge of $\sigma_w^2$, and ideal correlation properties of $\boldsymbol{s}_{\text{base}}$, the probabilities of False alarm (FA) and MD give [17]

$$\begin{aligned} P_{\text{FA}}(\lambda) &= \text{P}\left[H_1|H_0\right] = e^{-\lambda} \\ P_{\text{MD}}(\lambda) &= \text{P}\left[H_0|H_1\right] = \int_0^{\lambda} e^{-(a+\rho)} I_0(2\sqrt{a\,\rho})\, da \end{aligned} \tag{4}$$

where $I_0(x)$ is the zeroth order modified Bessel function, and where

$$\rho = \frac{|A|^2\, E_s}{\sigma_w^2} \tag{5}$$

is a Signal to noise ratio (SNR) measure.

### 3.4 Detection Algorithm

The basis of the detection algorithm is constituted by (3) which is applied in a serial search fashion for increasing values of $n_0$. When $\Lambda$ exceeds the threshold $\lambda$, then a packet is detected and digital demodulation begins. We will refer to this threshold check as *basic step*.

Incidentally, note that (3) requires that an estimate of the noise variance $\sigma_w^2$ is available. Such estimate can be derived by recalling that, in the absence of the signal $s_n$, the correlation $c_m$ is a Gaussian complex random variable with variance $E_s\sigma_w^2$. We thus set

$$\widehat{\sigma}_w^2 = \frac{\mathrm{E}\left[|c_m|^2\right]_{m<n_0}}{E_s} , \tag{6}$$

where $\mathrm{E}\left[\cdot\right]$ is a specific implementation of the expectation measure, typically a moving average, which results in a Maximum likelihood (ML) estimate.

In order to lower the FA probability, a further *verification step* (double dwell approach) can be foreseen after the basic step. Since a preliminary packet detection is available, this latter step can exploit samples all over the FH pattern. To comply with the ECMA standard, verification can be applied to at most 23 OFDM symbols, that is $V_{\max} = 23 \times 165 = 3795$ samples (here $165 = 128 + 37$ is the interval, at reception, occupied by the symbol), but, as will be shown in Sect. 4, a single OFDM symbol is sufficient to obtain a quasi optimum performance.

## 4 Performance Evaluation

### 4.1 Analytical Results

In this perspective, the reference state diagram for the packet detection algorithm is illustrated in Fig. 3. The black circle in figure represents the instant $W$ where the signal $s_{\text{base}}$ is placed, that is we expect that $c_m$ has its maximum for $m = W$. White circles represent instants where no packet is present, in which case $c_m$ is pure Gaussian noise.

The only absorbing states are: 1) FA when the threshold is exceeded but no signal $s_n$ has been sent; 2) MD when the threshold is not exceeded even if the signal has been sent; 3) Acquisition (ACQ) when the signal is correctly detected.

**FA** Transitions to the FA state are possible only from white circles, and with probability $P_{\text{FA}1}P_{\text{FA}2}$ where

$$P_{\text{FA}1} = P_{\text{FA}}(\lambda_1) , \quad P_{\text{FA}2} = P_{\text{FA}}(\lambda_2) , \tag{7}$$

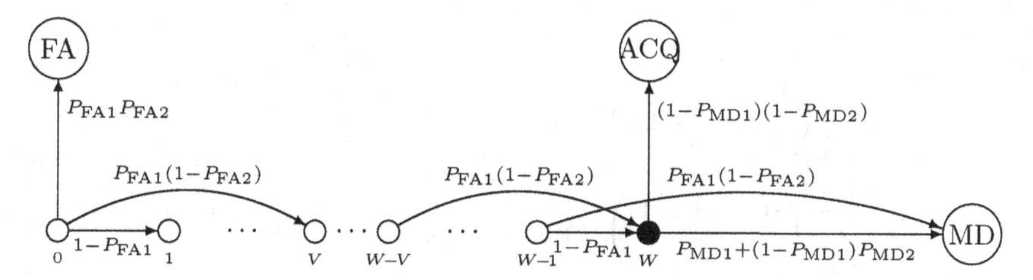


**Fig. 3.** State diagram for packet detection

with $\lambda_1$ the threshold in the basic step and $\lambda_2$ the threshold in the verification step. So, the FA state is reached only when both steps provide a false alarm. From each white circle we also have transitions to the next circle with probability $1-P_{FA1}$, which is the case when no false alarm occurs in the basic step. A final transition with probability $P_{FA1}(1-P_{FA2})$ is assured from each white circle to the circle which is $V$ positions to the right, $V$ being the length of time spent in the verification process, which corresponds to the case when the verification step rejects the FA activated by the basic step.

**ACQ** Transition to the ACQ state is instead possible *only from the black circle*, and only when both stages detect a packet, that is with probability $(1-P_{MD1})(1-P_{MD2})$, where

$$P_{MD1} = P_{MD}(\lambda_1) , \quad P_{MD2} = P_{MD}(\lambda_2) . \tag{8}$$

**MD** Finally, the transition to the MD state is again possible from the black state when the detection of either the basic step or of the verification step fails, that is with probability $P_{MD1} + (1-P_{MD1})P_{MD2}$. Transitions to the MD state are also possible when the verification step is activated (but not passed) in one of the states $W-V+1, \ldots, W-1$ (this is partially shown in figure).

Overall, we are interested in the probabilities of ending in the FA, ACQ, and MD states, which satisfy

$$P_{FA} + P_{ACQ} + P_{MD} = 1 .$$

So, our thresholds can be optimized in order to lower the probability of missed acquisition

$$P_{NACQ} = 1 - P_{ACQ} = P_{FA} + P_{MD} , \tag{9}$$

which provides an upper bound for both FA and MD probabilities.

A closed form analytical expression for $P_{NACQ}$ can be obtained by combinatorial arguments, giving

$$\begin{aligned} P_{NACQ} = P_{MD1} &+ (1-P_{MD1})P_{MD2} \\ &+ P_0\,(1-P_{MD1})(1-P_{MD2}) . \end{aligned} \tag{10}$$

where

$$\begin{aligned} P_0 = &-\sum_{n=1}^{W} \binom{W}{n} (-P_{\mathrm{FA1}})^n \\ &-\sum_{\ell=1}^{\lfloor W/V \rfloor} \binom{W-\ell(V-1)}{\ell} [P_{\mathrm{FA1}}(1-P_{\mathrm{FA2}})]^{\ell}(1-P_{\mathrm{FA1}})^{W-\ell V} \end{aligned} \tag{11}$$

We note that the results of [6] cannot be applied to the state diagram of Fig. 3. This is because of the presence of a transition from state $m$ to state $m+V$, and because transmission is non-continuous (in fact, it is packed-based).

## 4.2 Optimal Thresholds

By exploiting the analytical results, we can evaluate the thresholds $\lambda_1, \lambda_2$ giving the minimum probability $P_{\mathrm{NACQ}}$. To this end, we envision four different scenarios, depending on whether the verification step is applied or not, and whether we operate in reception or CCA mode.

Optimum thresholds in the four scenarios are shown in Fig. 4 as a function of $\rho$, and the resulting performance is given in plain lines in Fig. 5. The length $V$ of the verification stage is set to two OFDM symbols, that is $V = 330$ samples, since this is the optimum choice. This can be intuitively explained by the following rationale: although longer values of $V$ assure enhanced FA rejection, they imply longer pauses when a FA occurs in the basic step. In any case, the difference with longer choices of $V$ is negligible.

From Fig. 5 we see that verification can give an improvement both in reception and scanning mode, the improvement being more significant in the latter case. Also, the performance obtainable with verification is almost identical in both modes.

Figure 5 also shows in dashed-dotted lines the behavior of algorithms where fixed thresholds are employed in the whole range of $\rho$. Specifically, we used the thresholds which are optimum at $\rho = 17\,\mathrm{dB}$. In all cases, curves show a saturation for high values of $\rho$, and a misbehavior for low values of $\rho$. Moreover, also in this fixed threshold case, verification is highly recommended, providing a lower saturation value.

To conclude, in Fig. 6 we show the FA and MD probabilities (7) and (8). The solid and the dashed lines refer to the false alarm probability, respectively, $P_{\mathrm{FA1}}$ and $P_{\mathrm{FA2}}$. It can be seen that, although $P_{\mathrm{FA2}}$ is much higher than $P_{\mathrm{FA1}}$, the use of a verification stage allows a quite low false alarm probability as soon as the value of $\rho$ is larger than 15 dB.

The same remarks apply to the dash-dotted and dotted lines, which represent the missed detection probability, respectively, $P_{\mathrm{MD1}}$ and $P_{\mathrm{MD2}}$. Again, use of the verification stage allows a great reduction of MD probability.

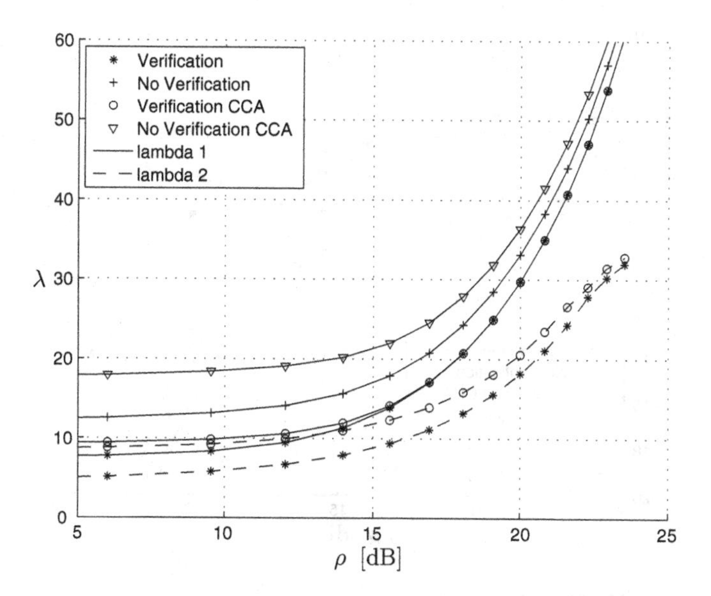


**Fig. 4.** Optimum thresholds $\lambda_1, \lambda_2$ as a function of $\rho$

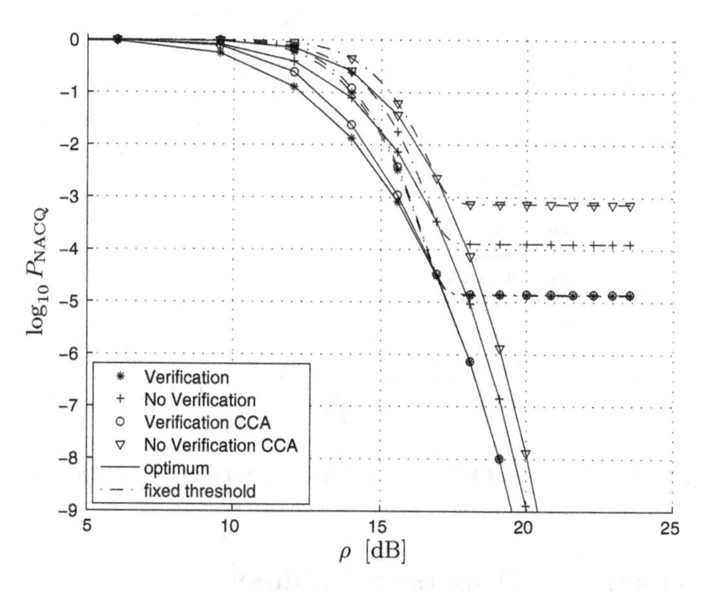


**Fig. 5.** Probability of incorrect acquisition $P_{\mathrm{NACQ}}$ as a function of $\rho$

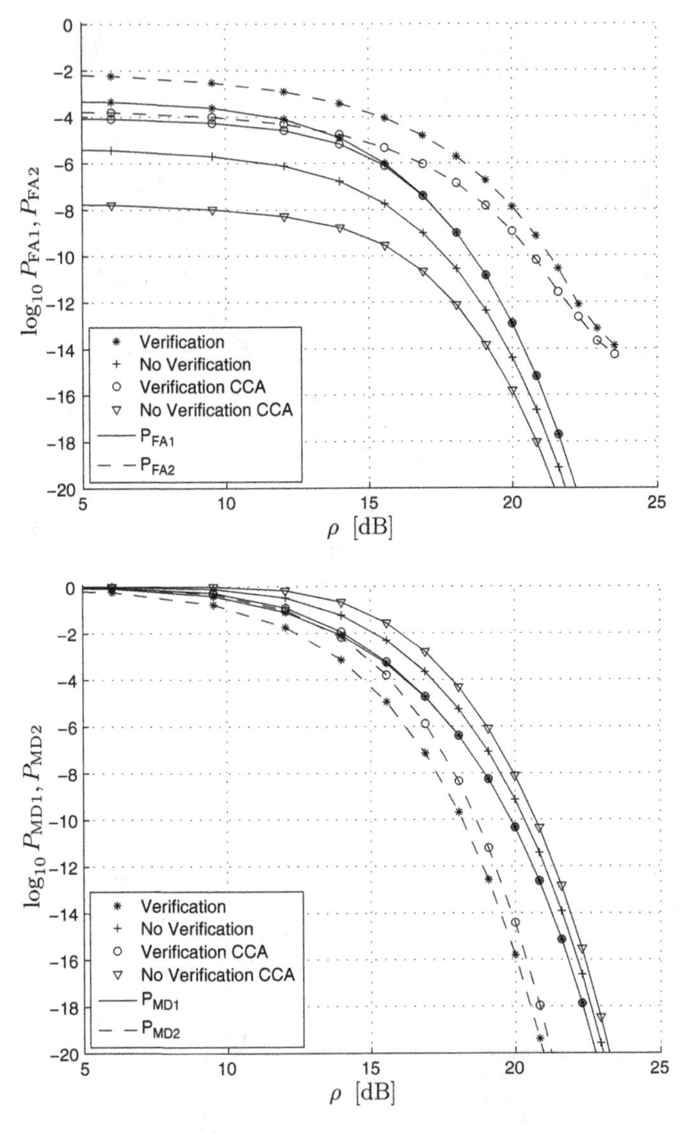


**Fig. 6.** Probabilities of FA (*top*) and MD (*bottom*) as a function of $\rho$

## 5 Performance in Dispersive Channels

In order to further verify the algorithm performance, we set up simulations in a dispersive channel environment using the reference indoor channel for UWB defined by the IEEE 802.15.4 group [18]. The simulation is carried out only in *reception mode*.

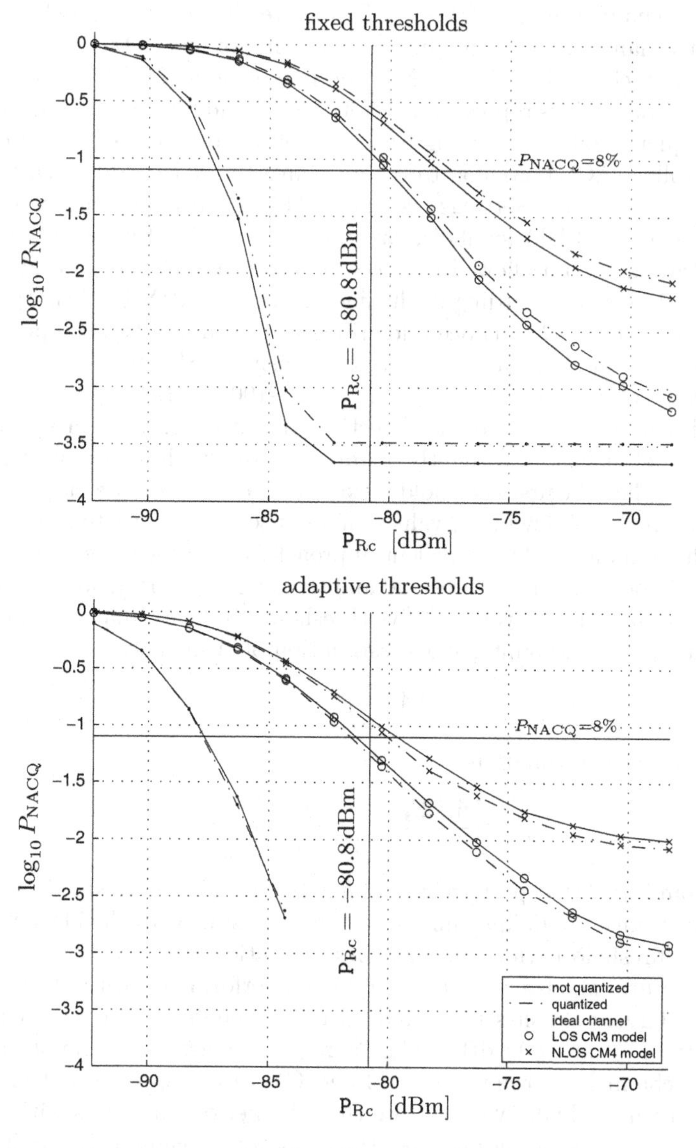


**Fig. 7.** System performance with dispersive channels: fixed threshold (*above*) and adaptive threshold (*below*)

Results are reported in Fig. 7 as a function of the received power $\mathrm{P_{Rc}}$, where, according to the standard, the noise model at reception assumes a 5 dB noise temperature.

Solid lines in figure represent a system taking into account filters at transmission and reception, so that dispersion effects are present also in the case of the ideal channel, $g(t) = \delta(t - t_0)$. Three are the channel models used for simulations, namely: the ideal channel (i.e. Line of sight (LOS) path only), the CM3 LOS channel, and the CM4 Non line of sight (NLOS) channel.

Dashed-dotted lines represent a system tested under typical effects, such as carrier frequency offset, sampling frequency offset, presence of an Automatic gain controller (ACG), Analog to digital converter (ADC) quantization and phase noise effects. The settings used are a 4 bit linear quantization, an ACG averaging over 2 OFDM symbols, sampling and carrier frequency offsets of 40 ppm (the maximum allowed by the standard), and an 8 bit internal representation. Moreover, according to the findings of the PRIN-UWB project, the phase noise process $\theta(t)$ has been described as the sum of two components, a Wiener process $\theta_W(t)$ with Power spectral density (PSD) of −85 dB at 1 kHz, plus a white component $\theta_0(t)$ with a PSD of −100 dB [19].

We also note that, since the detection algorithm assumes a single-path channel model, it has been slightly adapted to cope with dispersive channels. Specifically, when the first threshold is exceeded, a search on the maximum $\Lambda$ is performed over the following 32 values, in order to synchronize to the strongest multipath component. An equivalent approach is used for verification.

In Fig. 7, both the fixed threshold and the adaptive threshold performance is displayed. Specifically, the adaptive threshold can be estimated through (6) and by taking into account the ML estimation of $A$, namely

$$\widehat{A} = \frac{c_{n_0}}{E_s} . \tag{12}$$

In conclusion, the estimate is

$$\widehat{\rho} = \frac{|\widehat{A}|^2 E_s}{\widehat{\sigma}_w^2} = \frac{|c_{n_0}|^2}{\mathrm{E}\left[|c_m|^2\right]_{m<n_0}} . \tag{13}$$

We see from Fig. 7 that performance of fixed and variable threshold choices, is very similar, with the unique difference (clearly visible with the ideal channel) that fixed threshold performance reaches saturation.

In the figure, we also note that the system performance with ideal channel conditions largely satisfies the standard requirements of a missed detection probability of 8% at −80.8 dBm. The 8% request is instead not satisfied in the dispersive channel environments CM3 and CM4 by, respectively, 1 and 3 dB (with fixed thresholds). We also note that the performance loss with typical receiver inaccuracies, is very limited, thus making the chosen approach a very robust choice.

## 6 Conclusions

We have presented a new technique for packet detection that has been derived within the PRIN UWB project, for application to the Multiband OFDM UWB standard ECMA 368. Our algorithm is based on the generalized likelihood ratio test approach, which makes use of the correlation between the 128-sample zero auto-correlation transmitted sequence and the received signal, followed by a threshold-based decision.

The use of a double dwell procedure allows us to considerably lower the probability of false alarm. Unlike other applications of the GLRT in the literature we optimize threshold values in order to maximize the overall probability of correct packet acquisition.

The performance evaluation shows that our algorithm works by large on non dispersive channels, and meets the requirements of the standard (8% packet error rate at a receiver sensitivity of −80.8 dBm) even on the not very dispersive LOS CM3 model, with the use of adaptive thresholds. However, on more dispersive channels a way to collect the energy spread across several multipath components should be devised to ensure a more robust behavior, and satisfaction of the standard requirements.

## Acknowledgment

This work was supported by the Italian Ministry of University and Research within the framework of the PRIN project 2005-093524, "Enabling blocks for the integration in CMOS technology of a Multi-Band OFDM transceiver".

# References

1. ECMA International, *ECMA 368 Standard: High rate ultra wideband PHY and MAC standard*, Dec. 2005.
2. B. Muquet, Z. Wang, G.B. Giannakis, M. de Courville, P. Duhamel, "Cyclic prefixing or zero padding for wireless multicarrier transmissions?" *IEEE Trans. Commun.*, vol. 50(12), pp. 2136–2148, Dec. 2002.
3. A. Batra, J. Balakrishnan, G.R. Aiello, J.R. Foerster, A. Dabak, "Design of a multiband OFDM system for realistic UWB channel environments," *IEEE Trans. Microw. Theory Technol.*, vol. 52(9), pp. 2123–2138, Sept. 2004.
4. WiMedia alliance. [Online]. Available: http://www.wimedia.org
5. A.A. D'Amico, U. Mengali, "GLRT receivers for UWB systems," *IEEE Commun. Lett.*, vol. 9(6), pp. 487–489, June, 2005.
6. A. Polydoros, C.L. Weber, "A unified approach to serial search spread-spectrum code acquisition – Part I and II," *IEEE Trans. Commun.*, vol. COM-32(5), pp. 542–560, May 1984.
7. E. Brigant, A Mammela, "Adaptive threshold control scheme for packet acquisition," *IEEE Trans. Commun.*, vol. 46(12), pp. 1580–1582, Dec 1998.
8. J.H.J. Iinatti, "On the threshold setting principle in code acquisition of DS-SS signals," *IEEE J. Sel. Areas Commun.*, vol. 18(1), pp. 62–72, Gen. 2000.
9. T.W.S.R. Aedudodla, S. Vijayakumaran, "Acquisition of direct-sequence transmitted reference ultra-wideband signals," *IEEE J. Sel. Areas Commun.*, vol. 24(4), pp. 759–765, April 2006.
10. R.B. J. Ibrahim, "Two-stage acquisition for UWB in dense multipath," *IEEE J. Sel. Areas Commun.*, vol. 24(4), pp. 801–807, April 2006.

11. J.L.B.A. El Fawal, "A robust signal-detection method for Ultra-Wide-Band networks with uncontrolled interference," *IEEE Trans. Microw. Theory Technol.*, vol. 54(4), pp. 1769–1781, Apr. 2006.
12. G.G.Z. Tian, "A GLRT approach to data-aided timing acquisition in UWB radios – Part I and II," *IEEE Trans. Wirel. Commun.*, vol. 4(5), pp. 1638–1645, 2994–3004, Nov. 2005.
13. P.S.G. Lu, L.J. Greenstein, "A New Probability Density Function Enhancing Packet Detection Analysis for Low-SNR Links," *IEEE Trans. Vehicul. Technol.*, vol. 56(3), pp. 1230–1238, May 2007.
14. J.T. Lai, C.Y. Chu, A.Y. Wu, W.C. Chen, "A Robust Band-Tracking Packet Detector (BT-PD) in OFDM-Based Ultra-Wideband Systems," in *SIPS 06.* IEEE, Oct. 2006, pp. 165–170.
15. J. Iinatti, "Performance of ds code acquisition in static and fading multipath channels," *IEE Proc. Commun.*, vol. 147(6), pp. 355–360, Dec. 2000.
16. T.W.S. Vijayakumaran, "A search strategy for ultra-wide-band signal acquisition," *IEEE Trans. on Commun.*, vol. 53(12), pp. 2015–2019, Dec. 2005.
17. V. Rizzi, "Packet detection and frame timing acquisition in UWB MB-OFDM systems," Master's thesis, University of Padova, Italy, Oct. 2006.
18. J. Foerster, Ed., "IEEE802.15-02/490 – Channel modeling sub-committee report final," Tech. Rep. [Online]. Available: http://ieee802.org/15
19. J.J. Spilker, *Digital communications by satellite.* Prentice Hall, 1997.

# 6

# Low-Rate Predictive Feedback for the OFDM MIMO Broadcast Channel

Nevio Benvenuto, Ermanna Conte, Stefano Tomasin, and Matteo Trivellato

Department of Information Engineering, University of Padova, Via Gradenigo 6/B, 35131 Padova, Italy
{nb,ermanna.conte,tomasin,matteo.trivellato}@dei.unipd.it

**Summary.** For a frequency division duplexing (FDD) cellular system, the downlink channel state information (CSI) has to be fed back by mobile terminals (MTs) in order to perform beamforming and scheduling at base station (BS). In this paper, for an orthogonal frequency division multiplexing (OFDM) modulation format, we propose a novel CSI feedback strategy based on the prediction of channel variations. Indeed, quantization of CSI and feedback signalling are jointly designed in order to obtain a low rate feedback signal. Also, an user selection method is devised to allocate users across the subcarriers. Simulations for a 3G long term evolution scenario show a significant performance improvement achieved by the predictive strategy with respect to existing techniques.

## 1 Introduction

Future wireless cellular systems are expected to support high quality multimedia services; to this purpose, schemes with multiple transmit and receive antennas (multiple input multiple output, MIMO) have been considered. In fact, it has been shown that the network downlink capacity of a MIMO system with full channel state information (CSI) scales linearly with the number of transmit antennas [1].
As the capacity-achieving dirty paper coding (DPC) [2], based on interference subtraction at the base station (BS), has a high computational cost, many practical techniques have been proposed. Linear beamforming, although suboptimal, is a simpler approach that has been shown to achieve a large part of DPC capacity [3]. However, beamforming requires that BS has a full CSI, moreover the achievable throughput heavily depends on a reliable CSI. In a frequency division duplexing (FDD) system, CSI is provided to the BS by setting up an explicit feedback (FB) signalling from each mobile terminal (MT). Unfortunately, the number of bits required to describe the channel grows as product of the number of transmit and receive antennas, the channel delay spread and the number of MTs [4]. Then an optimization of the FB signalling

is essential to limit its impact on the network throughput. This optimization is particulary relevant for a broadband transmission where the channel is dispersive. Orthogonal frequency division multiplexing (OFDM) is considered a good candidate as modulation scheme for downlink of the long term evaluation (LTE) of 3G cellular systems [5], since il allows an efficient channel equalization. Until now, mainly single subcarrier MIMO systems for downlink cellular environment have been studied (e.g., [6–8]), while the investigation of MIMO OFDM systems is often left to future works.
In this paper, we consider a downlink cellular system based on MIMO OFDM with a low rate, error free, zero delay feedback signalling channel. Firstly, we propose an user strategy algorithm for the allocation of user data to the OFDM subcarriers. This strategy extends the greedy selection algorithm of [9] to a multicarrier scenario. Secondly, we investigate a predictive FB (PFB) technique, where both BS and MTs predict the CSI based on the past value of CSI, and the FB information given by MTs to the BS is a quantized version of the prediction error. The performance of the proposed FB technique, together with the user selection strategy, has been evaluated on the time-varying channels of the 3G-LTE cellular system [5], showing that the proposed methods allow to achieve a higher sum rate, with a lower FB rate, than existing techniques.
**Notation:** uppercase boldface letters denote matrices, lowercase boldface denote vectors and calligraphic denote sets. $^T$ denotes the transpose of a matrix or vector, $^H$ stands for the Hermitian transpose, and $^*$ stands for complex conjugate. $||\cdot||$ denotes the Euclidean norm of a vector, $|\mathcal{S}|$ indicates the cardinality of the set $\mathcal{S}$. $\mathbb{C}$ indicates the set of complex numbers.

## 2 System Model

We consider the downlink of a cellular system based on OFDM [10] with $N$ subcarriers; the BS has $M$ transmit antennas and $K$ MTs have one antenna each. Transmission is performed in time slots of $L$ OFDM symbols and in each time slot MTs feed back a partial CSI, which is used by the BS to schedule downlink transmissions.

For slot $n$, let $\mathcal{U}_c(n)$ be the set of MTs scheduled for downlink transmission on subcarrier $c \in \{1, \cdots, N\}$, while $\mathbf{x}_c(t)$, and $\mathbf{y}_c(t)$ are, respectively, the $M \times 1$ and $|\mathcal{U}_c(n)| \times 1$ transmitted and received column vectors on subcarrier $c$ of OFDM symbol $t$. We assume that the channel is quasi static, i.e., it can be considered invariant for the duration of one OFDM symbol. The frequency response of the MIMO channel on subcarrier $c$ is described by the complex $|\mathcal{U}_c(n)| \times M$ channel matrix $\mathbf{H}_c(t) = [\mathbf{h}_{1,c}(t), \dots, \mathbf{h}_{|\mathcal{U}_c(n)|,c}(t)]^T$, where the $M \times 1$ column vector $\mathbf{h}_{k,c}(t)$ collects the gains between the $M$ antennas of BS and the selected user $k$ on subcarrier $c$ of OFDM symbol $t$.

The discrete-time complex baseband model is given by

$$\mathbf{y}_c(t) = \mathbf{H}_c(t)\mathbf{x}_c(t) + \mathbf{n}_c(t)\,, \tag{1}$$

where $\mathbf{n}_c(t)$ is a $|\mathcal{U}_c(n)| \times 1$ complex Gaussian vector noise with independent and identically distributed (i.i.d.) components having zero mean and variance $\sigma_n^2$.

The transmit signal is subject to the average power constraint

$$\mathbb{E}\left[\|\sum_{c=1}^{N} \mathbf{x}_c(t)\|^2\right] \leq P\;, \tag{2}$$

where $P$ is the available power. From (2) and assuming that the channel has unitary average gain, the average signal to noise ratio per subcarrier is $SNR = P/(N \cdot \sigma_n^2)$.

Differently from most related works, we assume a realistic MIMO channel with time, frequency and spatial correlation among the elements of $\mathbf{H}_c(t)$. For a description of the considered channel model we refer to Sect. 6.

The channel matrix $\mathbf{H}_c(t)$ is not perfectly known at the BS while we assume that MT $k$ perfectly estimates the channel vectors (CVs) $\mathbf{h}_{k,c}(t)$, $c = 1, 2, \ldots, N$ once a slot. As in [8, 9] we adopt a double FB information and at each slot each MT feeds back (i) a channel direction information (CDI) on the direction of each CV, namely

$$\tilde{\mathbf{h}}_{k,c}(nL) = \frac{\mathbf{h}_{k,c}(nL)}{||\mathbf{h}_{k,c}(nL)||} \quad c = 1, \ldots, N, \tag{3}$$

and (ii) a channel quality information (CQI), based on the estimated signal-to-noise plus interference ratio (SNIR) at the receiver (see (13)). We assume that the FB channel has a finite rate of $N_b$ bits *per slot* and allows zero-delay error free transmissions. In this paper we focus on the quantization and FB of CDI and let the study of quantization and FB of CQI for future works. We observe also that the frequency responses of the channel on different subcarriers are correlated; hence feeding back CSI relative to an adequate subset of subcarriers provides enough information to the BS for interpolating CSI on the other subcarries. However, for sake of simplicity we consider FB per each subcarrier. The BS stores the partial CSI of selected users into the matrix

$$\overline{\mathbf{H}}_c(nL) = [\bar{\mathbf{h}}_{s_{1,c}}(nL), \ldots, \bar{\mathbf{h}}_{s_{|\mathcal{U}_c(n)|,c}}(nL)]^T, \quad s_i \in \mathcal{U}_c(n) \tag{4}$$

containing the unit-norm *reconstructed* CVs, which ideally should track $\{\tilde{\mathbf{h}}_{s_i,c}(nL)\}$. For ease of notation we drop both the slot $(n)$ and the OFDM symbol $(t)$ indices in the remaining of this section.

According to channel conditions and fairness criteria, at each slot the BS schedules which users are allocated on each subcarrier. Moreover, since a user can be allocated on more subcarriers, we denote as *stream* the user data block allocated on a specific subcarrier.

### 2.1 Zero-Forcing Beamforming

Once MTs have been selected for downlink transmission, the transmitted vector on subcarrier $c$ is obtained by beamforming, i.e.,

$$\mathbf{x}_c = \mathbf{G}_c \mathbf{u}_c \tag{5}$$

where $\mathbf{G}_c = \left[\mathbf{g}_{1,c}, \ldots, \mathbf{g}_{|\mathcal{U}_c|,c}\right]$ is the $M \times |\mathcal{U}_c|$ complex beamforming matrix, and vector $\mathbf{u}_c = (u_{1,c}, \ldots, u_{|\mathcal{U}_c|,c})^T$ contains the MT data symbols, which are assumed i.i.d with zero mean and unit variance. According to zero-forcing (ZF) criterion, it is

$$\mathbf{G}_c = \mathbf{F}_c \mathrm{diag}(\mathbf{p}_c)^{1/2} = \overline{\mathbf{H}}_c^H \left(\overline{\mathbf{H}}_c \overline{\mathbf{H}}_c^H\right)^{-1} \mathrm{diag}(\mathbf{p}_c)^{1/2}, \tag{6}$$

where $\mathbf{F}_c$ is the right pseudo-inverse of the reconstructed channel matrix $\bar{\mathbf{H}}_c$ and $\mathbf{p}_c$ is the vector of power normalization coefficients imposing the power constraint $P$ on the transmitted signal. Let $M_T = \sum_{c=1}^{N} |\mathcal{U}_c|$ be the number of streams allocated across all the subcarriers; under the assumption of equal power distribution among MTs, $\mathbf{p}_c$ has elements

$$p_{k,c} = \frac{P}{M_T \cdot ||\mathbf{f}_{k,c}||^2}, \quad k = 1, \ldots, |\mathcal{U}_c|, \tag{7}$$

where $\mathbf{f}_{k,c}$ is the $k$th column of $\mathbf{F}_c$.

## 3 Downlink Resource Allocation

The allocation procedure is split into two processes: a scheduling phase, where the BS provides priority for the allocation of users, and a user assignment phase, where user data streams are assigned to subcarriers.
We underline that we do not consider in this paper the power allocation problem and we equally divide the available power among all active streams; more on this in Sect. 3.

### 3.1 Scheduling Phase

In order to balance the opportunistic use of channel resources with fairness among users, we consider the multiuser proportional fair scheduling algorithm (MPFS) [11], which is an extension to the multi-user scenario of the proportional fair scheduling (PFS) algorithm [12].

Let $\mathcal{U}(n)$ be the set of couples of user index $k$ and subcarrier index $c$ such that user $k$ stream is scheduled on subcarrier $c$ at slot $n$ with rate $R_{k,c}(n, \mathcal{U}(n))$. Note that notation $R_{k,c}(n, \mathcal{U}(n))$ highlights the fact that rates achieved by users are mutually dependent, as simultaneous transmission on the same subcarrier yields interference.

For MPFS, the average throughput of user $k$ up to slot $n$ is denoted as $T_k(n)$ and is updated as follows:

$$T_k(n+1) = \left(1 - \frac{1}{\tau}\right) T_k(n) + \frac{1}{\tau} \sum_{c=1}^{N} R_{k,c}(n, \mathcal{U}(n)), \tag{8}$$

where $\tau$ is a parameter related to the time over which fairness should be achieved. In [12] it has been shown that proportional fairness, maximizing $\sum_k \log_2(T_k(n))$, is achieved by scheduling users according to the following criterion:

$$\mathcal{U}(n) = \arg\max_{\mathcal{U}} \sum_{(k,c)\in\mathcal{U}} \log_2 \left(1 + \frac{R_{k,c}(n,\mathcal{U})}{(\tau-1)T_k(n-1)}\right). \tag{9}$$

We observe that for $\tau >> 1$ we can approximate

$$\log_2 \left(1 + \frac{R_{k,c}(n,\mathcal{U})}{(\tau-1)T_k(n-1)}\right) \approx \frac{R_{k,c}(n,\mathcal{U})}{(\tau-1)T_k(n-1)} \tag{10}$$

and MPFS (9) coincides with the maximization of the weighed sum-rate

$$\mathcal{R} = \sum_{(k,c)\in\mathcal{U}(n)} w_k(n) R_{k,c}(n, \mathcal{U}(n)), \tag{11}$$

with weights

$$w_k(n) = \frac{1}{T_k(n-1)}. \tag{12}$$

### 3.2 User Assignment Phase

BS allocates data streams of different users to the subcarriers with the objective of maximizing the weighted sum rate (11). The allocation is based on the SINR relative to each user on each subcarrier, provided that each MT feeds back to the BS the CQI values

$$\xi_{k,c} \triangleq \frac{||\mathbf{h}_{k,c}||^2 |\tilde{\mathbf{h}}_{k,c}^T \bar{\mathbf{h}}_{k,c}^*|^2}{\sigma_n^2 + \frac{P}{M \cdot N} ||\mathbf{h}_{k,c}||^2 (1 - |\tilde{\mathbf{h}}_{k,c}^T \bar{\mathbf{h}}_{k,c}^*|^2)}, \tag{13}$$

for $c = 1, \cdots, N$, an estimate of the SNIR on subcarrier $c$ for MT $k$ is given by [9]

$$\gamma_{k,c} = \frac{P}{M_T \cdot ||\mathbf{f}_{k,c}||^2} \xi_{k,c}. \tag{14}$$

This estimate holds under the following assumptions: (a) CV with i.i.d. components, each Rayleigh fading with zero mean and unit variance; (b) equal power distribution among streams, (c) ZF beamforming at the BS, (d) $M-1$ interfering signal streams ; (e) the direction of error vector $\tilde{\mathbf{h}}_{k,c} - \bar{\mathbf{h}}_{k,c}$ isotropically distributed in the $M-1$ dimensional hyperplane orthogonal to $\bar{\mathbf{h}}_{k,c}$[8].

From (14), an estimate of $R_{k,c}(n, \mathcal{U}(n))$ is given by

$$R_{k,c}(n, \mathcal{U}(n)) = \log_2(1 + \gamma_{k,c}(n)) \ . \tag{15}$$

Note that the rate scheduled to user $k$ is

$$R_k(n, \mathcal{U}) = \sum_{c=1}^{N} \log_2(1 + \gamma_{k,c}(n)) \ . \tag{16}$$

For the choice of set $\mathcal{U}(n)$ we propose the greedy algorithm described in Table 1, which extends the algorithm described in [9] to OFDM. The algorithm operates iteratively, and at each iteration it selects the stream of user $k$ on subcarrier $c$ that maximizes the weighted sum rate. However, this new stream is accepted only if it increases the system weighted sum rate $\mathcal{R}$. In fact, we note that by adding a new user on subcarrier $c$ the weighted rate achieved by the set of selected users may be decreased due both to a new power distribution among all subcarriers and to the new beamformer on subcarrier $c$.

Indeed, for each MT selection, (15) requires to evaluate a new beamforming matrix (6) and the corresponding SNIR $\gamma_{k,c}(n)$ from (14). Overall, the algorithm requires the computation of $\mathcal{O}(N \cdot M \cdot K)$ SINRs, while an exhaustive search of all possible allocation would require $\mathcal{O}((K \cdot M)^N)$ SINR evaluations.

**Table 1.** Pseudo-code of the MT selection algorithm

1: INIT: $\mathcal{U} \leftarrow \emptyset$, $\mathcal{A} \leftarrow \{(k,c)|k = 1,2,\ldots,K; c = 1,2,\ldots,N\}$,
2: $\mathcal{R}(\mathcal{U}) \leftarrow 0$
3: **for** $m = 1$ to $M \cdot N$ **do**
4:
$$(\bar{k}, \bar{c}) \leftarrow \arg \max_{(k,c) \in \mathcal{A} \setminus \mathcal{U}} \mathcal{R}(\mathcal{U} \cup \{(k,c)\})$$
5: **if** $\mathcal{R}(\mathcal{U} \cup \{\bar{k}, \bar{c}\}) \le \mathcal{R}(\mathcal{U})$ **then**
6: $\mathcal{U} \leftarrow \mathcal{U}$
7: **break**
8: **else**
9: $\mathcal{U} \leftarrow \mathcal{U} \cup \{(\bar{k}, \bar{c})\}$
10: $\mathcal{U}_c \leftarrow \mathcal{U}_c \cup \{\bar{k}\}$
11: **end if**
12: **if** $|\{k|(k,\bar{c}) \in \mathcal{U}\}| = M$ **then**
13: $\mathcal{A} \leftarrow \mathcal{A} \setminus \{(k,\bar{c})\}$
14: **end if**
15: **end for**

## 4 Feedback Strategies

In order to perform beamforming and resource allocation, the BS needs the CSI relative to each MT. To this purpose, the MT first quantizes CSI relative to each subcarrier, and then feeds back a bit representation of the quantized value. In [8, 9] the codebook is a set of $2^b$ vectors randomly chosen from an isotropic distribution on the $M$-dimensional unit sphere. In other words, a random vector quantizer (RVQ) is used. For the feedback signalling, in [8, 9] it is proposed that MT $k$ quantizes the normalized subcarrier $c$ CV $\tilde{\mathbf{h}}_{k,c}$ into a codevector, whose index of $b$ bits is fed back at each slot. We denote this signaling technique as basic FB (RVQ-BFB). We consider as first refinement of RVQ-BFB a quantization technique where the codebook is designed according to the Linde Buzo and Gray algorithm with distance metric $d(\tilde{\mathbf{h}}_{k,c}, \mathbf{c}_i) = 1-|\tilde{\mathbf{h}}_{k,c}^H \cdot \mathbf{c}_i|^2$(see Appendix for details), while still FB provides the quantization vector index. We denote this technique as LBG-BFB. In both BFB strategies, the correlation in time of the MIMO channel is not exploited. In the next section, we propose a technique based on the prediction of channel variations, where the FB information is the subcarrier CV prediction error.

## 5 Predictive Feedback Technique

The predictive FB (PFB) strategy provides a different approach for CSI FB based on predictive vector quantization [13].

As depicted in Fig. 1, at slot $n$, both BS and MT obtain a prediction $\mathbf{h}_{k,c}^{(p)}(n)$ of the CV direction $\tilde{\mathbf{h}}_{k,c}(n)$, based on past reproduced values $\{\bar{\mathbf{h}}_{k,c}(m), m<n\}$. For example, a simple first order linear predictor yields $\mathbf{h}_{k,c}^{(p)}(n) = \bar{\mathbf{h}}_{k,c}(n-1)$ where only the previous CSI value is used for prediction. Next, each MT quantizes the prediction error $\mathbf{e}_{k,c}(n) = \tilde{\mathbf{h}}_{k,c}(n) - \mathbf{h}_{k,c}^{(p)}(n)$ and feeds back to BS $\mathbf{i}_{k,c}(n)$, a binary representation of the quantized vector error $\hat{\mathbf{e}}_{k,c}(n)$ using $b$ bits. Both BS and MT update the reproduced CV $\bar{\mathbf{h}}_{k,c}(n)$ by combining the predicted with the quantized prediction error, i.e.,

$$\bar{\mathbf{h}}_{k,c}(n) = \frac{\mathbf{h}_{k,c}^{(p)}(n) + \hat{\mathbf{e}}_{k,c}(n)}{||\mathbf{h}_{k,c}^{(p)}(n) + \hat{\mathbf{e}}_{k,c}(n)||}, \tag{17}$$

denoted as $+/||.||$ in Fig. 1.

In this case, the codebook of the prediction error quantizer is designed by the LBG algorithm, using the mean square error, $d(\mathbf{e}_{k,c}, \mathbf{c}_i) = \mathrm{E}[||\mathbf{e}_{k,c} - \mathbf{c}_i||^2]$, criterion. We follow the open loop approach, hence from a training sequence (TS) $\{\tilde{\mathbf{h}}_{k,c}(n)\}$ we first obtain the set of channel predictions and channel prediction errors $\{\mathbf{e}_{k,c}(n)\}$, which are then used to design of the codebook by the LBG algorithm, as described in the Appendix.

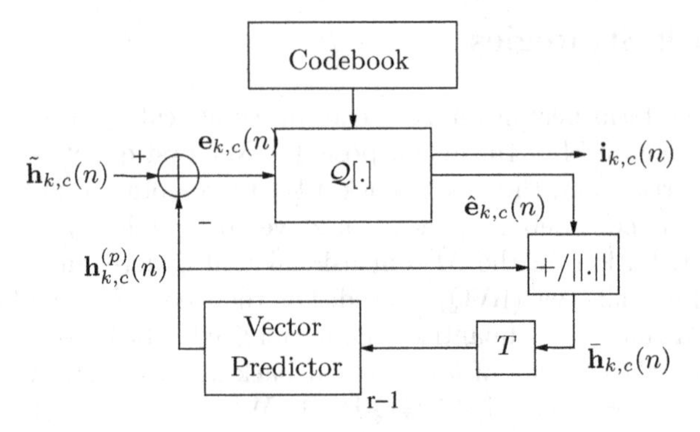


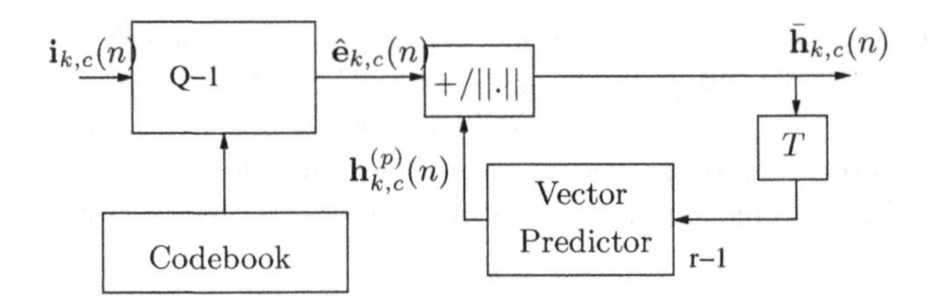


Base Station

**Fig. 1.** Predictive FB (PFB)

## 6 Performance Results

We consider a BS equipped with $M = 4$ antennas and a set of $K = 20$ users; the subcarrier frequency is 2 GHz, the transmission bandwidth is 3.84 MHz with $N = 8$ subcarriers, and the distance between two adjacent transmit antennas is 10 wavelength. The channel is modeled as time-variant, frequency selective Rayleigh fading according to the spatial channel model (SCM) [14]. The time slot duration is $T = 0.5$ ms and each MT transmits the FB once per slot. We define the average sum rate (SR) as

$$\mathrm{E}\left[\sum_{k=1}^{|\mathcal{U}(n)|} \bar{\mathcal{R}}_k(t)\right] \tag{18}$$

with $\bar{\mathcal{R}}_k(t)$ the achievable rate of user $k$ at OFDM symbol $t$ of slot $n$, i.e.,

$$\bar{\mathcal{R}}_k(t) = \sum_{c=1}^{N} \log_2\left[1 + SNIR_{k,c}(t)\right] \tag{19}$$

and [9]

$$SNIR_{k,c}(t) = \frac{p_{k,c}|\mathbf{h}_{k,c}^T(t)\mathbf{f}_{k,c}^*(n)|^2}{\sigma_n^2 + \sum\limits_{i \in \mathcal{U}_c(n) \backslash \{k\}} p_{i,c}|\mathbf{h}_{k,c}^T(t)\mathbf{f}_{i,c}^*(n)|^2} \,. \tag{20}$$

The codebook for predictive error quantization is designed from a TS composed of CVs of SCM for MT moving at 3, 50 and 130 km h$^{-1}$ with equal probability and a first order linear predictor is adopted. The value chosen for the window length in PFS is $\tau = 0.1$ s. We compare the proposed PFB strategy with

- LBG quantizer designed according to Sect. 4 and BFB strategy;
- RVQ quantizer designed according to Sect. 4 and BFB strategy;
- Perfect CSI (PCSI): perfect CSI is available at the BS at each time slot. The corresponding SR provides an upper bound.

In Fig. 2 we compare the FB strategies in terms of average sum rate as a function of the number of FB bits where the average SNR is 15 dB. PFB outperforms both RVQ-BFB and LBG-BFB, for every FB rate and MT speeds. This is due to the time correlation of the channel, which is exploited by PFB

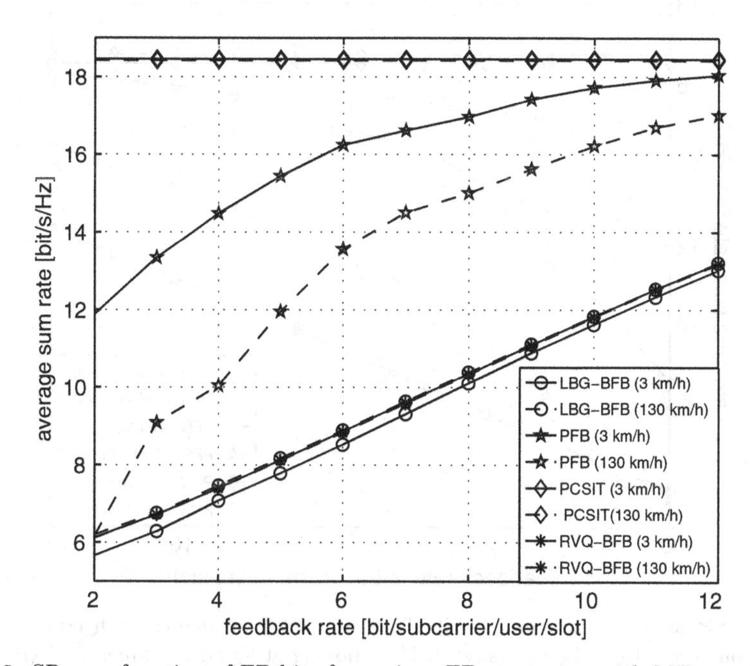


**Fig. 2.** SR as a function of FB bits for various FB strategies, with MTs moving at 3 and 130 km h$^{-1}$ and $b = 6$ feedback bits per subcarrier per user. SNR = 15 dB, $M = 4$ and $K = 20$

and disregarded by the other FB methods; in fact, PFB performance is very close to that of PCSI even with a low number of FB bits and high speeds.

We modify now the power allocation criterion, and instead of equal stream power we adopt an equal subcarrier power, i.e., we force equal power allocation across the subcarriers, $\sum_{k \in \mathcal{U}_c} P_{k,c} = P/N, \quad c \in \{1, \cdots, N\}$. In this case, users are allocated indipendently on each subcarrier and the scheduler is noteworthly simplified. However, as it can be seen in Fig. 3, this strategy yields a slightly lower sum rate. In fact, imposing all the subcarriers to be active and transmitting at the same power is not the optimal choice.

In Fig. 4, we evaluate the sum rate as a function of the system average SNR; we observe that PFB still outperforms LBG-BFB and RVQ-BFB methods for all MT speeds; the gain grows with the system SNR, as at high SNR the interference dominates noise. Again, we note a slight worsening in performance when imposing equal power allocation across the subcarriers (see Fig. 5). Anyway, Figs. 2 and 4 underline that the LBG quantization method provides better performance than RVQ for every average SNR, FB rate and MT speeds.

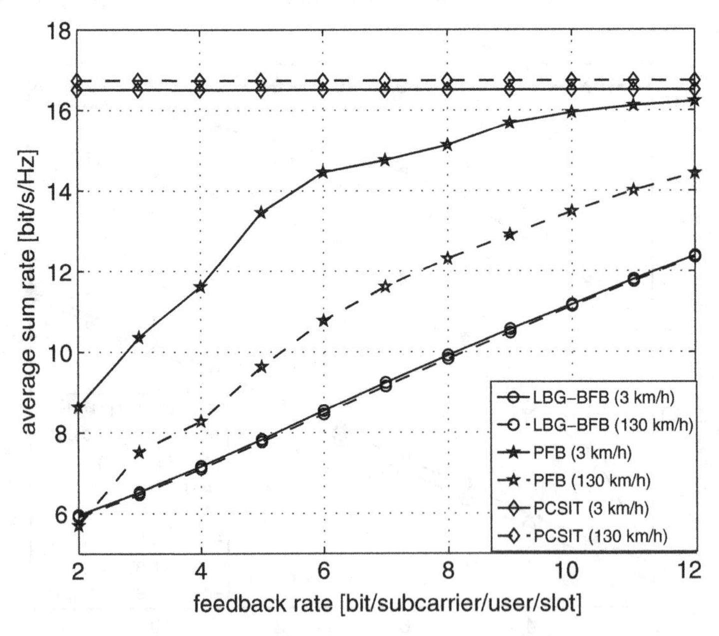


**Fig. 3.** SR as a function of FB bits for various FB strategies with equal power allocation across the subcarriers, with MTs moving at 3 and 130 km h$^{-1}$. SNR = 15 dB, $M = 4$ and $K = 20$

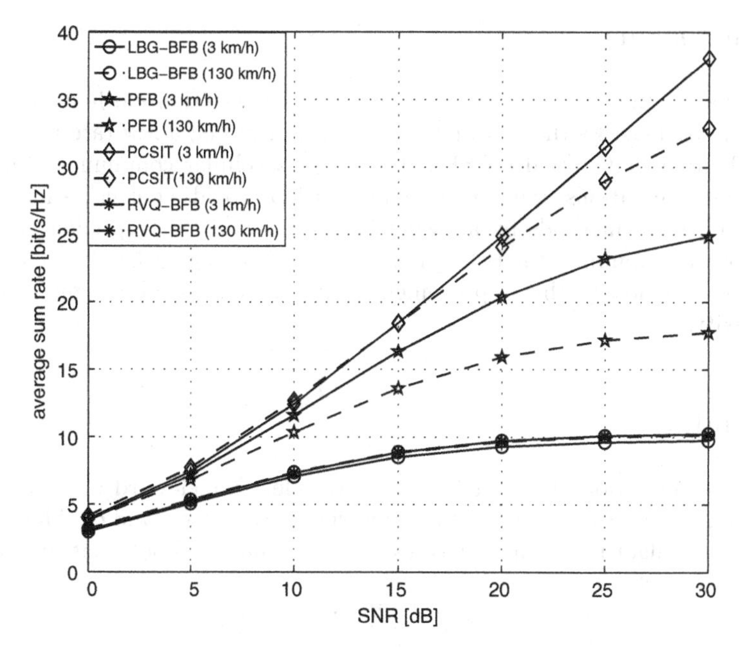


**Fig. 4.** SR as a function of SNR for various FB strategies, with MTs moving at 3 and 130 km h $^{-1}$ and 6 FB bits per user. SNR = 15 dB, $M = 4$ and $K = 20$

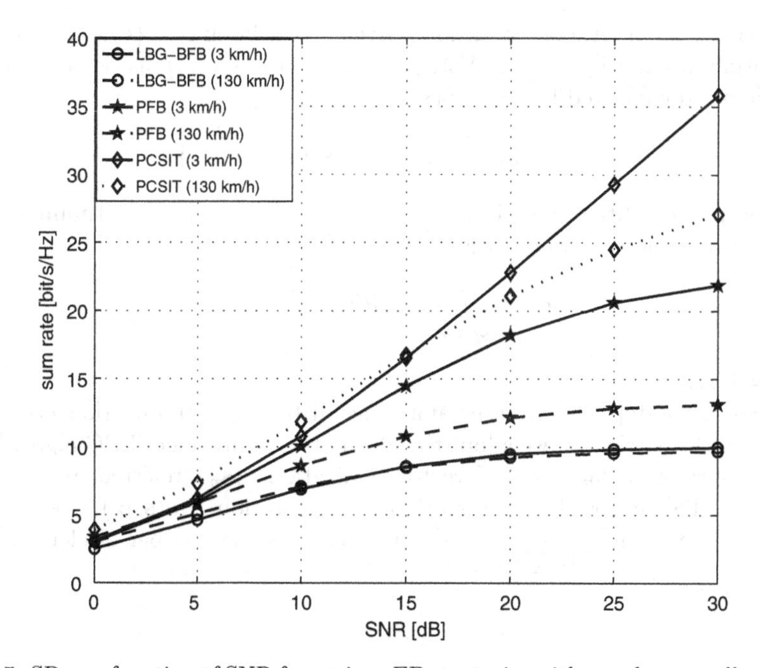


**Fig. 5.** SR as a function of SNR for various FB strategies with equal power allocation across the subcarriers, with MTs moving at 3 and 130 km h$^{-1}$. SNR = 15 dB, $M = 4$ and $K = 20$

## 7 Conclusions

In this paper, in conjunction with a suitable user selection mechanism which allocates users across the subcarriers, we have proposed a low rate predictive FB technique that provides CSI at the BS of a cellular communication system. From simulations results it is seen that PFB provides better performance than traditional methods by allowing both MTs and BS to predict channel variation, and minimize the FB signaling rate, by feeding back only the prediction error. Obviously, this improvement comes at the cost of a slightly higher complexity.

## Appendix

For the quantization of vectors $\mathbf{s} \in \mathbb{C}^M$ we select a codeword $\mathbf{c}_i \in \mathbb{C}^M$ in a codebook $\mathcal{C}$. Let $d(\mathbf{s}, \mathbf{c}_i)$ be a distance metric between the $\mathbf{s}$ and $\mathbf{c}_i$. The generalized Lloyd algorithm [15], generates the optimum codebook that minimizes the average distortion,

$$\min_{\mathcal{Q}[\cdot]} \mathrm{E}\left[d(\mathbf{s}, \mathbf{c}_n)\right] \tag{21}$$

where $\mathbf{c}_n = \mathcal{Q}[\mathbf{s}]$ is the quantized vector at minimum distortion.

The algorithm comprises two steps:

- Nearest neighborhood condition. Given a codebook $\mathcal{C} = \{\mathbf{c}_1, \ldots, \mathbf{c}_\nu\}$, the optimum partition region (Voronoi cell) $\mathcal{V}_i$, $i = 1, \ldots, \nu$ associated to the codevector indexed by $i$ satisfies

$$\mathcal{V}_i = \{\mathbf{s} : d(\mathbf{s}, \mathbf{c}_i) \geq d(\mathbf{s}, \mathbf{c}_\ell), \forall \ell \neq i\} . \tag{22}$$

- Centroid condition. For given regions $\{\mathcal{V}_i, i = 1, \ldots \nu\}$, the optimum quantization code-vectors $\mathbf{c}_i$ satisfy

$$\mathbf{c}_i = \arg\min_{\mathbf{c}_\ell \in \mathbb{C}^M} \mathrm{E}\left[d(\mathbf{s}, \mathbf{c}_\ell) \middle| \, \mathbf{s} \in \mathcal{V}_i\right] \tag{23}$$

for $i = 1, \ldots, \nu$.

These two steps are iterated until the distortion minimization criterion converges. In particular, we adopt the alternative approach of LBG algorithm, which considers a large set of vector realizations $\{\mathbf{s}_n\}$ referred as training sequence (TS) and replace the statistical expectation $\mathrm{E}\left[d(\mathbf{s}, \mathbf{c}_i) \middle| \, \mathbf{s} \in \mathcal{V}_i\right]$ by the sample average $\frac{1}{m_i} \sum_{\mathbf{s}_n \in \mathcal{V}_i} d(\mathbf{s}_n, \mathbf{c}_i)$, where $m_i$ is the number of elements of TS inside $\mathcal{V}_i$.

## References


1. D. Tse and P. Viswanath, *Fundamentals of Wireless Communication*, Cambridge University Press, 2005.
2. G. Caire and S. Shamai, "On the achievable throughput of a multiantenna broadcast Gaussian channel", *IEEE Trans. Inform. Theory*, vol. 49, pp. 1691–1706, July 2003.
3. T. Yoo and A. Goldsmith, "On the optimality of multiantenna broadcast scheduling using zero-forcing beamforming", *IEEE J. Select. Areas Commun.*, vol. 24(3), pp. 528–541, Mar. 2006.
4. D. J. Love, R. W. Heath, W. Santipach, and M. L. Honig, "What is the value of limited feedback for MIMO channels?", *IEEE Commun. Mag.*, pp. 54–59, Oct. 2004.
5. H. Ekstrom, A. Furuskar, J. Karlsson, M. Meyer, and S. Parkvall, "Technical solutions for the 3G long-term evolution", *IEEE Commun. Mag.*, pp. 38–45, Mar. 2006.
6. P. Xia and G. B. Giannakis, "Design and analysis of transmit-beamforming based on limited-rate feedback", *IEEE Trans. Signal Processing*, vol. 54(5), pp. 1853–1863, May 2006.
7. K. Mukkavilli, A. Sabharwal, E. Erkip, and B. A. Aazhang, "On beamforming with finite rate feedback in multiple antenna systems", *IEEE Trans. Inform. Theory*, vol. 49(10), pp. 2562–2579, Oct. 2003.
8. T. Yoo, N. Jindal, and A. Goldsmith, "Multi-antenna broadcast channels with limited feedback and user selection", *IEEE J. Sel. Areas Commun.*, (in press).
9. M. Trivellato, F. Boccardi, and F. Tosato, "User selection schemes for MIMO broadcast channels with limited feedback", in *Proc. IEEE Vehic. Tech. Conf. 2007 (VTC '07)*, Dublin (Irleand), Apr. 2007.
10. L. Hanzo, M. Münster, B.J. Chei, and T. Keller, *OFDM and MC-CDMA for Broadband Multi-User Communications, WLANS and Broadcasting*, Wiley, 2003.
11. M. Kountouris and D. Gesbert, "Memory-based opportunistic multi-user beamforming", in *Proc. Int. Symp. Info. Theory (ISIT)*, Sept. 2005, pp. 1426–1430.
12. A. Jalai, R. Padovani, and R. Pankaj, "Data throughput of CDMA-HDR a high efficiency-high data rate personal communication wireless system", in *Proc. Vehic. Tech. Conf. (VTC)*, May 2000.
13. A. Gersho and R. M. Gray, *Vector Quantization and Signal Compression*, KAP, 1992.
14. J. Salo, G. Del Galdo, J. Salmi, P. Kyösti, M. Milojevic, D. Laselva, and C. Schneider, "Matlab implementation of the 3gpp spatial channel model", Tech. Rep., 3GPP TR 25.996, Jan. 2005, Available on-line : http://www.tkk.fi/Units/Radio/scm/.
15. C. R. Murthy and B. D. Rao, "Quantization methods for equal gain transmission with finite rate feedback", *IEEE Trans. Signal Processing*, vol. 55(1), pp. 233–245, Jan. 2007.

# References

# Session 2

## Ad-Hoc and Cellular Networks

7

# Interferer Nulling Based on Neighborhood Communication Patterns

Robert Vilzmann[1], Jörg Widmer[2], Imad Aad[2], and Christian Hartmann[1]

[1] Technische Universität München, Institute of Communication Networks, Munich, Germany lastname@tum.de
[2] DoCoMo Euro-Labs, Future Networking Lab, Munich, Germany lastname@docomolab-euro.com

**Summary.** Beamforming using smart antennas can improve the performance of wireless devices. Beyond the increase of antenna gain in a desired direction, beam patterns can also be modified to have very small gains in the direction of interference and noise sources (nulling). In this paper we propose nulling algorithms based on the long-term communication patterns and routing tables in ad hoc networks. Simulation results show that this approach considerably improves spatial reuse and increases SNR, without necessarily increasing path length.

## 1 Introduction

Smart antennas have a number of useful properties that help to improve the performance of wireless devices. Beamforming allows to emit power in a given sector around a device, by phase shifting the signals sent by different antenna elements so that they constructively add in a specific direction. Similarly, antenna patterns can be modified to have a very small antenna gain in the direction of interferers or source of noise (nulling). Prior work has shown that smart antennas positively affect network properties such as connectivity, path length, interference, etc. in comparison to omni-directional antennas [1, 2]. The increase in hardware complexity for smart antennas and the increase in protocol complexity to exploit their properties vary significantly depending on the specific solution, but smart antennas have even been used successfully for very resource constrained devices such as sensor nodes [3].

Existing approaches can be divided into two categories. In work that falls into the first category, the beamforming direction is either fixed [3] or random [2]. Slightly more sophisticated algorithms may select the beamforming direction based on aggregate information collected from the neighborhood, e.g., setting the beamforming direction based on the number of neighbors found in specific directions [4]. This results in low MAC protocol complexity and a relatively stable topology, since the beamforming direction changes slowly (if at all). Higher performance gains are possible with schemes that fall into the second category: communication-based beamforming. Here, the beamforming direction may be changed on a per packet basis. This approach, however, substantially increases MAC layer and routing complexity, having to deal with directional deafness and very quick changes in the communication topology that may be hard to control. It also requires that nodes frequently infer the direction in which the communication partner lies through direction of arrival (DoA) estimation [5], or have very accurate information about the positions of their neighbors [6, 7].

In this chapter we explore an alternative solution intending to reduce interference and increase the spatial reuse of radio resources. It can be used without prior beamforming, but also on top of fixed or random direction beamforming. Its complexity is still limited compared to communication-based beamforming on a per packet basis. The proposed algorithms adapt the beamforming pattern to the communication pattern in a given neighborhood. This communication pattern is inferred from the forwarding table of a node, as well as the overheard traffic. We investigate a simple algorithm where nodes start out with an omni-directional antenna pattern. A node distinguishes between neighbors with which it communicates and "undesired" neighbors by looking at the local next hop information. When the node overhears a certain number of packets from a neighbor with which it does not communicate, the node places a "null" in the antenna pattern in the direction of this neighbor. To allow establishment of new routes between nodes where a null exists, it is possible to either periodically send and receive control packets in an omni-directional manner, or let the nulls time out after a certain period.

This chapter is organized as follows. Sect. 2 explains the proposed nulling approach in more detail. Modeling assumptions and the considered beamforming method are summarized in Sect. 3. Simulation results are shown and discussed in Sects. 4 and 5. We conclude with Sect. 6.

## 2 Algorithms

The goal of the algorithms is to adapt the shape of a node's antenna pattern in order to increase the SINR of communicating nodes. To this end, antenna nulls can be placed toward other nodes to attenuate interference. Furthermore, placing the antenna main lobe toward communicating nodes can enhance the signal reception. Nulling can be combined with both directional and omni-directional antenna patterns. This work will, however, only consider antenna nulls on top of omni-directional patterns.

For a given node, a neighbor is called a *desired* node if the node either transmits to or receives from that neighbor in connection with any of the data flows in the network. All nodes that are not desired nodes are called *undesired*.

## 2.1 The Basic Approach

Each node in the network annotates its neighborhood table with information from the PHY layer about signal strength. Neighbors are divided into *desired* (communicating nodes) and *undesired* (potential interferers), based on the forwarding table and previous communication events. In order to reduce the interference and thus increase the SINR, a node places a null toward the undesired neighbor with the highest signal strength. With a reasonable number of antenna elements, the remaining degrees of freedom of the antenna array can then be used to maintain the original (omni-directional) pattern as much as possible. No further topological dynamics are thus introduced, and desired communication links can be preserved.

## 2.2 Advanced Nulling Approaches

Placing a null in one interferer's direction is usually sub-optimal and a node may place further nulls (on several interferers) to increase its signal quality. However, in practice, placing a null in a given direction also changes the shape of the whole antenna pattern. Therefore, placing a null toward an interferer does not necessarily increase the SINR, since it may decrease the antenna gain toward desired nodes.

Finding the optimal solution requires a node to check all combinations of placing nulls toward interferers to attenuate their signals without decreasing the gain of the antenna main lobes, to keep a satisfactory SINR level to communicating nodes. Due to the complexity of such an approach, we adopt a simple greedy algorithm here. It may also be possible to use different beamforming techniques than the one described in Sect. 3.3. However, we use the latter in this work since it has desirable properties from a networking perspective, such as explicit null directions and transparent behavior of the antenna response, relatively low complexity, and applicability to both omni-directional patterns and gain-maximizing beamforming prior to placing antenna gain nulls.

In our greedy approach as used in the simulations section, a node first starts establishing a list of undesired and desired neighbors, and nulls the one with the highest interference power. The node then checks whether the link budget of desired neighbors is still satisfactory (i.e., no connections to communicating nodes are lost so as to maintain the existing routes), due to the possible changes in the antenna beam pattern. If the signal quality degrades below a given threshold the node removes the null. Irrespective of whether or not the null is placed, the node subsequently places a null toward the (second) next highest interferer, and the same procedure is repeated. The algorithm ends after nulling all undesired neighbors in the list, or upon reaching the nulling limitations at the node, which depend on the number of antenna elements.

An example nulling scenario is shown in Fig. 1. After nulling as many undesired nodes as possible, the links toward the desired nodes are still present.

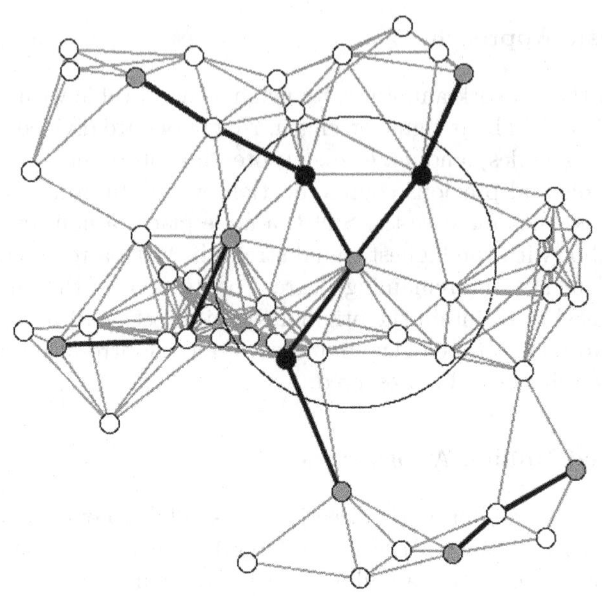

(a) Before nulling. *Dark lines* indicate data flow routes

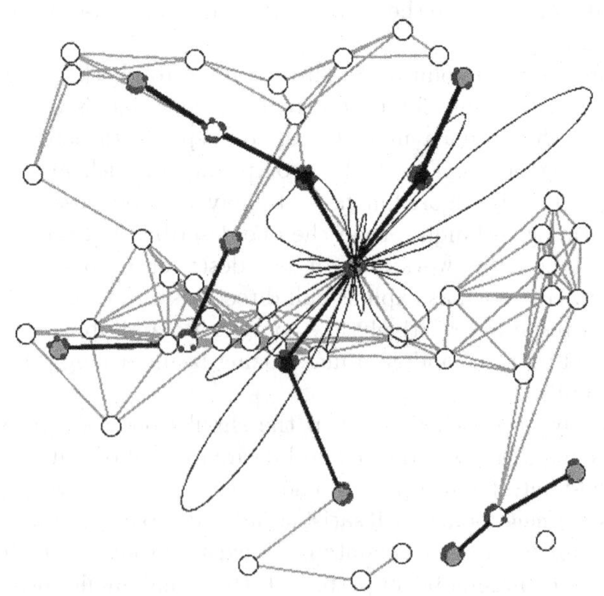

(b) After nulling. *Small gray dots* indicated antenna null directions

**Fig. 1.** Sample scenario with four data flows (*gray nodes* are end points). For the node under consideration (center of the indicated antenna pattern), the four data flows result in three desired nodes (*black*). All remaining nodes (*gray and white*) are undesired nodes, in case they are close enough to be detected

While all active nodes carry out the nulling procedure, only the pattern of the sample node is illustrated for the sake of clarity. Small gray dots indicate null directions.

### 2.3 Updating Antenna Nulls

Several factors mandate recomputing null directions:

- Changes in network topology: when old communicating nodes (respectively interferers) move out of and new ones move into a node's neighborhood.
- Changes in communication pattern: when communication sessions end and communicating nodes become potential interferers (and vice versa).
- Changes in channel conditions: when fading and shadowing may change the signal strength from communicating nodes and interferers.

The frequency of recomputing null-directions depends on the application scenario and must take the above factors into consideration.

## 3 Antenna and Network Modeling

In this chapter we apply a nulling technique to a realistic antenna model, and estimate to what extent interference can be eliminated by applying nulling on the physical layer. The corresponding modeling assumptions are summarized in this section, including the channel and link model.

### 3.1 Channel and Link Model

In order to model the wireless channel, we apply the modified free-space model with an attenuation exponent $\alpha$. By applying this model, we implicitly assume line-of-sight propagation. That is, no explicit multipath propagation is considered. With this model we can describe the received power $p_r$ at a node located at distance $s$ from a sending node transmitting with power $p_t$ as:

$$p_r = p_t \cdot g_t \cdot g_r \cdot \left(\frac{\lambda}{4\pi s}\right)^2 \left(\frac{s_0}{s}\right)^{\alpha-2}, \qquad (1)$$

where $\lambda$ is the carrier wavelength, $s_0$ is a reference distance, $g_t$ is the antenna gain of the transmitter in direction toward the receiver, and $g_r$ is the antenna gain of the receiver in direction toward the transmitter. The directive antenna gains at the nodes, which are effective for transmission as well as for reception, are given by the beam pattern of the adaptive antennas. How these beam patterns are generated is explained in Sect. 3.3.

Now that we have the channel model in place we can describe the link model, i.e., we can determine when two nodes can establish a communication link between each other. For this purpose we define a threshold power value $p_{r0}$, referred to as the receiver sensitivity. If the received power $p_r$ according to (1) is not lower than the sensitivity $p_{r0}$, we assume that a communication link can be established. According to the channel model and considering that the antenna gain patterns are effective both for transmission and reception, links are always bi-directional. Note that this does not guarantee that data packets are always received successfully on such a link. Rather, packets can still be lost when packet collisions (severe interference from parallel transmissions in the neighborhood) occur.

In addition to the receiver sensitivity $p_{r0}$, we apply another power threshold, the detection threshold $p_{rd}$: If the received power $p_r$ according to (1) is lower than the receiver sensitivity $p_{r0}$ but not lower than the detection threshold $p_{rd}$, the nodes cannot establish a communication link, since the received power would be too weak for decoding. However, the signal can still be detected and the received power is assumed to be sufficiently high in order to perform DoA estimation. Hence, in such a scenario, nodes can detect each other as interferers and are able to estimate the direction from which the interference is coming, which can then be used to advise the beamformer to produce an antenna pattern that eliminates this interference.

The actual detection threshold used in our simulations ($p_{rd} = -91\,\mathrm{dBm}$) together with the applied noise figure (temperature 295 K, bandwidth 22 MHz, noise factor 1.0) yields an SNR threshold of 9.5 dB, which should be sufficient for DoA estimation (cf. e.g., [8]), assuming that interference from other nodes does not harm the DoA estimation.

### 3.2 Antenna Model

We assume that nodes are equipped with $m$-element uniform linear array (ULA) antennas. That is, $m$ antenna elements are placed in one line with a fixed distance $\Delta = \lambda/2$ between neighboring elements, where $\lambda$ represents the carrier wavelength. Each element is modeled as an isotropic radiator. The overall transmit power of the array is fixed to $p_t$, however the power can be distributed unevenly among the antenna elements, such that $\sum_1^m p_{t,i} = p_t$, where $p_{t,i}$ represents the power radiated by the $i$th element. The power-to-element assignment is determined by the antenna steering vector $\mathbf{a} = (\mathbf{a_1}, \ldots, \mathbf{a_m})^\mathbf{T}$. The steering vector comprises $m$ complex factors, determining the amplitude and phase shift for each antenna element. The actual beamforming method, i.e., the choice of the steering vector depending on the requested number and directions of nulls in the pattern is explained in Sect. 3.3. Throughout this chapter we assume ULA10 antennas, i.e., $m = 10$ antenna elements.

### 3.3 Nulling Method

Our goal is to investigate the potential of improving the performance of a multi-hop network by nulling out interferers. However, nulling out interferers requires changes in the beam pattern of nodes, which in turn can change the connectivity of the network significantly. Obviously, we would like to be able to control these changes in topology in the following ways:

- The link to the undesired interferer should be removed, while
- The links to the other neighbors in the connectivity graph should remain. This is to avoid rapid changes in the topology and thus in the routing tables.

In order to approach these goals, we apply a beamforming method which places nulls in a desired direction while minimizing the deviation (in terms of the mean square difference) of the new pattern from the pattern before the null was added. In particular, we adopt the beamforming method described in [9]. It is briefly sketched in the following.

Since we are assuming that the antenna elements comprising the array are isotropic radiators, the antenna pattern can be represented by the *array factor*. In the case of an ULA with a spacing of $\lambda/2$ which we are considering here, we can express the array factor $\kappa(\theta)$ as a function of the angle $0 \leq \theta < 2\pi$

$$\kappa(\theta) = \sum_{n=1}^{m} a_n e^{-jn\pi \sin\theta} , \tag{2}$$

when the steering vector $\mathbf{a}$ is applied. Assume the current array factor is $\kappa_0(\theta)$, determined by the steering vector $\mathbf{a_0}$. This pattern might either be omnidirectional or it might already have some desired nulls at angles $\theta_1, ..., \theta_{\mu-1}$. Now we want to place an additional null in the pattern at angle $\theta_\mu$. Since we want the new pattern to be as similar as possible to $\kappa_0(\theta)$, we have to choose the new steering vector $\mathbf{a_a}$, yielding the new array factor $\kappa_a(\theta)$, which solves the following optimization problem:

$$\min\left\{\epsilon(\kappa_a) = \int_0^{2\pi} |\kappa_0(\theta) - \kappa_a(\theta)|^2 d\theta\right\} , \tag{3}$$

subject to

$$\kappa_a(\theta_i) = 0 , \qquad i = 1, \ldots, \mu . \tag{4}$$

This problem can be solved in close form as detailed in [9], as long as $\mu < m$. That means, we can place up to $m - 1$ distinct nulls in our pattern. Exemplifying beamforming patterns for the considered antenna configuration are illustrated in Figs. 2–4. Note that the antenna pattern of linear arrays is always symmetric. Each antenna gain null implies a null in the direction symmetric to the array.

### 3.4 Network and Traffic Model

For our simulations we generate random scenarios in which 50 nodes are independently placed according to a uniform distribution in an area of

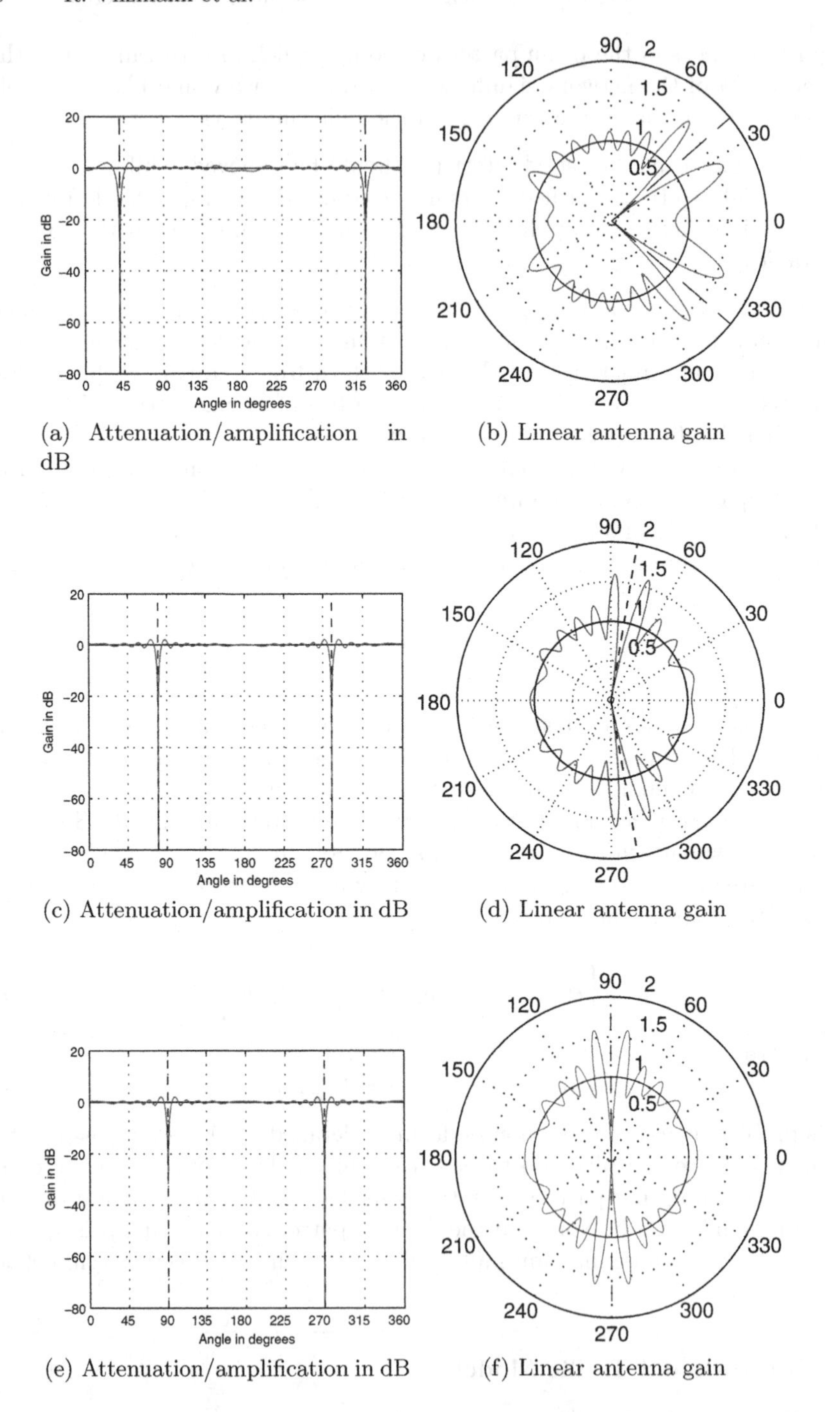


(a) Attenuation/amplification in dB

(b) Linear antenna gain

(c) Attenuation/amplification in dB

(d) Linear antenna gain

(e) Attenuation/amplification in dB

(f) Linear antenna gain

**Fig. 2.** Sample antenna gain pattern of a ULA10 with varying directions (40°, 80°, 90°) of one antenna null. Other nulls (at 320°, 280°, and 270°) result from the symmetry of linear array antennas

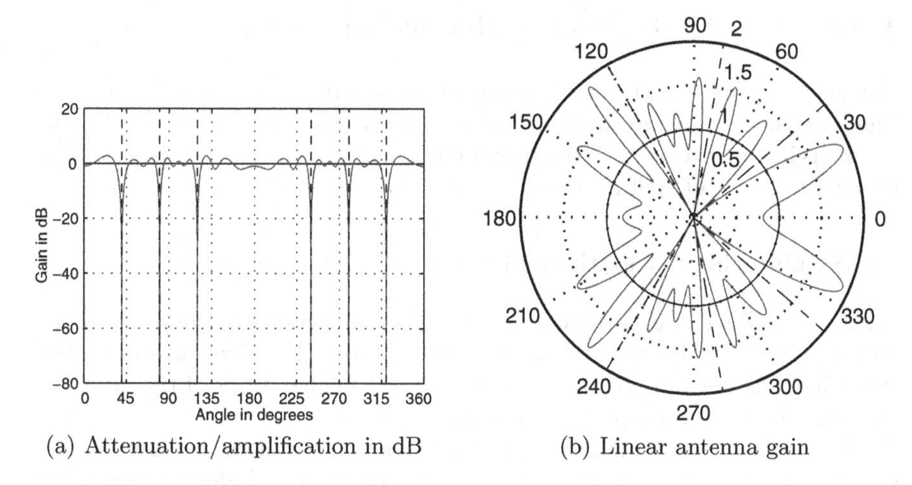


(a) Attenuation/amplification in dB (b) Linear antenna gain

**Fig. 3.** Antenna gain pattern of a ULA10 with three antenna nulls in the directions 40°, 80°, 120°

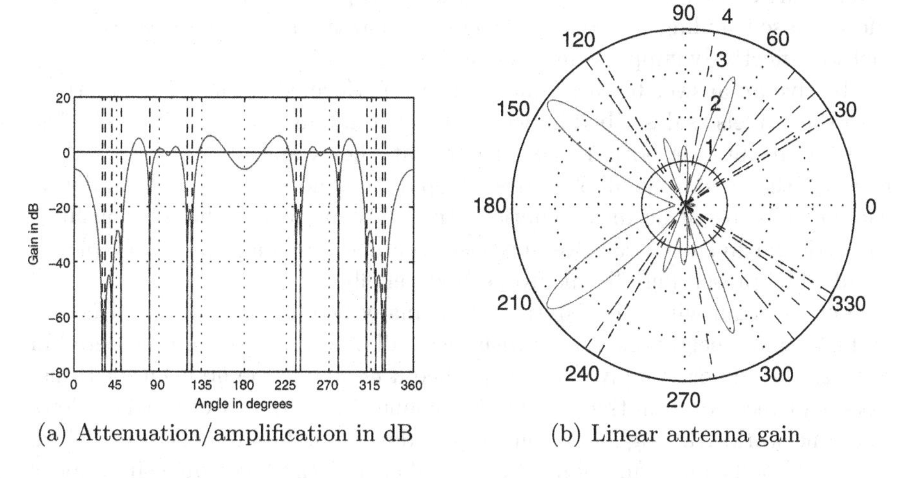


(a) Attenuation/amplification in dB (b) Linear antenna gain

**Fig. 4.** Antenna gain pattern of a ULA10 with seven antenna nulls in the directions 30°, 33°, 40°, 50°, 80°, 120°, 125°

$500 \times 500\,\mathrm{m}^2$. For each scenario multi-hop traffic is generated. Active nodes, i.e., nodes generating traffic, randomly choose one of the 49 remaining nodes as sink, and determine the shortest path to this sink in terms of the number of hops. We will look at two different setups: in the first setup, 10 nodes out of the 50 nodes are active. In the second, all 50 nodes are active and generate traffic. For all figures in this chapter, statistics were averaged over 1,000 random networks, and all contemplable nodes in these networks.

## 4 Analysis of Interference Reduction

This section provides an analysis of the above-described approach by means of simulations. We analyze the interference cancellation capabilities of adaptive antennas in connection with the proposed nulling algorithm, and the resulting benefits in terms of spatial reuse.

### 4.1 Neighborhood Relationships and Null Placement

With our nulling approach, not all contiguous interference sources can be nulled in arbitrary traffic relationship constellations. Possible gains in spatial reuse are thus not only depending on the nulling capabilities of the adaptive antennas, but also on topology and traffic scenarios.

When placing antenna gain nulls based on neighborhood communication patterns several questions arise: How many desired and undesired neighbors does a node typically see? Does the presence of desired nodes inhibit null placement, in particular since the limited number of antenna elements of the antenna array do not allow for arbitrary antenna gain shaping (cf. Fig. 4)? Can the proposed nulling approach in connection with the chosen beamforming method effectively suppress interference?

To give an answer to these questions, we performed extensive simulations with the models as described in Sect. 3. In the random network with 50 nodes, we show results for both 10 concurrent data flows and 50 concurrent data flows. In the latter case all 50 nodes set up a multi-hop data flow to randomly chosen sinks, resulting in a scenario with highly meshed traffic relationships. In particular in such a case we must ask whether antenna nulls can be placed at all while preserving all links to desired neighbors.

Let us begin with an analysis of how many neighbors are desired nodes, and how many neighbors a node wishes to null. As described in more detail in Sect. 2, a neighbor of a given node is called a *desired* if the node either transmits to or receives from that neighbor in connection with any of the data flows in the network. All neighbors that are not desired nodes are called *undesired*. The probability mass functions of the number of desired and undesired nodes for a node in the network are shown in Fig. 5.

It is now interesting the analyze how many nulls a node actually forms under the constraints of preserving desired links and limited degrees of freedom of the antenna array. The probability mass function of the number of placed nulls is shown in Fig. 6. Interestingly, even in the case of 50 active data flows, many nodes fully use the degrees of freedom (= 9) of the ULA10 for nulling.

### 4.2 Node Decoupling and Interference Suppression

Each antenna gain null is placed explicitly toward an undesired node. However, we expect that each placed null will on average suppress more than one

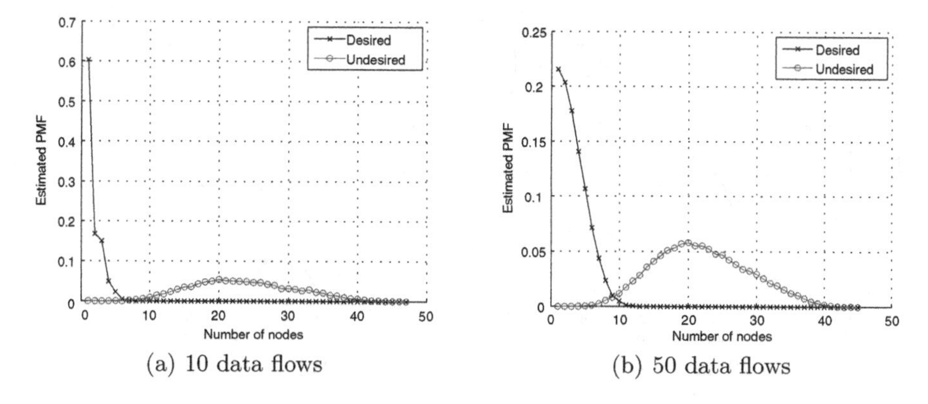


**Fig. 5.** Probability mass function of the number of desired and undesired nodes

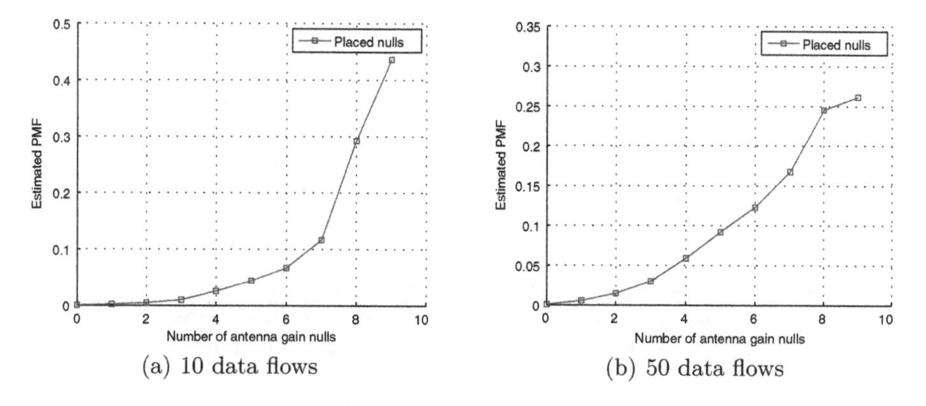


**Fig. 6.** Probability mass function of the number of placed nulls

interferer lying in the angular direction of the null. Fig. 7 shows the probability mass function of the number of nodes in sensing range before and after nulling.

As can be seen, this number is reduced dramatically. From these results we expect a significant reduction in MAC layer blocking, by explicit control messages or by carrier sensing, and a corresponding increase in the spatial reuse of radio resources.

We are also interested in the interference reduction, change in signal level of the desired signal, and SIR improvement provided by nulling. As can be seen in Fig. 8 and Fig. 9, nulling both decreases interference and increases the signal level of the desired signal, on average. Here, the notion of interference is the worst-case interference level, which occurs when all remaining nodes which are part of at least one of the multi-hop paths transmit simultaneously. As a result of increased signal level and reduced interference, the worst-case SIR (Fig. 10) benefits significantly from nulling.

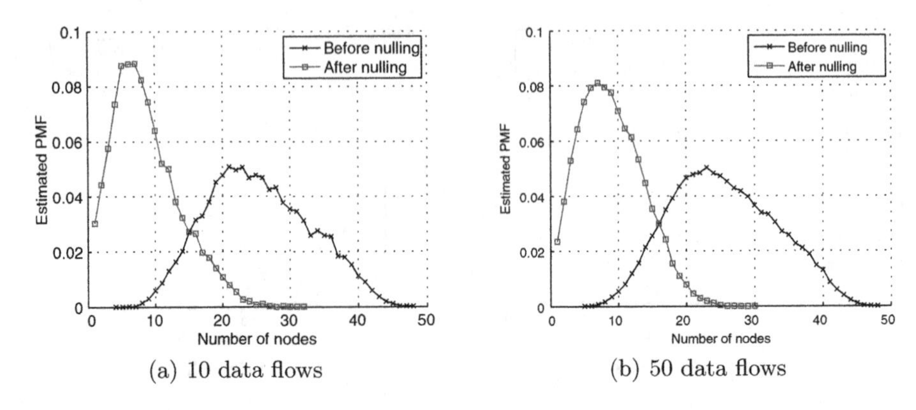


(a) 10 data flows (b) 50 data flows

**Fig. 7.** Probability mass function of the number of nodes in sensing range

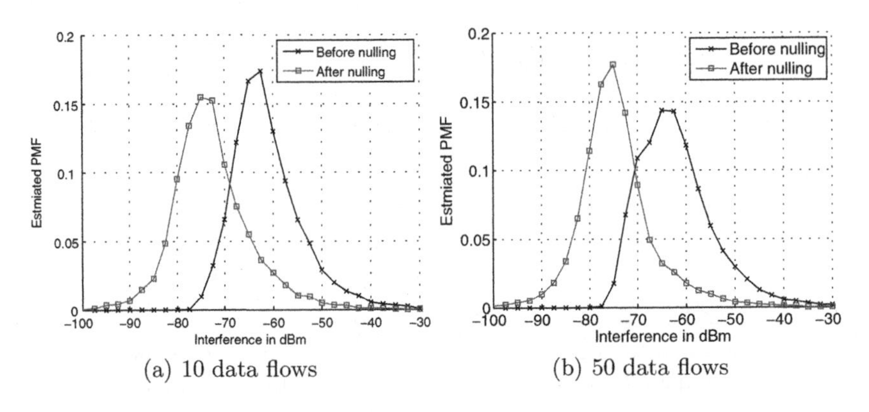


(a) 10 data flows (b) 50 data flows

**Fig. 8.** Probability mass function of the worst-case interference level

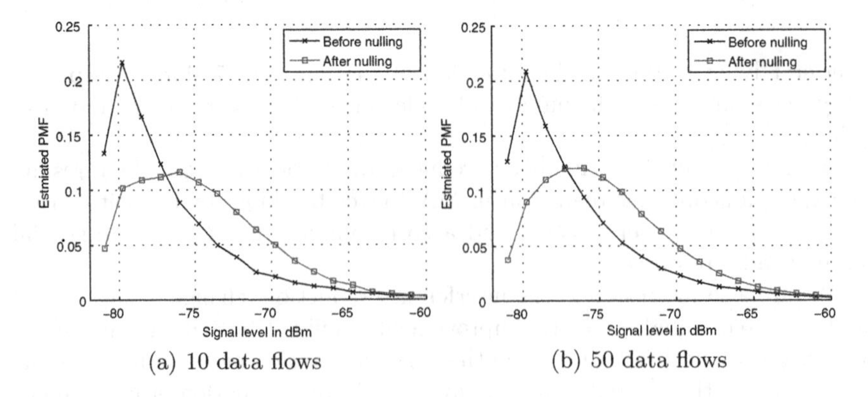


(a) 10 data flows (b) 50 data flows

**Fig. 9.** Probability mass function of the signal level of the desired signal

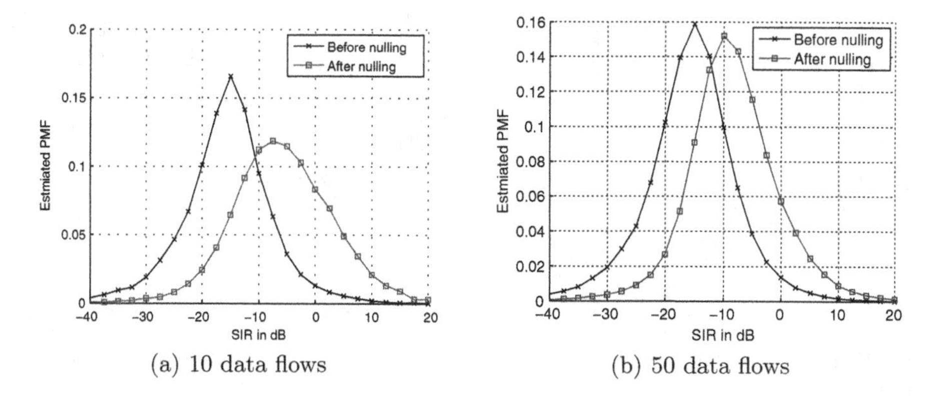


(a) 10 data flows

(b) 50 data flows

**Fig. 10.** Probability mass function of the worst-case signal-to-noise ratio

# 5 Aspects of Connectivity and Routing

The previous sections showed how placing antenna gain nulls can drastically reduce interference. Thereby, our nulling approach tries to ensure that desired links persist. Antenna nulls may still prohibit desired links in case of movement or rotation of mobile nodes, changes in communication relationships, and other reasons for necessary changes of communication paths, such as shadow fading. The involved loss of desired links might induce losses in terms of connectivity, and result in an increase of path lengths. Means to counteract these shortcomings could be as follows:

- Updating antenna nulls by repeating the process of neighborhood exploration and DoA estimation. Factors mandating such a re-computation of null directions have been summarized in Sect. 2.
- Regular or tentative removal of some or all antenna nulls, as a conservative means to avoid long-term disconnection and routing detours.
- Analysis of routing protocol control packets: with certain routing protocols, a node may infer that a (control) packet has traversed one of its nulled neighbors. Removing the corresponding antenna null can eliminate the obviously existing route detour.

In the following, we will go one step back by asking the following question: With our nulling approach, do connectivity and path lengths deteriorate at all for the network scenario considered in this work? If so, how much?

We answer this question by dynamically changing traffic relationships in the network. In particular, upon adapting antenna nulls to desired and undesired nodes in the neighborhood, we let each node *reselect* its chosen sink. While the prevailing nulls have been placed in response to the initially chosen sinks and the resulting communication paths, the new sinks now have to be reached via the links that remained after this null placement.

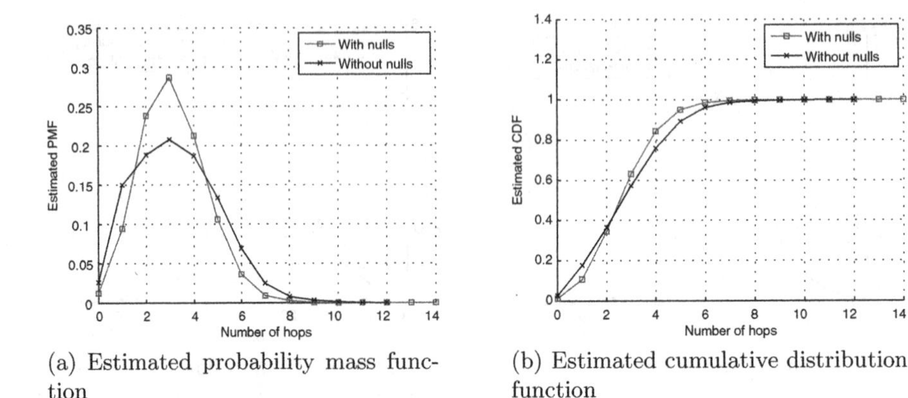


(a) Estimated probability mass function

(b) Estimated cumulative distribution function

**Fig. 11.** Path length distribution from source to reselected sink

Upon reselecting sinks and performing routing to these sinks, we will numerically analyze the resulting path length distribution. We then compare the results to the path length distribution we would have obtained if no nulls would have been placed at all.

In the same network setup as considered above, all 50 nodes set up a data flow toward a randomly chosen sink, separately. The results for the path length distribution for the reselected sinks are shown in Fig. 11. On the horizontal axis, a number of hops equal to zero means that the sink cannot be reached (disconnected), and a number of hops equal to one means that the sink is a one-hop neighbor. The curve labeled "Without sinks" represents the path length distribution that would be obtained if the antenna nulls placed upon selecting the initial sinks would be removed before routing to the reselected sinks.

Surprisingly, it can be observed that nulling does not deteriorate the path length distribution. Less long paths exist when the nulls based on the initially chosen sinks/routes are maintained. This means that although the antenna nulls have been placed for different traffic relationships, the same antenna gain patterns let the new sinks often be reached in a smaller number of hops.

This result can be explained by looking at the antenna patterns depicted in Sect. 3, and relating to previous work on so called *Random Direction Beamforming* [2] in multi-hop networks. Antenna gains exceeding a factor of one can lead to very long links in the network. These long links can provide shortcuts toward the sink which are not available with omni-directional antennas. On the other hand, the number of neighbors (node degree) typically decreases both with gain maximizing beamforming and interference nulling. For this reason, the number of one hop paths is shorter with nulling than with omni-directional patterns (cf. Fig. 11, data point for one hop).

The means to counteract disconnection listed above may still be necessary in practice. Nevertheless, the results shown here indicate that for many scenarios the proposed nulling scheme does not inhibit a well-connected multi-hop network, even when nulls are not up-to-date with respect to current communication paths.

## 6 Conclusions and Future Work

In this chapter we proposed algorithms to adapt the antenna patterns of nodes with smart antennas based on the long-term communication pattern observed in a node's neighborhood. This provides a useful trade-off between complexity and performance, and may be more useful in practice than per-packet beamforming in connection with per-packet DoA estimation.

The provided simulation results indicate that the proposed scheme is beneficial in terms of interference and the decoupling of nodes contending for radio resources. In case of outdated neighborhood information, spuriously placed antenna gain nulls do not affect the connectivity of the wireless network.

For future work, we intend to extend our algorithms to also support nulling of interference from neighbors outside the communication range, and external interference from other systems. Further, we are interested in topological effects of using nulling on top of gain maximizing beamforming as used in previous work. In our current work, we go beyond topological analysis and evaluate the MAC layer performance when using the proposed nulling scheme. Initial results show that the benefits described here also translate into remarkable improvements in terms of the end-to-end throughput.

**Acknowledgments**

The authors would like to thank Diego Herrero for implementing the adaptive beamforming algorithms used in this work.

## References

1. R. Choudhury, X. Yang, R. Ramanathan, and N. Vaidya, "On designing MAC protocols for wireless networks using directional antennas," *IEEE Transactions on Mobile Computing*, vol. 5, May 2006
2. R. Vilzmann, C. Bettstetter, D. Medina, and C. Hartmann, "Hop distances and flooding in wireless multihop networks with randomized beamforming," in *Proc. ACM International Workshop on Modeling, Analysis, and Simulation of Wireless and Mobile Systems (MSWiM)*, (Montreal, Canada), Oct. 2005
3. N. Malhotra, M. Krasniewski, C. Yang, S. Bagchi, and W. Chappell, "Location estimation in ad hoc networks with directional antennas," in *IEEE International Conference on Distributed Computing Systems (ICDCS)*, June 2005
4. R. Vilzmann, J. Widmer, I. Aad, and C. Hartmann, "Low-complexity beamforming techniques for wireless multihop networks," in *Proceedings of IEEE Communications Society Conference on Sensor, Mesh and Ad Hoc Communications and Networks (SECON)*, Sept. 2006

5. M. Takai, J. Martin, A. Ren, and R. Bagrodia, "Directional virtual carrier sensing for directional antennas in mobile ad hoc networks," in *Proceedings of MobiHoc*, 2002
6. T. Korakis, G. Jakllari, and L. Tassiulas, "A MAC protocol for full exploitation of directional antennas in ad-hoc wireless networks," in *Proceedings of MobiHoc*, 2003
7. S. Roy, D. Saha, S. Bandyopadhyay, T. Ueda, and S. Tanaka, "A network-aware MAC and routing protocol for effective load balancing in ad hoc wireless networks with directional antenna," in *Proceedings of MobiHoc*, 2003
8. Z. Shan and T.-S. Yum, "A conjugate augmented approach to direction-of-arrival estimation," *IEEE Transactions on Signal Processing*, vol. 53, pp. 4104–4109, Nov. 2005
9. H. Steyskal, "Synthesis of antenna patterns with prescribed nulls," *IEEE Communications Magazine*, no. 2, pp. 273–279, 1982

# 8

# On the Beneficial Effects of Cooperative Wireless Peer-to-Peer Networking

L. Militano[1], F.H.P. Fitzek[2], A. Iera[1], and A. Molinaro[1]

[1] Università Mediterranea di Reggio Calabria, Italy,
leonardo.militano@unirc.it, antonio.iera@unirc.it,
antonella.molinaro@unirc.it
[2] Aalborg University, Denmark, ff@kom.aau.dk

**Summary.** Conventional cellular systems efficiency is currently being challenged by the continuously changing and more demanding world of mobile services and communications. The major problems in conventional cellular communication systems are the limited autonomy of battery-powered mobile devices and the low data rate of currently available services. In this chapter a viable solution is presented to overcome the aforementioned limitations by proposing the concept of peer-to-peer cooperation among mobile phones. Cooperation with instantaneous pay-off is the key to break the trade-off between complexity and energy consumption. In this chapter we highlight one possible scenario of cooperative data reception among users of a cellular network which benefits both the end user in terms of energy consumption, data rate, and transmission delay, and the network provider with better resource sharing and revenues coming from exploitation of new service paradigms.

## 1 Introduction

Fourth generation (4G) wireless communications systems are requested to give an answer to all crucial needs raised by emerging applications and services conceived for mobile devices. Most of the new applications, such as TV-on-mobile and video streaming, are quality driven and put very demanding constraints both on the system and on the end-user device performance. A thorough analysis of service levels offered by the system is indeed of utmost importance when planning the launch of innovative services.

Recent studies on novel quality of service (QoS) paradigms advise to focus on the quality the user perceives. Usually, users are only aware of either how clear an image appears to them or how often they need to recharge the battery of the device. Clearly, high-level service requirements can be matched by addressing low-level target performances, such as data rate increase, transfer delay reduction, and energy saving. Currently deployed cellular systems are limited by low data rate and high terminal energy consumption; this is why unconventional forms of user communications and network synergies are needed to meet highlighted exigencies.

In this chapter a peer-to-peer *cooperation* paradigm is conceived among cellular subscribers cooperating over a short-range radio network (such as WiFi or Bluetooth) with the aim of achieving mutual beneficial effects.

The chapter is organized as follows. In Sect. 2 we give a brief description of cooperation as communication paradigm in wireless networks. The reference scenario for our analysis is introduced and discussed in Sect. 3, while the test-bed set-up implementation will be described in Sect. 4. The experimental results of the implemented scenario are presented in Sect. 5. Concluding remarks are given in Sect. 6.

## 2 Cooperative Communications Paradigms

Cooperation in wireless networks is a communication paradigm of increasing interest that is nowadays testified by intense and very different research. A well-known approach to cooperation includes multi-hop or relaying techniques, cooperative diversity and antennas, cooperative coding, and so on. Another emerging approach that is particularly relevant to 4G networks, requires cooperation between heterogeneous networks (e.g., cellular and short-range wireless communications). This latter approach, whose main potentialities and characteristics have been discussed in [1], [2] and [3], is followed in this chapter.

Several ideas behind the concept of cooperation are inspired by nature and animal behavior. Cooperation is defined in [3] as the strategy of a group of entities to achieve either a common or an individual goal. Furthermore, cooperation can be seen as the action of obtaining advantages by giving, sharing or allowing something.

In principle, cooperation has the potential to enhance key capabilities and performance figures of wireless communications, such as increasing the data rate, limiting power consumption, enhancing spectrum efficiency and capacity. On the other hand, cooperative communications among users and synergies among networks require new business and cost models to be studied by network and service providers.

A good classification of cooperative communication, first introduced in [3], distinguishes three levels of cooperation, namely *implicit cooperation*, *explicit macro cooperation* and *explicit micro cooperation*. Implicit cooperation is characterized by fairness and mutual respect among entities in a passive way. Involved entities share common resources in a fair way without any further gain (an example is given by a medium access control policy). According to explicit macro cooperation, the cooperating entities are wireless system terminals, access points, router, which directly interact with each other to reach a common or individual goal (an example is given by the multi-hop relaying scenario). The explicit micro cooperation level, instead, involves parts of the entities, such as antennas, batteries, CPUs, or other terminal hardware components. The energy saving strategies reported in this chapter refer to an architecture implementing an explicit micro cooperation model. This latter exploits the potentialities of peer-to-peer communication.

Users typically have a natural disincentive to cooperate in a peer-to-peer content-sharing network [4], which often has the "free riding" problem as a common consequence. Free riders are users who benefit from the resources of others without giving their own resources in exchange [5]. Notwithstanding, effective incentive mechanisms for aforementioned architectures have been introduced in literature. Based on a Evolutionary Prisoner's Dilemma (EPD) model in [6] and on a Generalized Prisoner's Dilemma (GPD) model in [7], incentives for cooperation are proposed and analysed. In [8] a new mechanism in a BitTorrent file distribution system is proposed being more robust against free riders.

## 3 The Reference Scenario for Cooperation: Motivations and Benefits

In Fig. 1 the cooperative multi-network communications architecture, that is used as a reference in this chapter, is represented. Multimode terminals are provided with both cellular and short-range communication capabilities. Cellular technologies can be either General Packet Radio Service (GPRS) or

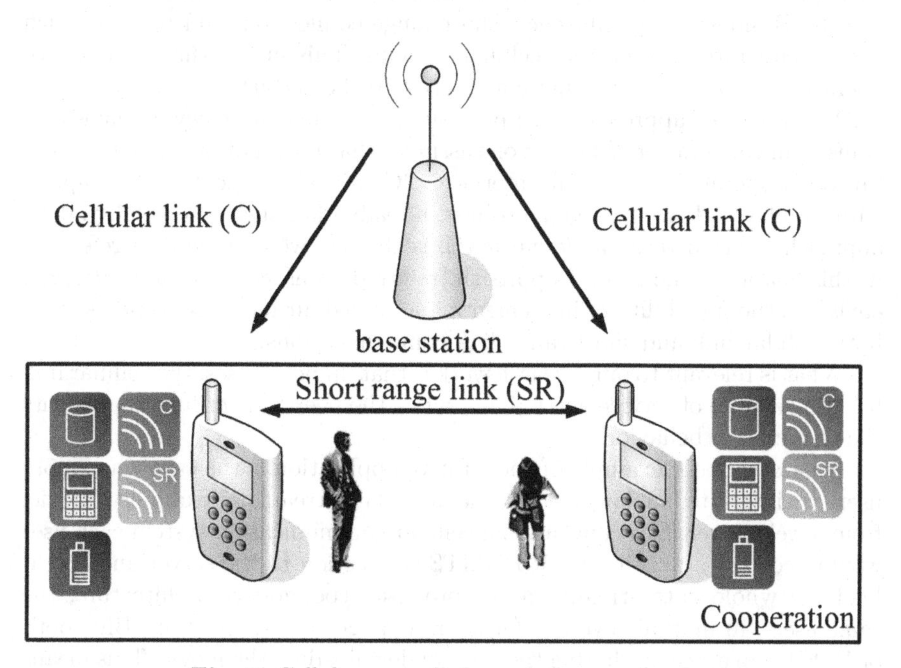


**Fig. 1.** Cellular controlled peer-to-peer network

Universal Mobile Telecommunication System (UMTS); Bluetooth and Wireless Local Area Network (WLAN) technologies are a viable solution to short-range connection needs.

The depicted scenario implements micro cooperation based on the synergy between long range and short-range involved networks. Each terminal is connected to the base station (or access point) by means of a cellular link and, simultaneously, with other terminals on a short-range link communication basis. As briefly introduced in Sect. 2, free riding is one of the main problems in peer-to-peer communication paradigms. Nevertheless, in this chapter we assume a *fair* cooperative behavior of the user terminals. This means that each user is available both to benefit from the resources of others and to offer his own resources to others in exchange.

Complementarities and capabilities of cellular and short-range radio networks can be beneficially exploited by cellular subscribers operating in this cooperative scenario. Cellular link has the key role of service entry point, while the peer-to-peer short-range link implements the cooperative user's behavior. Enabling cooperative peer-to-peer communication between terminals over short-range links while simultaneously downloading data over cellular links can achieve energy saving and data rate improvements. It is worth recalling that cellular links between base station and user devices need high power to cover some kilometer distances, while device-to-device short-range connections need low power to cover tens of meters only. This feature is the basis of all energy consumption related benefits offered by the novel architecture. Besides energy efficiency, short-range connections can provide much higher data rate than current cellular systems. This makes the reference cooperative architecture even more attractive to the end-user.

Two different approaches are possible in the view of achieving envisaged gains from cooperation. The first one assumes that terminals are able to simultaneously operate over both air interfaces. 2G and 3G mobile devices equipped with cellular and short-range interfaces, already offer such a possibility. This approach referred to as *multi modality*, is the one we assume as a reference in this chapter. The second approach, referred to as *common air interface*, envisages the possibility of implementing a unified air interface to be used for both cellular link and short-range link communications.

What is relevant to our research work is that, in both cases, the cellular link has the key role of service entry point, while the peer-to-peer link implements the cooperative behavior.

A paradigmatic example of cooperative application is the case when more users are interested in downloading the same data stream, for example a movie, from a web server. In a conventional cellular communication system each user would need to establish a GPRS/UMTS connection to the server and download the whole data stream. In the proposed cooperative architecture, the same users, in mutual coverage for a short-range connection (e.g., Bluetooth or WiFi), can agree on sharing the costs of downloading the movie. This means that they will download different fractions of the movie from the server over the GPRS/UMTS connection and, by setting up their short-range connections, they will cooperatively exchange the movie portions.

All users will achieve their personal goal of receiving the complete amount of data belonging to the desired service. Noteworthy, the objective is reached by downloading only a fraction of the data over the expensive, slow, and high energy demanding cellular link.

## 4 The Cooperation Environment Test-Bed Setup

The study and performance evaluation of the envisaged *micro cooperation architecture* has to pass through the choice of a reference test-bed scenario. Attention of our research is on a situation in which two sample users are interested in downloading the same data, for example an mp4 video file, from a web server. The two terminals will download different shares of the same video and, by setting up a short-link connection, they will exchange the correspondent shares of data in a cooperative fashion.

Both terminals will achieve their personal goals of receiving the complete amount of data belonging to the desired service. Consequently, an "egoistic" reasoning driven by the instantaneous pay-off will motivate each terminal to cooperate with a possible peer entity.

The set-up of a test-bed for a GPRS-Bluetooth cooperative architecture has been deployed. The achievement of the objective passed through the study and design of the overall novel cooperative architecture that required:

- The accomplishment of the cooperative terminal behaviors
- The development of graphical user interfaces for the cooperative application
- The integration of the Bluetooth connection set up into the application
- The implementation of the HTTP transaction management on cellular links
- The exploitation of HTTP/1.1 Range Request [9] feature for a dynamic, server independent and peer-failure tolerant file-sharing

Two cooperative mobile phones equipped with Bluetooth network cards have been used for a first measurement campaign. The test-bed scenario also foresees a remote server, where the two mobile terminals can download an mp4 file from. Two different data delivery approaches have been implemented: a *static server-aware* approach and a *dynamic server-unaware* approach.In both cases the server will deliver a complete file to the non-cooperating clients, while partial files will be delivered to cooperating clients.

### 4.1 The Static Server-Aware Approach

We refer as *static sever-aware* approach the case in which the server needs to be provided beforehand with the necessary data to support the cooperative-willing entities. In the specific implemented case that the cooperating devices are two, two files are available to the cooperating clients from the server, which are exactly the first and the second half of the whole file. An 2.546 kB mp4 file has been used for testing purposes; it corresponds to two files of 1.273 kB in the static cooperative case. When a mobile is not cooperating with the other one, it will download the complete file. When the two devices are cooperating, the master of the Bluetooth link will download the first half of the whole file and the slave of the same link will download the second half of it.

Each device, before starting any transaction, can take the decision of either downloading the complete file by itself, or first trying to establish a Bluetooth connection with another available device. After the Bluetooth connection is set up, the device will know which half of the file to download. At the end of the cooperative transaction, both devices will have received both parts of the file. They will then build the complete movie file by simply appending data received over the cellular link to data received over the Bluetooth connection. The complete mp4 file can eventually be displayed and watched by each user on his own terminal.

This really simplified scenario, presenting many limitations and assumptions, can be extended to more devices. This requires the upload of a proportional number of shares of the file to the server in a static way. Moreover, problems can arise in case of Bluetooth link failure, caused for example by the decision of one of the two peers to stop the transaction. The user would need to start a new non-cooperative transaction to the server to fetch the complete file and will be loosing all previously made efforts. To answer these issues and to obtain a better performing cooperative scenario, a second alternative approach based on 1.1 version of HTTP protocol has been studied.

### 4.2 The Dynamic Server-Unaware Approach

The exploitation of HTTP/1.1 protocol features fits perfectly our cooperative *dynamic server-unaware* scenario. HTTP/1.1 *Range Request* is one of the main features introduced to improve previous protocol version. The Range Request header is intended to support dynamic data delivery. As an example of exploitation of this feature, we mention a parallel-access scheme proposed in [10]. In the proposed architecture, end-users access multiple servers at the same time, getting by means of Range Request calls, different portions of a file from different servers and reassembling them locally. The end result is that users experience significant speedups. Moreover, one of the main advantages stated in [10], is the dynamic aspect of the access scheme which does not require any modifications to servers or to content.

In the implemented test-bed, the two cooperating peers will perform an HTTP HEAD call to the server to obtain meta-information about the file to access. This call will be made over the cellular link, but will not cause the transfer of the entity-body itself. The data size information, retrieved from the information obtained after the HTTP HEAD call, will be used to set the bytes Range-Request of the subsequent HTTP GET call.

In the test-bed scenario the master and the slave will fetch, respectively, the first and the second half of the data. In a scenario involving more terminals, a valid alternative algorithm needs to be studied.

The main advantage of implementing HTTP/1.1 *Range Request* feature is that we are not forced to upload additional files to the server to support the cooperative service. Moreover, in case of a Bluetooth link failure, the peer will not loose previously made efforts, but will simply perform a new GET call to the server to fetch the missing bytes over the cellular link. For these reasons we refer to this approach as a *dynamic server-unaware* approach.

The main limit for a wide-spread use of this algorithm, is that old HTTP/1.0 servers present on the Web do not support the Range Request header. Moreover, as stated in [11], HTPP/1.1 servers, when a client makes a Range Request specifying one or more contiguous ranges of bytes, can either return one or more ranges in the response or ignore the Range header.

Clearly, the described scenario is a simplified representation of a real architecture, in which reasonably more than two terminals will cooperate. Notwithstanding, it is an effective test-bed for experimental measurements aiming at testifying the benefits introduced by cooperation in terms of energy consumption, data rate and transfer time. The power saving potentialities of the proposed communication architecture are easily predictable. The performance comparison between the *stand-alone* and the *cooperative* model, which is the subject of our simulation campaign, will show the exact measure of achievable gains.

## 5 Implementation Issues and Measurement Results

Mobile phones used for the performance evaluation campaign (Nokia N70) are based on Symbian operating system (OS). Efficient management of simultaneous events behaving asynchronously and the good interaction level with other services on the Symbian device, are Symbian OS features offering the opportunity to effectively support the desired cooperative behavior.

A mobile application has been developed for aforementioned devices, to implement the test-bed scenario described in previous sections. For the set-up of the Bluetooth connection a Bluetooth module, implemented by using Symbian OS Application Programming Interfaces (API), has been integrated into the application. The module was developed by the Mobile Device Group of Aalborg University [12] and has been the basis for the implementation of the transactions on the short-range link and for the accomplishment of the cooperative terminal behavior. As for the cellular data transmission, we have implemented an HTTP transaction on a GPRS link. Timer implementations within the mobile application have been required to detect elapsed time and transfer rate during transactions. The test-bed application has been provided with the design of a graphical interface depicting the instantaneous and

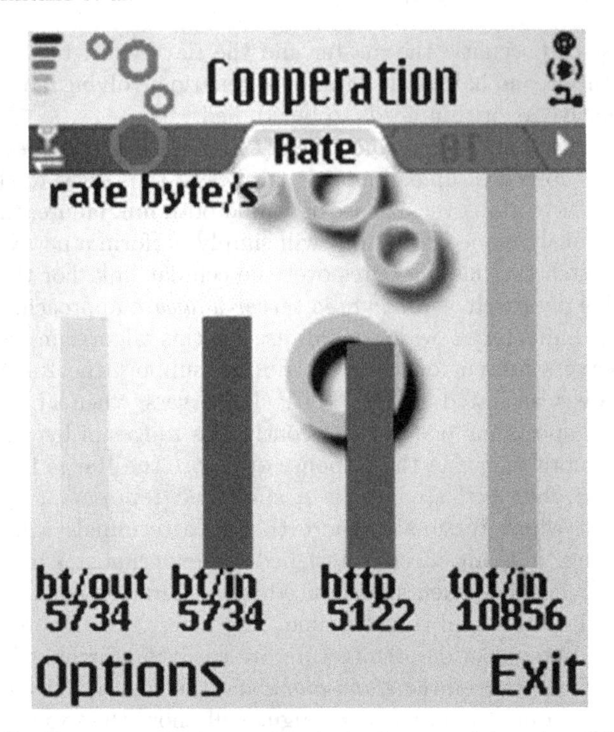


**Fig. 2.** Application screen-shot reporting instantaneous data rates of the cooperating terminals on the two simultaneously working networks

total resulting values of transmission time and download rate. The application screen-shot in Fig. 2 refers to one of the four implemented views.

During the cooperative transaction the download rate is displayed, split into four different terms: *Bluetooth Out*, *Bluetooth In*, *HTTP In*, and *Total In*, respectively. All values are instantaneous values in bytes $\mathrm{s}^{-1}$, considered on a 10 s interval. The first three values are representative of the download rate over the short-range (sending rate and receiving rate) and over the cellular link connection, respectively. The *Total In* value is defined as the *virtual data rate*, representative of all data instantaneously incoming at the mobile device.

The application screen-shots in Fig. 3 compare a cooperative and a non-cooperative scenario. Specifically, results refer to two complete data downloading processes with the same data size. The two screen-shots on the left side of the figure, refer to the process performed according to the cooperative behavior, while the two screen-shots on the right side of the figure are representative of the non-cooperative behavior. From this glint of comparison between the two cases, the enormous benefit each cooperating device gains, in terms of transmission time and download rate, clearly appears. In the specific case, the upper part of the figure shows a comparison of the instantaneous data rate at the devices on the different involved networks. Comparing the

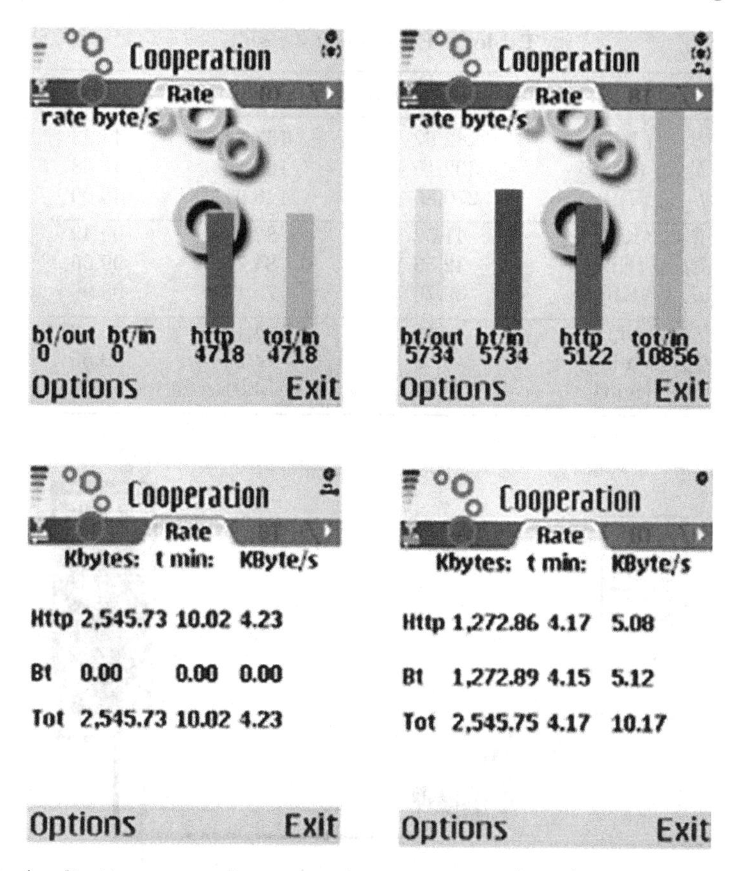


**Fig. 3.** Application screen-shots reporting cooperative (*right*) and non-cooperative (*left*) instantaneous and final download rate, and transmission time results

instantaneous total at device incoming rate(tot/in value in the screen-shots), the gain in the cooperative case is clearly depicted. The lower part of the figure shows an overall average data rate of $10.2\,\mathrm{kB\,s^{-1}}$ registered by the cooperative device (on the right side) when transaction is completed. This value, compared to the $4.2\,\mathrm{kB\,s^{-1}}$ of the non-cooperative device (on the left side), is more than double, and consequently the transmission time is more than halved. Even if this is a very optimistic case, the overall average results of the measurements campaign are reported in Table 1.

As for the energy consumption evaluation, the circuit showed in Fig. 4 has been used, where the 3.7 V battery of the mobile phone is directly connected to a multimeter. The multimeter is set to read the average value of current used by the phone downloading the desired data flow over the chosen measurement period.

To obtain the best comparison is possible between traditional and cooperative scenarios, measurements have been taken in the same physical position,

**Table 1.** Experimental results

| | No cooperation | Cooperation | Gain (%) |
|---|---|---|---|
| $E_{avg}$ (J) | 282.62 | 157.34 | 44.33 |
| $E_{max}$ (J) | 302.15 | 167.15 | 45.68 |
| $E_{min}$ (J) | 270.86 | 146.01 | 46.09 |
| $R_{avg}$ (Kbps) | 41.52 | 82.8 | 99.42 |
| $R_{max}$ (Kbps) | 42.56 | 84.76 | 99.06 |
| $R_{min}$ (Kbps) | 39.76 | 79.12 | 98.99 |
| $T_{avg}$ (min) | 8.18 | 4.10 | 49.88 |
| $T_{max}$ (min) | 8.52 | 4.29 | 49.65 |
| $T_{min}$ (min) | 7.97 | 4.01 | 49.67 |

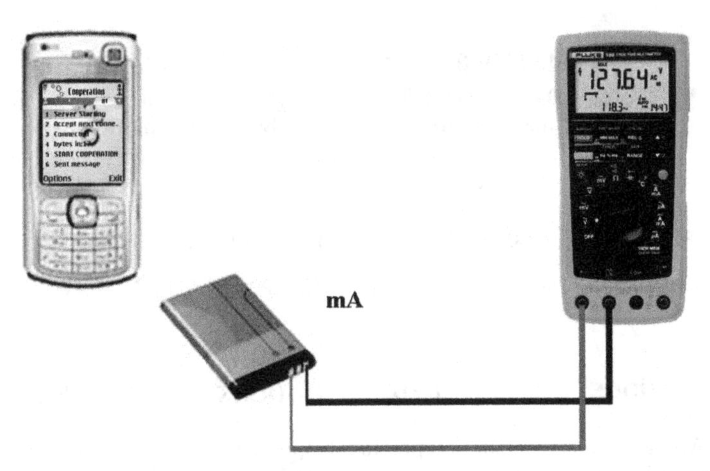


**Fig. 4.** Energy consumption measurement circuit. Multimeter is connected to one of the two cooperating terminals

alternating a non-cooperative with a cooperative download from two devices. Measurements have been repeated in different days, at about the same time in the evening. This is because during that period of the day the lower cell load gave more stable and better comparable results. According to (1), the energy value has been obtained by multiplying the voltage by the current and by the time

$$\mathbf{Energy(J) = Voltage(V) \times Current(I) \times Time(s)} \qquad (1)$$

The main benefits obtained by cooperation in the simplified reference test scenario are detailed in Table 1:

- *Energy consumption.* Energy consumption results are the most interesting ones among those obtained and, maybe, the less obvious as well. By comparing the energy consumption values of the cooperative and the non-cooperative scenarios, a manifest gain in the cooperative case has been obtained. Energy consumption gain reached peaks of 46%, having an average value of 44.33%. This means, by using a practical example, that the mobile terminal end-user, will, for instance, need to recharge his phone almost one time every second day, instead of each day. The reason for this gain, is that the short-range Bluetooth connection consumes very much less energy than the long range cellular GPRS connection. Each device, downloading half of the data over GPRS and exchanging the received data with another device over Bluetooth connections, will experience the cited energy consumption gain.

- *Virtual data rate* (total at terminal incoming rate, deriving from data simultaneously received over GPRS and Bluetooth). The results in terms of downloading rate in the non-cooperative scenario, showed (during a complete download of the 2.546 kB mp4 file) an average value of 41.52 Kbps, a maximum rate of 42.56 Kbps, and a minimum rate of 39.76 Kbps. This shows that a reasonable variation range of download rates for the GPRS connection is 39–43 Kbps. Differently, by considering the cooperative scenario, a gain of about the 100% has been registered on the average. Compared to the simple GPRS case, this shows a doubled achievable download rate. This is an expected result as the two devices accumulate the cellular links. The average rate in the cooperative scenario (on a complete download of the 2.546 kB mp4 file) is 82.8 Kbps; the maximum rate is 84.72 Kbps, while the minimum rate is 79.12 Kbps. This shows that the virtual data rate range is 79–85 Kbps.
- *Transmission time.* In terms of transmission time, an average gain of about 50% is reached, which means a much lower transmission delay. In the specific test-bed scenario it means that the waiting time for the end-user to finally see the result of his download (a movie in our test-bed) is reduced to one half. This is not surprising as the virtual data rate has doubled.

The aforementioned results are quite promising. Noteworthy, if more than two devices would be involved downloading even less than half the data over the cellular link and exchanging it over the short-range, proportional increasing gains can be achieved.

Two more benefits of the cooperative scenario have to be underlined. These do not need any experimental measurements to be supported and are especially interesting for network and service providers:

- *Cell capacity.* In the cooperative scenario the devices will download only half of the data over GPRS, which means halved usage of GPRS resources, and consequently increased cell capacity for the network.

- *Service cost.* A device downloading, let's say, half of a file, will stay connected half of the time; this allowing a proportional reduced billing. As this is attractive for the end-user interest, service and network providers can design new business scenarios and cost models to incentive support for cooperative networking and enlarge the subscribers basin.

## 6 Conclusions and Perspectives

In this chapter we have presented promising energy saving strategies for a novel cooperative communication architecture for wireless and mobile networks. The cooperative behavior between cellular terminals provides great benefits not only related to energy consumption. Using a peer-to-peer short-range link for cooperation among terminals, combined to a service entry point cellular network, data rate, transmission time, cell capacity and service cost registered significant improvements.

Supported by the relevant and significant results obtained, the introduced concepts need to be considered as a first step towards further research and implementations of micro cooperative scenarios and networks. As perspective, we intend to extend the cooperative scenario to more than two devices. A dynamic algorithm needs to be studied to fairly share the data amount to be downloaded from the devices over the cellular link. The proposed *dynamic server-aware* approach, can be considered as a good starting point for further improvements.

A second perspective is to consider other networks for the cooperative scenario. For instance, an implementation of a cooperative scenario based on a WLAN short-range network, opens up the opportunity to compare and analyse different energy saving behaviors in a dynamic and multiple network scenario. Moreover, it is expected that involving more terminals and considering different combinations of cellular and short-range networks, better performing and even more interesting benefits can be achieved for all network architecture entities.

## References

1. S. Frattasi, H. Fathi, F.H.P. Fitzek, M.D. Katz, and R. Prasad, "Defining 4G Technology from the Users Perspective" in *IEEE Network*, Jan./Feb. 2006
2. S. Frattasi, F.H.P. Fitzek, B. Can, and R. Prasad, "Cooperative Services for 4G" in *14th IST Mobile and Wireless Communications Summit*, Dresden, Germany, June 2005
3. F.H.P. Fitzek and M. Katz, Eds., *Cooperation in Wireless Networks: Principles and Applications–Real Egoistic Behavior is to Cooperate!*, ISBN 1-4020-4710-X. Springer Berlin Heidelberg New York, April 2006

4. J. Chuang, M. Feldman, K. Lai and I. Stoica, "Quantifying disincentives in peer-to-peer networks" in *Workshop on Economics of Peer-to-Peer Systems. Berkeley, CA*, June 2003
5. M. Feldman and J. Chuang, "Overcoming free-riding behavior in peer-to-peer systems" in *ACM Sigecom Exchanges, 6*, 2005
6. I. Stoica, K. Lai, M. Feldman and M. Chuang, "Incentives for Cooperation in Peer-to-Peer Networks" in *Workshop on Economics of Peer-to-Peer Systems. Berkeley, CA*, June 2003
7. Michal Feldman, Kevin Lai, Ion Stoica, and John Chuang, "Robust incentive techniques for peer-to-peer networks" in *EC '04: Proceedings of the 5th ACM conference on Electronic commerce*, New York, NY, USA, 2004, pp. 102–111, ACM Press
8. Seung Jun and Mustaque Ahamad, "Incentives in bittorrent induce free riding" in *P2PECON '05: Proceeding of the 2005 ACM SIGCOMM workshop on Economics of peer-to-peer systems*, New York, NY, USA, 2005, pp. 116–121, ACM Press
9. R. Fielding, J. Gettys, J. Mogul, H. Frystyk and T. Berners-Lee, "Hypertext Transfer Protocol – HTTP/1.1" Jan. 1997. RFC 2068
10. P. Rodriguez and E.W. Biersack, "Dynamic parallel access to replicated content in the Internet" in *IEEE/ACM Trans. Networking, vol. 10,* pp. 455–465, Piscataway, NJ, USA, Aug. 2002
11. B. Krishnamurphy, J.C. Mogul and D.M. Kristol, "Key differences between HTTP/1.0 and HTTP/1.1" in *WWW '99: Proceeding of the eighth international conference on World Wide Web, Toronto, Canada,* May, 1999
12. Mobile Device Group, "http://mobiledevices.kom.aau.dk"

9

# Relay Quality Awareness in Mesh Networks Routing

Claudio Casetti, Carla Fabiana Chiasserini, and Marco Fiore

Dipartimento di Elettronica, Politecnico di Torino, C.so Duca degli Abruzzi 24, Torino, Italy, casetti@tlc.polito.it, chiasserini@tlc.polito.it, fiore@tlc.polito.it

**Summary.** We consider a wireless mesh network whose nodes use the IEEE 802.11 technology. We address the problem of designing an efficient traffic forwarding scheme, which takes into account communication links quality and MAC issues. Through experimental measurements and simulation results, we derive some guidelines on designing an efficient routing scheme. Based on these findings, we define a novel relay selection algorithm which is implemented as an extension of the OLSR routing protocol, and show through simulation that it significantly improves system performances, in terms of both throughput and fairness.

## 1 Introduction

Mesh networking [1] is an emerging wireless technology that aims at providing robust and reliable wireless broadband service access at relatively low cost. Mesh networks are composed of nodes which operate both as hosts and routers, forwarding packets on behalf of other nodes that may not be within direct transmission range of their destinations; in addition, a small percentage of the mesh nodes implements gateway/bridge functionalities.

In this work, we consider a mesh network based on the IEEE 802.11 technology. We focus on a network scenario in which multiple mesh clients wish to connect to a mesh gateway, through either direct or multihop communications. The problem we address is how to transfer traffic between the wireless nodes and a gateway in a fair, efficient manner.

Routing protocols for wireless networks are usually designed considering that all nodes within transmission range of a transmitter are equivalent. However, this is often false, as the quality of the channel toward (and from) different one-hop neighbors may significantly vary with distance, presence of obstacles and interfering transmissions. Also, since 802.11 off-the-shelf devices implement rate-adaptation techniques, the link quality directly determines the data transmission rate to be used between pairs of nodes. Thus, measuring routing distances in terms of number of hops may be misleading, as routing through a larger number of high-rate hops may lead to a higher network

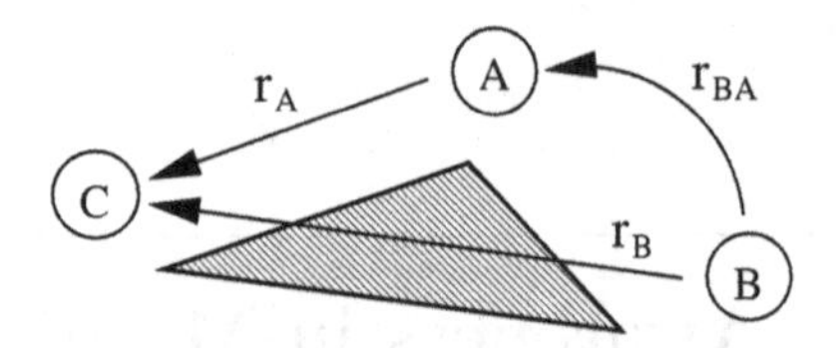


**Fig. 1.** Single-hop anomaly scenario

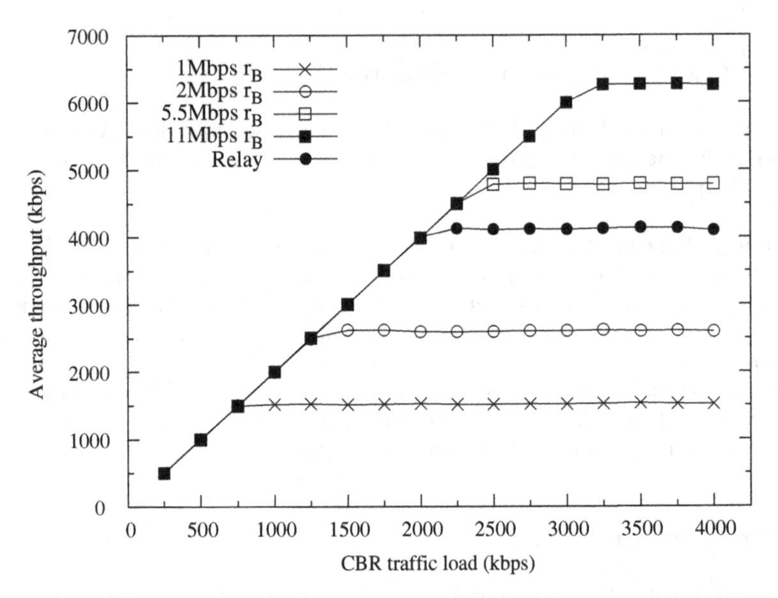


**Fig. 2.** Network throughput vs. per-node offered load, with $r_{AC}$ and $r_{BA}$ equal to 11 Mbps and varying $r_{BC}$

throughput with respect to performing fewer low-rate forwards [5]. An example of this behavior can be observed in the simple scenario of Fig. 1. Here, all nodes are within receive range of each other, and both $A$ and $B$ send UDP traffic to $C$ using the 802.11 distributed coordination function (DCF) to access the radio channel. Consider $A$ enjoying an optimal channel quality toward $C$, while $B$ experiences a low quality channel due to distance and/or obstacles. The resulting decrease in $B$s data transmission rate ($r_B$), is the cause of the well-known 802.11 anomaly phenomenon [7], which reduces the overall network throughput, as shown in Fig. 2. However, if the channel quality between $B$ and $A$ is good enough and $A$ relays data from $B$ toward $C$, then the resulting overall network throughput is better than that obtained with a 2 Mb s$^{-1}$ direct data transfer between $B$ and $C$. Even higher improvements are obtained when employing TCP, which introduces reverse acknowledgments flows from $C$ to $A$ and $B$.

Our goal is to design a routing protocol for ad hoc networks accounting for medium access control (MAC) and physical layer performance in the route computation, by making multiple fast hops preferable to single slow ones. The joint design of MAC and routing schemes is however a challenging issue. After reviewing some related work in Sect. 2, we outline some guidelines on designing an efficient, cross-layer, relay selection scheme that accounts for quality and transmission rate of the available links (Sect. 3). Furthermore, we define a relay-quality aware routing as an extension of the optimized link state routing (OLSR) [3] protocol. Section 4 describes its implementation and shows some performance results. Section 5 concludes the chapter.

## 2 Related Work

Routing in wireless networks has received a great deal of interest, and several schemes have been proposed which exploit various metrics for route selection.

In [2] it has been observed that a routing scheme using the hop count as a metric for route selection may not be the best choice. Indeed, while this scheme may be appropriate in single-rate networks, in a multi-rate environment it tends to select short paths composed of maximum length links. Since long distance links operate at low rates, poor throughput performances are likely to be obtained. To select high-throughput paths in multihop networks, the use of the expected transmission count (ETX) metric is proposed in [5]. Based on the ETX metric, the route featuring the fewest expected number of transmissions (including retransmissions) to deliver a packet is chosen.

The solutions presented in [6, 8] design multirate-aware routing schemes to increase utilization of 802.11-based, multihop networks. In particular, the key idea in [8] is to change the next-hop node to another node where higher data rates are available. The work in [6] presents a modified routing metric with respect to ETX, which accounts also for the bandwidth available at the 802.11 nodes.

Similar issues are addressed in [9], where a MAC-layer, relay-enabled point coordination function (PCF) protocol is presented. The scheme allows packet delivery through a relay node if the direct link has low quality and low rate.

## 3 Principles of Relay Quality-Aware Routing

The results derived in our previous work [4] allow us to draw a set of observations that might be used as a springboard to design a fair relay selection algorithm (FRSA). Below, we summarize the main conclusions that simulation has provided:

- The 802.11 performance anomaly can be solved through the use of relays.
- Single relays offer a good compromise between the use of higher bit rates and lower numbers of packet replicas.
- Best performance are achieved if relay(s) uses similar rates between gateway and mesh nodes.

- If multiple relays are available, it is best to distribute far nodes among them.
- Recursive relaying should be avoided.

The definition of FRSA leverages existing routing algorithms for ad hoc networks; we chose to refer to a proactive routing scheme, such as OLSR, since it seemed more suitable for including FRSA (see the following section).

The main requirement of the algorithm is the identification and ordering of the available paths between a node and the gateway. The paths providing the best throughput will be ranked first.

Following common notation, the resulting topology information stored by each wireless node can then be mapped into a graph $G(V, A)$, where $V$ is a set of vertices, representing the network nodes ($n_i$), and $A$ is a set of arcs, representing links between pairs of nodes ($l_{ij}$); each link is associated to a cost $c_{ij}$, a function of the link feasible rate $r_{i,j}$, i.e., the highest bit rate that can be used on that link. We then select the minimum-cost path to each destination, that can be computed by a shortest path algorithm, such as Dijkstra's or Bellman–Ford's.

## 4 FRSA Implementation as an Extension of the OLSR Scheme

In this section, we briefly summarize the main features of the OLSR protocol, and then describe how we extend the OLSR scheme to apply our FRSA, leading to a relay-quality aware routing. Finally, we show some performance results obtained via simulation.

### 4.1 Background on OLSR

OLSR is a proactive link state protocol, which involves regular exchange of topology information among the network nodes. It employs designated nodes called multi point relays (MPRs) to facilitate controlled flooding of topology information. MPRs are also the sole constituent nodes in the route between any source–destination pair in the network.

*HELLO message broadcast and processing.* Every OLSR node periodically broadcasts heartbeat HELLO messages, with information about its neighbors and the corresponding link states. A link state can be symmetric, asymmetric or MPR. An MPR link state with a neighbor indicates that the neighbor has been selected by this node as an MPR; MPR links are symmetric. The HELLO messages are broadcast to all one-hop neighbors, but are not relayed to nodes which are further away. A (*two-hop*) *Neighbor Set* at each node stores the information about (two-hop) neighbors, and is updated on reception of a HELLO message, which causes a node to create or refresh the neighbor entries

corresponding to the node which sent the message (Neighbor Set) and to its one-hop neighbors (two-hop Neighbor Set).
*Multipoint relays.* Based on the information obtained from the HELLO message, each node in the network selects a set of nodes amongst its symmetrically linked neighbors, that help in controlled flooding of broadcast messages. This set of nodes is called the Multipoint Relay set of the node. The neighbors of the node which are not in its MPR set, receive and process broadcast messages from the node, but do not retransmit them. The MPR set is selected such that it covers all the nodes that are two hops away.
*Topology control message broadcast and processing.* The *Topology Control (TC)* messages are broadcast by MPRs in the network to declare their *Multipoint Relay Selector Set (MPRSS).* The MPRSS of an MPR $x$ includes the nodes that have selected $x$ as an MPR. A node obtains MPRSS information from periodic HELLO messages received from its neighbors. Each node in the network maintains a *Topology Table*, in which it records the information about network topology as obtained from the TC messages. An entry in the topology table contains the destination address, $T_{dest}$, and the address of the last hop to the destination, $T_{last}$. Each such entry means that node $T_{dest}$ has selected node $T_{last}$ as an MPR and that node $T_{last}$ has announced this information through a TC message.
*Routing table calculation.* The routing table is evaluated based on the connectivity information in the neighbor table and topology table. Shortest path algorithm is employed for route calculation. Each resulting route entry consists of the destination entry, the next node from the sender, and number of hops to the destination.

## 4.2 Fair Relay Selection Algorithm

In principle, in order to perform a relay quality-aware routing, a node would need a complete knowledge of the network, in terms of nodes, connected pairs and relative rates. Since acquiring such knowledge is unrealistic, we will describe a scalable approach below.

The measurement of the data transmission rate requires cross-layer interaction between the routing protocol and the MAC layer, which is common to many recent proposals. However, the MAC layer evaluation of the maximum rate between nodes raises a further issue. Since broadcast transmissions are performed by 802.11 at $1\,\mathrm{Mb\,s^{-1}}$, determining the actual maximum achievable rate between node pairs under a trial-and-error approach requires unicast transmissions. Sending unicast data to every node within transmission range would introduce complexity, overhead and unreliability (e.g., latency in detecting topology changes). Our idea, instead, exploits the signal to noise plus interference ratio (SINR) to estimate the available data transmission rate. This information can be obtained at the MAC layer at each packet reception, no matter the data transmission rate employed by the sender. Thus, a single broadcast transmission allows all neighbors to estimate the quality of the channel from the originating node, at a time. Note that this approach is especially fit to proactive routing such as OLSR: indeed, the topology control messages each node is required to send at short, regular time intervals allow a comprehensive and frequently updated SINR estimation for each neighbor. Recorded SINR information is handed over to the routing protocol, which smooths it through an exponentially weighted moving average (EWMA) filter to avoid short-term effects, and extrapolates the corresponding data rate by looking at SINR thresholds, which separate the working intervals of the different channel coding techniques [8].

The rate resulting from the previous computation refers to the link from the neighbor node to the node which receives the data, i.e., the *reverse* link. Since link symmetry is not guaranteed in a wireless environment, this value could not correspond to the rate achievable from the current node to its neighbor. Thus, the information about the reverse link quality must be communicated back to the neighbor that generated the transmission.

Once a node receives the reverse link rate information from a neighbor (corresponding to its *forward* rate to the neighbor), it can add this information to its knowledge base, and notify it along with other topology information in the control messages that are necessary to the proactive routing functioning.

Referring to the notation introduced in Sect. 3, we implemented the cost function as $c_{ij} = K + 1/(r_{ij} \cdot w_i)$ where $K$ is a constant, $r_{ij}$ is the data transmission rate from node $i$ to node $j$, and $w_i$ is the *relay willingness* of node $i$, i.e., a measure of the willingness of the node to act as a relay for the data coming from an additional node. The additive constant $K$ weighs each extra hop by an empirically determined value of 0.25, so as to fulfill the guidelines listed in Sect. 3. The relay willingness is locally set by each node considering factors such as the locally generated traffic, the already relayed traffic, the level of mobility, the selfishness of the user. As an example, a highly mobile node should not act as a relay, since its link will often break. The relay willingness information must be advertised by nodes, and, again, this can be easily done through topology control messages.

Next, we describe how OLSR can be easily extended so that each node performs a relay quality-aware routing limited to its two-hop neighborhood, and not applied to the whole network. This means that, for nodes that are farther than two hops away, a node uses the standard OLSR routing table computation. This choice is justified by the fact that, once the forwarded data exits the two-hop neighborhood of the originator, the node that relayed the data last can apply the relay quality-aware routing over the next two hops, and so on.

Our design for a relay quality-aware routing extension to OLSR does not require any change in the structure of control messages. However, *reserved* fields of the HELLO message format are used to exchange link quality and relay willingness information, as described below and shown in Fig. 3.

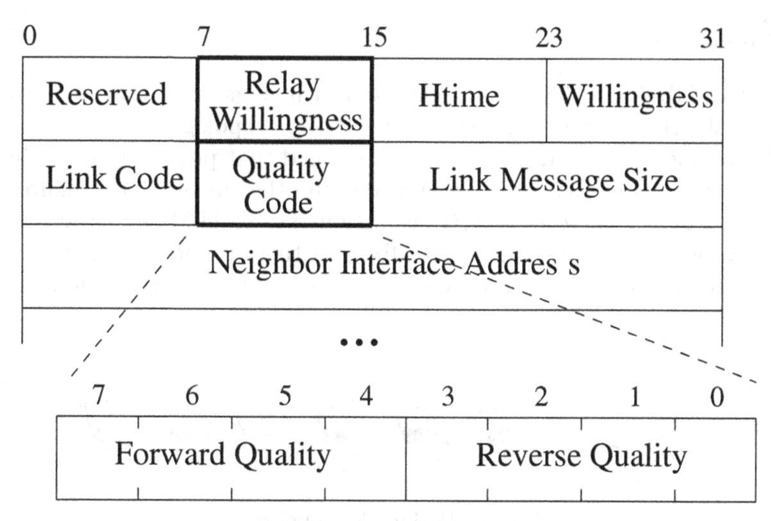


**Fig. 3.** Modified HELLO message format

- An 8-bit *Relay Willingness* field, advertising the willingness of the node to act as a relay for data flows. This information is used to compute the link cost, as described before.
- An 8-bit *Quality Code*, advertising the forward and reverse link quality to and from the neighbor nodes, characterized by the given Link Code, listed below. The 4-bit forward link quality is used to compute the cost of the link, while the 4-bit reverse link quality lets each neighbor know its forward quality to the node originating the message, as previously discussed. Four bits allow a wide range of choices: in the simplest implementation, five codes are allowed, corresponding to the four data rates provided by 802.11b.

The following changes to the record structure defined by the standard are also needed.

- Two fields, *L_fwd_quality* and *L_rev_quality*, are added to the Link Tuple format, storing forward and reverse link quality. The reverse quality value is computed from the SINR observed for the current link, while the forward link quality is obtained from HELLO messages.
- A *N_fwd_quality* field is added to the two-hop Neighbor Set, storing the quality of the link from the node neighbor to the two-hop node neighbor, and is used to compute the cost of the link.

Clearly, this scheme, limiting the relay quality-aware routing to two times the transmission range, could generate sub-optimal results. An extension of the relay quality-aware routing to the whole network is also possible, by operating an MPR selection based on measured and advertised link quality scores, similar to [3], and then introducing link quality information on OLSR TC messages.

### 4.3 Simulation Results

We tested FRSA on three topologies, shown in Figs. 4–6, that we dubbed, respectively, the *parking lot*, the *fork* and the *fan* topology.

Where not specified differently, it is assumed that propagation conditions are such that nodes that are adjacent to each other in the above figures can achieve *on average* a reciprocal transmission rate of 11 Mb s$^{-1}$; the rate between two wireless nodes falls to 5.5 Mb s$^{-1}$ if there is one intermediate node among them, and to 2 and 1 Mb s$^{-1}$ in case of two or three intermediate

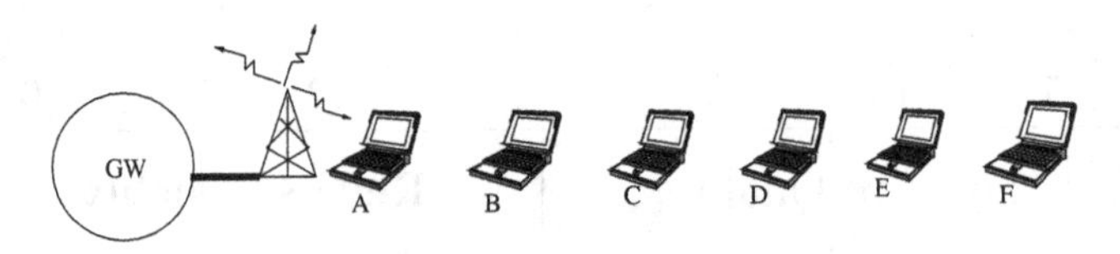


**Fig. 4.** Parking lot topology

**Fig. 5.** Fork topology

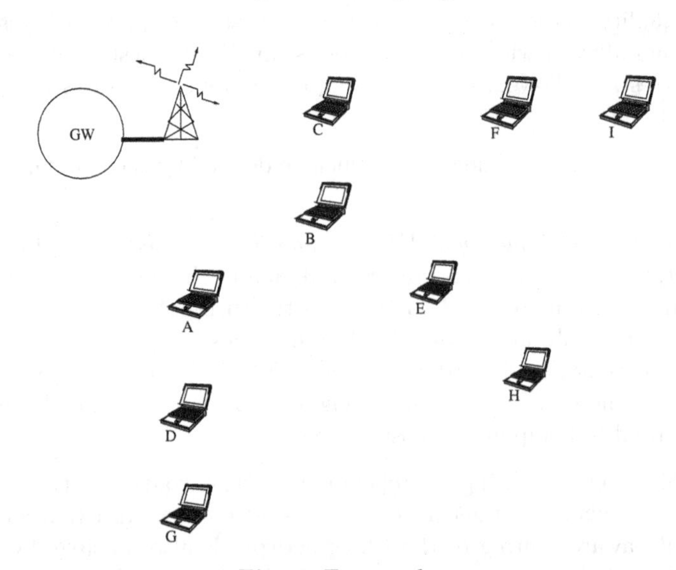


**Fig. 6.** Fan topology

nodes, respectively. Nodes further than that cannot decode each other's transmissions, but they still sense each other, thus all of the presented scenarios identify a *clique* when the relative interference graph is taken into account.

Therefore, as an example, node B in the fork topology can communicate at $11\,\mathrm{Mb\,s^{-1}}$ with both C and D, and at $5.5\,\mathrm{Mb\,s^{-1}}$ with both E and F. As already pointed out, these are average values, since the coupled simulation of ARF techniques at nodes and propagation-dependent channel errors force the nodes to vary their rates over time. All traffic is supposed to be of CBR over UDP nature and flowing in the downlink direction, i.e., from the mesh gateway (GW) to the mesh nodes.

The first set of results refers to the parking lot topology and it shows the average throughput achieved by every node as a function of the source transmission rate. Simulations are performed using standard OLSR (Fig. 7) and FRSA (Fig. 8), and they show that a significant throughput increase (over 50%) can be achieved by FRSA, while maintaining fair access. Figure 9 provides additional proof of our claims by reporting the aggregate throughput.

The fork topology illustrates a special case: due to their distance from the AP, no node can achieve $11\,\mathrm{Mb\,s^{-1}}$ on its direct link from the AP, the node that enjoys the highest rate from the AP being node A, with a $5.5\,\mathrm{Mb\,s^{-1}}$ transmission rate. Standard OLSR (Fig. 10) has nodes C, D communicate directly with the AP at $1\,\mathrm{Mb\,s^{-1}}$, while it has nodes E, F use D as their relay, achieving a combined $11\,\mathrm{Mb\,s^{-1}}+1\,\mathrm{Mb\,s^{-1}}$ rate. This is clearly a suboptimal solution, since it lowers all transmission rates, hence decreasing the overall performance due to the anomaly. FRSA, on the contrary has C, D use A

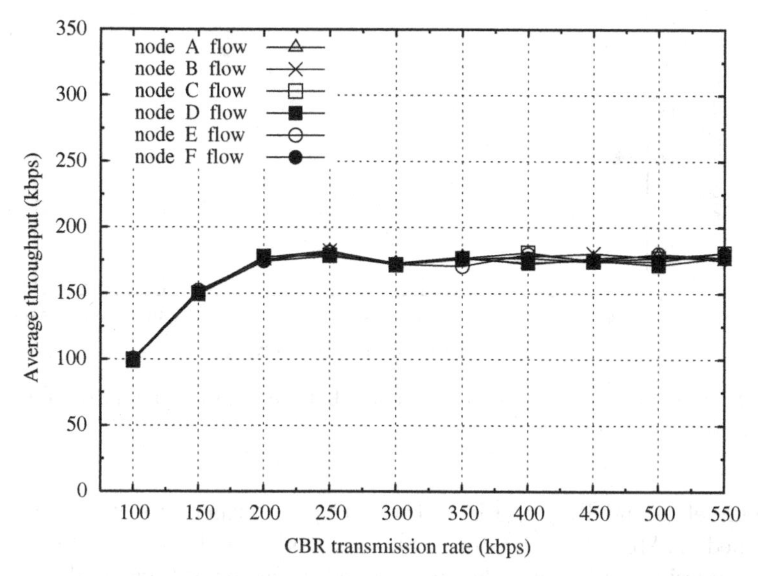


**Fig. 7.** Per-flow throughput with standard OLSR (*parking lot topology*)

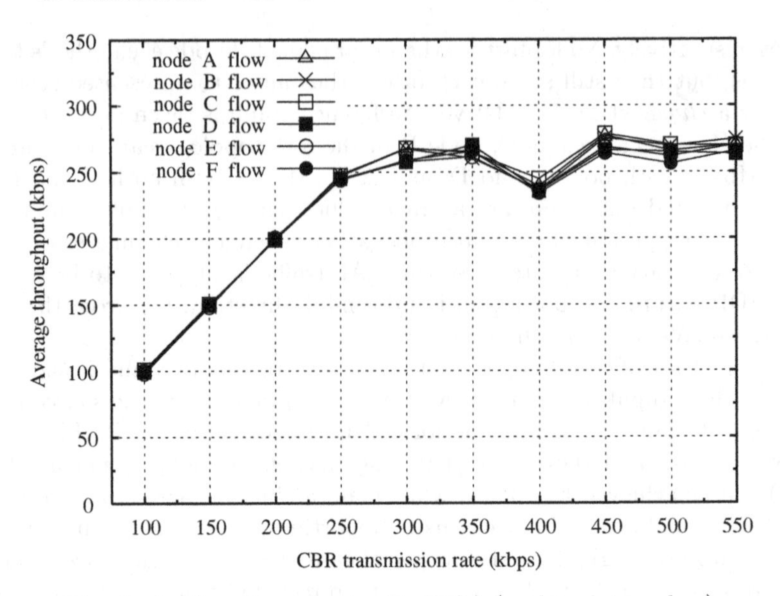


**Fig. 8.** Per-flow throughput with FRSA (*parking lot topology*)

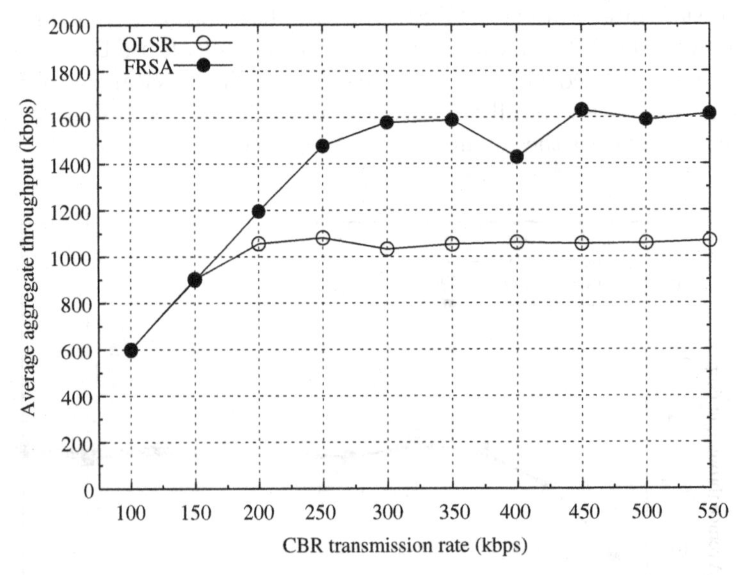


**Fig. 9.** Aggregate throughput comparison between standard OLSR and FRSA (parking lot topology)

as their relay, and E, F use B as their relay; in either case, the rates are a combined $5.5\,\mathrm{Mb\,s^{-1}} + 5.5\,\mathrm{Mb\,s^{-1}}$. Since no node has the upper hand in terms of transmission rates, the balance is tipped in favour of those nodes, namely A and B, who receive their traffic through fewer hops (Fig. 11). The other

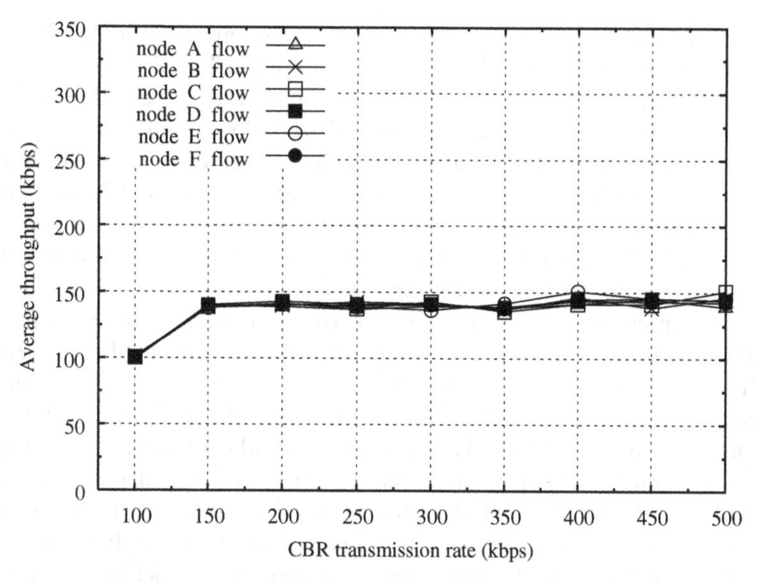


**Fig. 10.** Per-flow throughput with standard OLSR (fork topology)

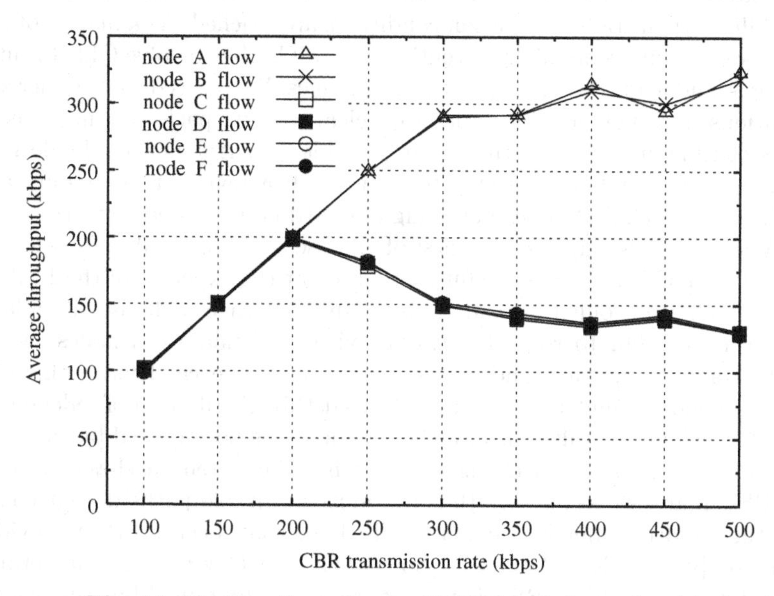


**Fig. 11.** Per-flow throughput with FRSA (fork topology)

nodes perform similarly to the OLSR case, and the overall throughput of the system again experiences a 50% rise. Note that the unfairness experienced by the relayed nodes, absent in the other topologies examined, is determined by the fact that in this configuration two nodes have to share a single relay, and thus achieve a halved throughput.

The fan topology defines three concentric areas around the AP. Nodes on the same branch are referred to as *relative* nodes. The first rim, including nodes A, B and C, can achieve a 5.5 Mb s$^{-1}$ direct rate to the AP; the second rim, including nodes D, E and F, has 1 Mb s$^{-1}$ direct rate to the AP, or a 5.5 Mb s$^{-1}$ rate toward their relative inner rim node; finally, no direct link can be established between nodes on the third, outer rim and the AP. These nodes can instead establish a 1 Mb s$^{-1}$ link toward their relative first rim node, or a 5.5 Mb s$^{-1}$ link toward their relative intermediate rim node. Note that these rates only refer to relative nodes, i.e., two nodes on separate branches can experience different rates with respect to those mentioned above, depending on their distance. In this case, standard OLSR can route traffic to nodes on the third rim through any of the six nodes in the inner rims, and, since it just considers the hop count as the cost metric, all the six inner rim nodes are considered as equivalent. This can bring standard OLSR to select paths which are trivially sub-optimal. As an example, in one of our simulations standard OLSR initially routed traffic directed to node G through node B, while node A would have been a better relay in any case, its distance from G being clearly lower than Bs. Moreover, standard OLSRs lack of preference in the relays toward the outer rim nodes brings the network to very high route instability, when traffic saturation conditions are reached. As a matter of fact, the losses deriving from high contention and full buffers involve OLSR routing messages and lead to very frequent path changes. We noticed that, under such conditions, it is not possible to identify clear routing paths anymore, as the nodes continuously modify their routing tables, struggling for reliable links. Obviously, it would be desirable to avoid such a behavior, as it only adds overhead and definitely does not bring any advantage to the network.

When FRSA is employed, we first of all observe an optimal route selection, as the fast 5.5 Mb s$^{-1}$ links are fully exploited at the expenses of the 1 Mb s$^{-1}$ links, which are avoided instead. This means that each node uses its relative inner rim neighbor to route the traffic, with the middle rim nodes reached through a two-hop path and the outer rim nodes reached via a three-hop route. Secondly, simulation results show that this distribution of independent data flows along the different branches of the fan is not affected by the traffic load. The routes are rarely changed, even when the system reaches saturation, as FRSA manages to identify that the most profitable paths do not change with the uniform traffic load increase. The beneficial effect of FRSA is evident when comparing Figs. 12 and 13, showing the per-flow throughput obtained in the two cases: when FRSA is used, fairness is maintained but the network performances are noticeably improved, leading to an aggregate throughput, depicted in Fig. 14, nearly doubled at high loads.

In a fourth scenario, we tested a mobile environment where a node is moving on a straight line toward the AP, along the parking lot topology. The mobile node starts outside radio range of node F and inches toward the AP, receiving traffic as soon as it can establish a link with any node in the neighborhood. All other nodes are supposed to have no traffic of their own, but

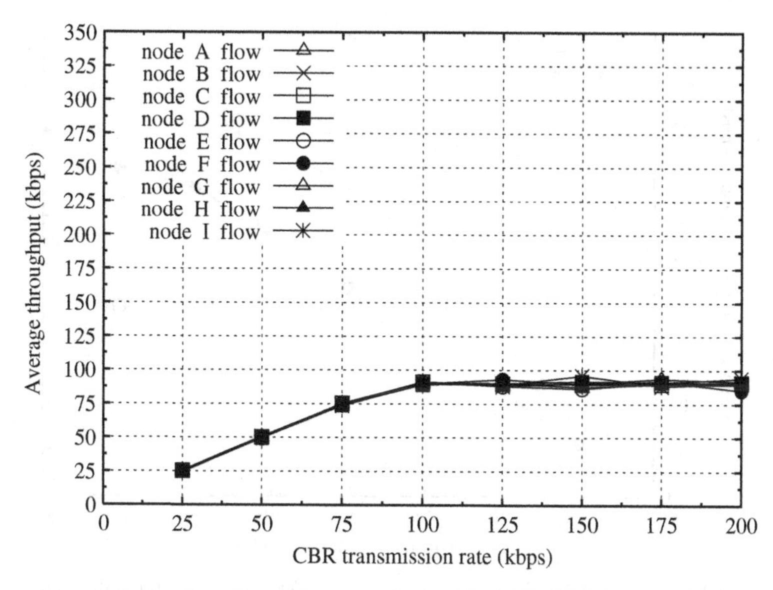


**Fig. 12.** Per-flow throughput with standard OLSR (*fan topology*)

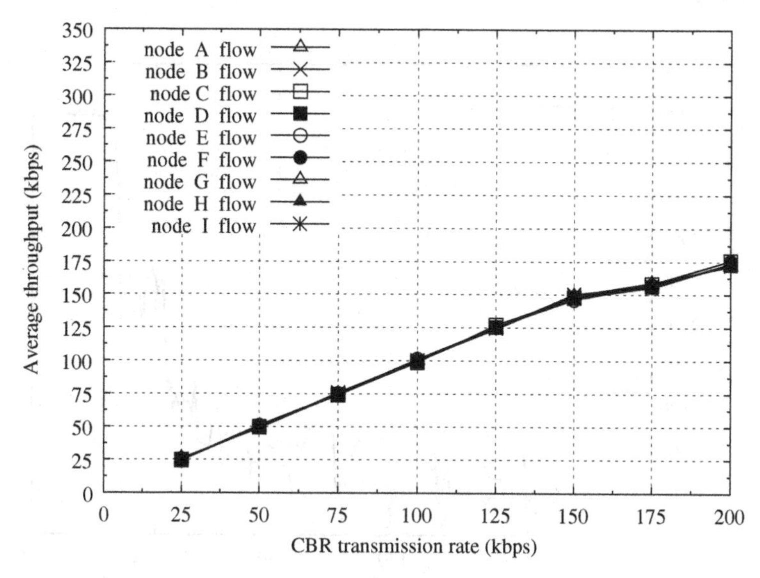


**Fig. 13.** Per-flow throughput with FRSA (fan topology)

they relay traffic to the mobile node, if needed. Figure 15 shows the achievable throughput over time as the node cruises along the topology toward the AP, again showing the higher throughput achieved by FRSA. The histograms in Figs. 16 and 17 show the distribution of relay choices with standard OLSR and FRSA, respectively; each bar reports the fraction of packets routed through the mesh node at each achievable rate.

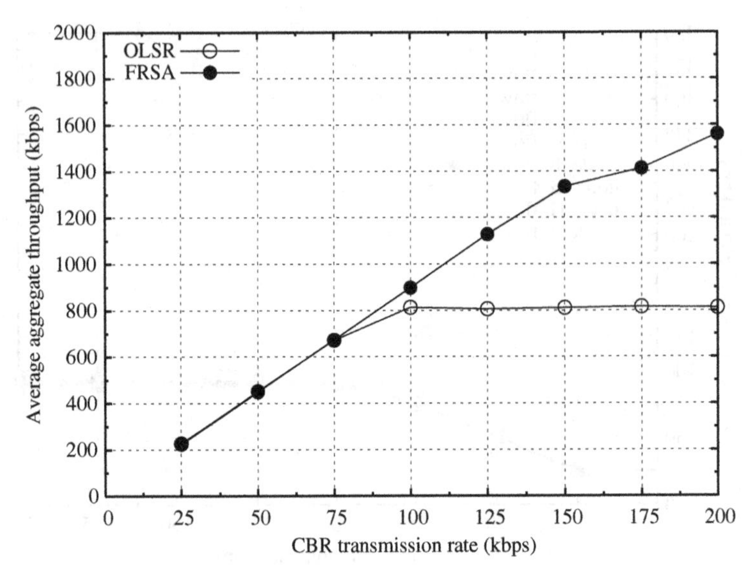


**Fig. 14.** Aggregate throughput comparison between standard OLSR and FRSA (fan topology)

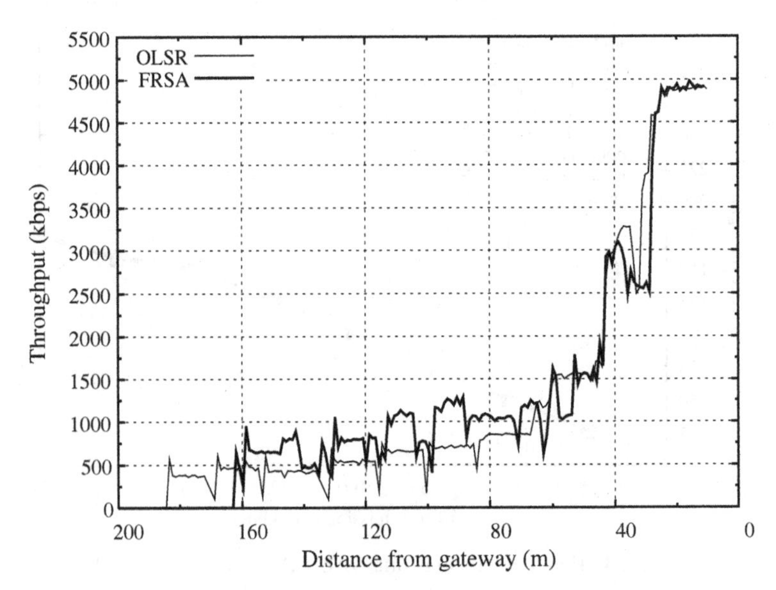


**Fig. 15.** Temporal diagram of aggregate throughput of standard OLSR and FRSA (*parking lot topology with mobile node*)

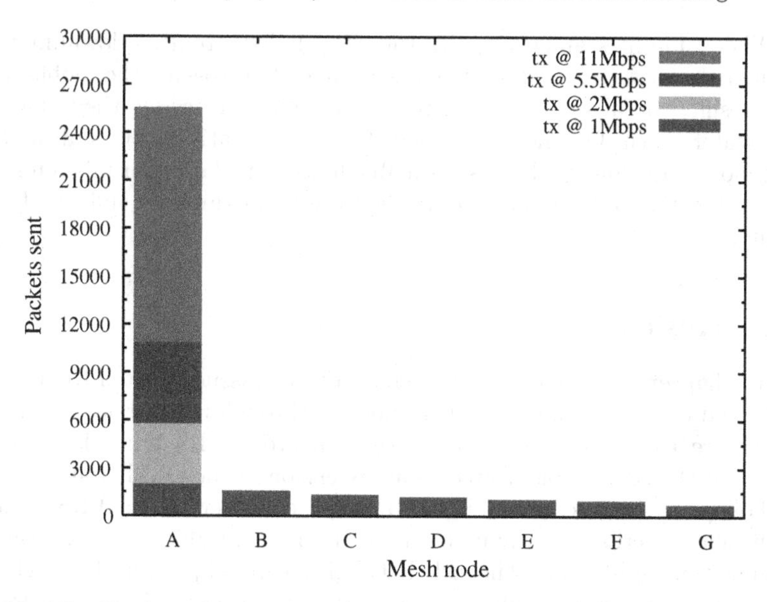


**Fig. 16.** Distribution of relay choices with standard OLSR (*parking lot topology with mobile node*)

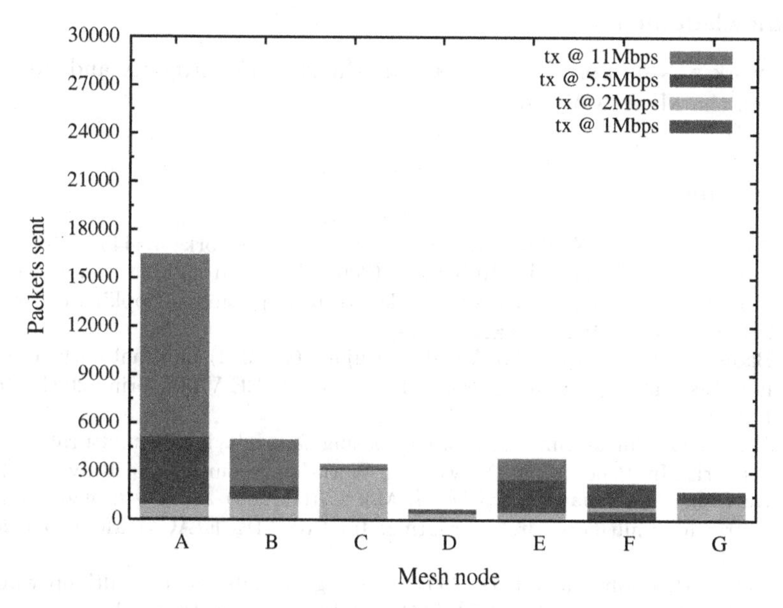


**Fig. 17.** Distribution of relay choices with FRSA (*mobile topology*)

When running standard OLSR, the AP privileges routing downlink traffic through the closest neighbor that has a direct link toward the mobile node. As a result, the algorithm tends to use the lowest rates, as they guarantee the greater reach. On the other hand, FRSA alternatively picks all nodes of the topology exploiting the best available links. The data transmission is only routed directly to the mobile node when the link exhibits a sufficiently high quality.

## 5 Conclusions

In this chapter, we provided some perspectives on issues related to mesh networks and considered multihop transmission through traffic relays as a countermeasure to the anomaly effect. We then proposed a fair relay selection algorithm (FRSA), implemented as an extension of the OLSR routing protocol. Following the guidelines laid out by empirical observations, FRSA aims at an efficient selection of relay nodes in an automated fashion. An extensive set of simulation results have showed that higher throughput and fairer channel access among all mesh nodes are achievable when FRSA is used instead of standard OLSR in both static and mobile mesh scenarios.

### Acknowledgment

This work was supported by the Italian Ministry of University and Research through the MEADOW project.

## References

1. Akyildiz I, Wang X, Wang W (2005) Computer Networks 47:445–487
2. Awerbuch B, Holmer D, Rubens H (2003) High throughput route selection in multi-rate ad hoc wireless networks. Tech. Rep., Johns Hopkins University, `http://www.cs.jhu.edu/archipelago/`
3. Badis H, Munaretto A, Al Aghal K, Pujolle G (2004) Optimal path selection in a link state QoS routing protocol. In: Proc. IEEE VTC Spring 2004, Milan, Italy
4. Casetti C, Chiasserini C, (2006) Choosing fair relaying in multi-rate 802.11 networks. In: Proc. Italian Networking Workshop, Courmayeur, Italy, Jan. 2006
5. De Couto D, Aguayo D, Bicket J, Morris R (2003) A high-throughput path metric for multihop wireless routing. In: Proc. IEEE/ACM MobiCom 2003, San Diego
6. Draves R, Padhye J, Zill B (2004) Routing in multi-radio, multihop wireless mesh networks. In: Proc. IEEE/ACM MobiCom 2004, Philadelphia
7. Heusse M, Rousseau F, Berger-Sabbatel G, Duda A (2003) Performance anomaly of 802.11b. In: Proc. IEEE Infocom'03, San Francisco
8. Seok Y, Park J, Choi Y (2003) Multi-rate aware routing protocol for mobile ad hoc networks. In: Proc. IEEE VTC 2003-Spring, Jeju, Korea
9. Zhu H, Cao G (2004) Mobile Networks and Applications 9:423–434

# 10

# Fundamental Bound on the Capacity of Ad Hoc Networks with Conventional Hop-by-Hop Routing

Anthony Acampora[1], Michael Tan[2], and Louisa Ip[3]

[1] Department of Electrical and Computer Engineering and Center for Wireless Communications, University of California, San Diego, La Jolla, CA 92093, USA
acampora@ece.ucsd.edu

[2] Department of Electrical and Computer Engineering and Center for Wireless Communications, University of California, San Diego, La Jolla, CA 92093, USA
mytan@ucsd.edu

[3] Department of Electrical and Computer Engineering and Center for Wireless Communications, University of California, San Diego, La Jolla, CA 92093, USA
cepstrum@gmail.com

**Summary.** Peer-to-peer wireless networks have recently attracted a great deal of attention. In this paper, we establish two fundamental bounds on the capacity of such networks with conventional hop-by-hop routing. One of these bounds is based on fundamental information theory, but applied to switching systems instead of transmission systems. This bound scales well with network size. The second bound, based on more traditional network flow optimization, is somewhat tighter, but doesn't scale well to large networks. Both bounds present a yardstick against which the capabilities of cooperation in peer-to-peer wireless networking may be compared.

## 1 Introduction

In this chapter, we consider fundamental upper bounds on the throughput capacity of any ad hoc network consisting of $N$ nodes which communicate among themselves via conventional hop-by-hop routing. By this we mean that no wireless ad hoc network that operates in the absence of cooperation among its nodes can possibly produce a higher throughput than indicated by these bounds. Thus, these bounds serve two purposes:

1. They provide a yardstick against which the throughput of a conventional hop-by-hop ad hoc network may be gauged, that is, any network architected such that its throughput is nearly equal to these best case bounds is essentially the best that can be designed; and
2. They provide a second yardstick against which the throughput of a particular wireless ad hoc scheme with cooperation may be compared, that is, one can assess the potential benefits of a particular scheme by comparing

its throughput against that of the (possibly unachievable) best that any hop-by-hop scheme can produce.

Neither of the upper bounds that we have produced are constructive, that is, neither provides an approach to actually achieve the bound. Nonetheless, these bounds have utility, as noted above.

As will be described, the first of these bounds is produced by a flow deviation method. Although providing a tighter bound than our second technique, the bound becomes difficult to compute as the number of nodes grows large and, as a practical limit, we have not applied this technique to networks containing more that 100 nodes. The second bound is based on a unique application of basic Shannon Theory, but applied to switching networks rather than noise-corrupted digital links. Although not as tight, this bound is particularly simple to compute, and can be applied to arbitrarily large networks. Both bounds account for path loss, shadow fading, and multipath fading.

## 2 Starting Point

Input to both bounds are two matrices, to be defined below, which we refer to as the link capacity matrix and the relative traffic matrix:

$$\underline{C} = \begin{bmatrix} C_{1,1} & C_{1,2} & \dots & C_{1,N} \\ C_{2,1} & C_{2,2} & \dots & C_{2,N} \\ \vdots & & & \\ C_{N,1} & C_{N,2} & \dots & C_{N,N} \end{bmatrix}, \tag{1}$$

$$\underline{T} = K \begin{bmatrix} t_{1,1} & t_{1,2} & t_{1,3} & \dots & t_{1,N} \\ t_{2,1} & t_{2,2} & t_{2,3} & \dots & t_{2,N} \\ \vdots & & & & \\ t_{N,1} & t_{N,2} & t_{N,3} & \dots & t_{N,N} \end{bmatrix}. \tag{2}$$

As its name would suggest, element $C_{j,k}$ of the link capacity matrix is the maximum rate at which information may flow between node $j$ and node $k$. To insure that an upper bound is produced, this number is taken as the Shannon limit for link $(j, k)$, given the interference-free signal-to-noise ratio (SNR) produced at node $k$ when node $j$ sends at its maximum power level. The capacity of link $(j, k)$ does, however, reflect path loss caused by the physical separation between nodes $j$ and $k$, as well as any shadow and multipath fading present along path $(j, k)$. Note that use of the Shannon limit in defining the elements of the link capacity matrix is separate and distinct from the application of basic Shannon Theory to find the second of our bounds.

Element $t_{j,k}$ of the relative traffic matrix represents the relative exogoneous traffic originating at node $j$ and destined for node $k$; convenient units are packets per second, each of which contains a fixed number of bits. For this exogenous demand, and subject to the link capacity constraints, our bound seeks to maximize the scalar $K$ by which the relative traffic matrix may be scaled such that all traffic can be delivered. Then, for this maximum value of $K$ and since the link capacities may be expressed in bits/s/Hz, we can apply the appropriate normalization and sum over the elements of the relative traffic matrix to find the maximum capacity, in bits/s/Hz, for the $N$ node wireless ad hoc network. Of course, this bound applies for a particular relative traffic matrix and a particular set of propagation impairments as reflected in the link capacity matrix. Averages over propagation impairments are then easily found, and the resulting network capacity is plotted vs. the normalized SNR, defined as that produced between two nodes separated by normalized distance with no shadow fading and no multipath fading.

## 3 Characteristics and Constraints of Wireless Peer-to-Peer Networks

A primary distinction between wireless ad hoc networks and wired networks is that in the former, the signal sent by any node is, in general, broadcast to all other nodes. Thus, a connectivity diagram would show all nodes as fully connected: any node can send directly to any other node, although the maximum allowable data rate between any pair of nodes would certainly be constrained by the physical separation between those nodes and any intervening propagation impairments. Since some of these paths may be quite poor, multi-hopping via intermediate nodes is normally used to send information between any pair. Thus, at any given moment, a transmitting node sends only to one intended receiver, and a receiving node listens to only a single transmitter. Further, a given node may not send and receive at the same time. By contrast, for wired networks, a node can send only to those nodes to which it has a direct connection. In general, the connectivity diagram is not fully connected and, again, multihopping is necessary. However, a given transmitting node may simultaneously send independent information streams to those nodes to which it is connected, and a given receiving node can simultaneously accept independent information streams from those nodes to which it is connected. Further, nodes may send and receive at the same time.

Accordingly, fundamental constraints on the capacity of any network of wireless ad hoc nodes are:

1. At any given time, no node can receive a packet from more than one transmitting node.
2. At any given time, no node can send a packet to more than one receiving node.
3. No node can send and receive simultaneously.

## 4 Summary: Flow Deviation Bound

The flow deviation method to bound the capacity of a wireless ad hoc network seeks to find the flow of exogeneous traffic streams over the network links such that the maximum utilization of any node is as small as possible, where utilization is defined to be the sum of the fraction of time that a node is sending and the fraction of time that the same node is receiving. By minimizing the maximum utilization of any node, the constant $K$ in (2) may then be scaled such that the utilization of the maximally utilized node equals 100%. From this, the maximum throughput for that network is readily found.

The time taken to flow a given amount of traffic along a particular link $(j, k)$ is inversely proportional to the capacity of that link, $C_{j,k}$. For a given flow, one can then readily compute the sum of the transmit plus receive times for each network node. We have shown that the problem of finding the flow that minimizes the utilization of the maximally utilized node is an unconstrained, non-linear, differentiable, multi commodity flow problem, and that the utilization of the maximally utilized node is a convex function of the flow variables. Thus, the flow deviation method may be applied to recursively find the solution.

## 5 Summary: Information-Theoretic Bound

Our information-theoretic bound is found by computing the maximum number of fixed-length packets that can be switched per unit time by each network node. For the $j$th node, this is found by considering capacities of the links leading from that node to each of the remaining $N - 1$ nodes. At node $j$, three events are possible. First, a fixed length packet containing $M$ bits may arrive via node $j$s receiver that needs to be relayed to node $k$. The minimum time $T_1$ needed to accomplish this is given by:

$$T_1 = \left(\frac{M}{C_{j,k}}\right) + \left(\frac{M}{C_M}\right), \tag{3}$$

where $C_M$ is the capacity of the fastest link leading to node $j$.

Second, a packet may arrive at node $j$s receiver that needs to be delivered to the user attached at node $j$. The minimum time needed to accomplish this is given by:

$$T_2 = \left(\frac{M}{C_M}\right). \tag{4}$$

Third, a packet may originate at node $j$ that needs to be routed to node $k$ as the next hop. The time needed to accomplish this is given by:

$$T_3 = \left(\frac{M}{C_{j,k}}\right). \tag{5}$$

For a packet that arrives at node $j$ via node $j$s receiver, let $p_{j,k}$ be the probability that the packet must be relayed to node $k$, $k = 1, 2, \ldots, N$; $k \neq j$. Also, let $p_{j,u}$ be the probability that a packet arriving over the radio interface at node $j$ is intended for the user attached at node $j$. Similarly, let $q_{j,k}$ be the probability that a packet generated by the user at node $j$ is to be relayed to node $k$, $k = 1, 2, \ldots, N$; $k \neq j$. Note that:

$$\sum_{k=1, k \neq j}^{N} p_{j,k} + p_{j,u} = 1, \tag{6}$$

$$\sum_{k=1, k \neq j}^{N} q_{j,k} = 1. \tag{7}$$

Thus, we see that the maximum average rate at which switching decisions can be made by node $j$ for a fixed length packet arriving at its receiver is given by:

$$R_j^{Radio} = - \left[ \sum_{k=1, k \neq j}^{N} \frac{p_{j,k} \log p_{j,k}}{\left( \frac{M}{C_{max,j}} + \frac{M}{C_{j,k}} \right)} \right] - \frac{p_{j,u} \log p_{j,u}}{\left( \frac{M}{C_{max,j}} \right)}. \tag{8}$$

Also, the maximum average rate at which switching decisions can be made by node $j$ for a fixed length packet generated by the user attached at node $j$ is given by:

$$R_j^{User} = - \sum_{k=1, k \neq j}^{N} \frac{q_{j,k} \log q_{j,k}}{\left( \frac{M}{C_{j,k}} \right)}. \tag{9}$$

Overall, the maximum rate at which switching decisions can be made by node $j$ is found by maximizing (8) with respect to the output probabilities, subject to (6), and separately maximizing (9) with respect to the output probabilities, subject to (7), and choosing the larger. These maximizations are readily performed with the help of LaGrange multipliers. Finally, summing over all $N$ nodes, one obtains the maximum number of switching decisions per unit time that the network can possibly perform. Call this number $D$.

Next, one can define the traffic entropy of the $j$th source node

$$\epsilon_j = - \sum_{k=1}^{N} r_{j,k} \log r_{j,k}, \tag{10}$$

where $r_{j,k}$ is the probability that a packet originating at node $j$ is intended for node $k$. Let us define:

$$R_j = \sum_{l=1}^{N} t_{j,l}. \tag{11}$$

Then,

$$r_{j,k} = \frac{t_{j,k}}{R_j} = \frac{t_{j,k}}{\sum\limits_{l=1}^{N} t_{j,l}}, \tag{12}$$

The traffic entropy of node $j$ corresponds to the minimum number of switching decisions per packet that must be performed by the network to route node $j$s traffic to the intended destinations. Summing over all $N$ network nodes, and comparing this with the maximum number of switching decisions per second that the network is capable of performing, one can the bound the maximum capacity of the $N$ node network in units of total bits per second. The result is:

$$C \leq \frac{DM}{W} \frac{\sum\limits_{j=1}^{N}\sum\limits_{k=1}^{N} t_{j,k}}{\sum\limits_{j=1}^{N}\sum\limits_{k=1}^{N} t_{j,k} \left[\log\left(\sum\limits_{l=1}^{N} t_{j,l}\right) - \log t_{j,k}\right]} \tag{13}$$

where $C$ is the capacity of the network, with respect to the relative traffic matrix, expressed in bits/s/Hz.

## 6 Results

Results for both bounding approaches were found by randomly placing the $N$ nodes over a rectangular service region; one such placement is called a constellation. For a pair of nodes separated by distance $R$, the path loss was assumed to be $1/R^4$. For the flow deviation method, only path loss is considered. For the information-theoretic approach, for each pair of nodes, a shadow loss factor was assigned; assuming log-normal shadow fading, this corresponds to a factor of $10^{g/10}$, where $g$ is Gaussian with zero mean and standard deviation 6 dB. Also, for the information-theoretic bound, a multipath fading factor was assigned to each pair, assuming flat Rayleigh fading.

From these factors, and assuming some normalized SNR corresponding to two nodes at unit separation with no shadowing or multipath, the capacity matrix $C$ can be found. Various types of traffic matrices were considered, corresponding to different types of traffic non-uniformities and different types of traffic statistics. For each bounding technique, results were then obtained for 100 constellations at each SNR.

Sample results for our upper bounds are as shown below, plotted vs. network size, $N$. Uniform random traffic is assumed, that is, each element of the traffic matrix is a random variable drawn fro the same probability distribution. For the flow deviation method, we show the mean and standard deviation of the upper bound over the 100 simulation trials. Also shown is a an achievable lower bound, assuming the absence of interference. This lower bound is constructively achieved by superimposing a set of non-conflicting flows onto the nodes. For the information theoretic bound, we show only mean results since, for the relatively large number of nodes considered, the standard deviation was observed to be small (Figs. 1 and 2).

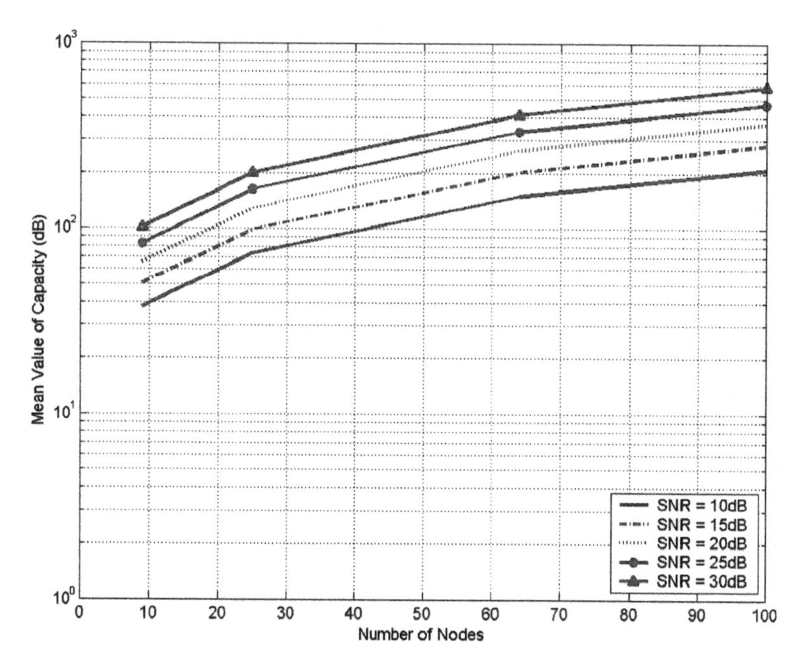


**Fig. 1.** Sample capacity upper bound for information-theoretic approach

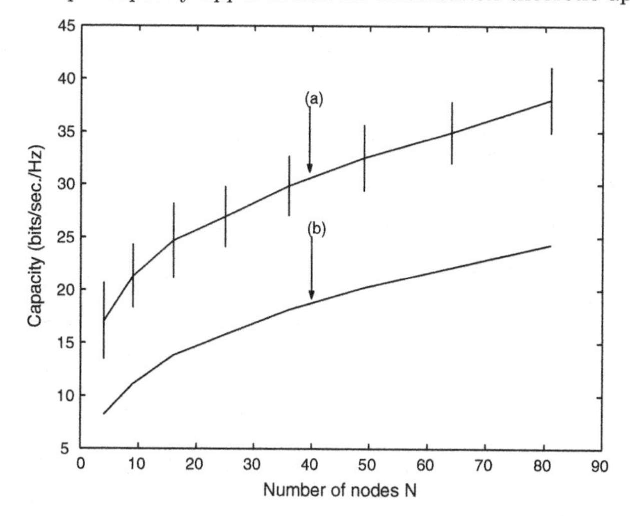


**Fig. 2.** Sample capacity upper and lower bounds for flow deviation approach with SNR = 20 dB

[illegible] of interference. [illegible] nodes [illegible] Figs. 1 and 2.

Fig. 1 [illegible]

Fig. 2 Sample [illegible] upper and lower [illegible] 20 dB [illegible]

# 11

# A Stochastic Non-Cooperative Game for Energy Efficiency in Wireless Data Networks

Stefano Buzzi[1], H. Vincent Poor[2], and Daniela Saturnino[1]

[1] Università degli Studi di Cassino
03043 Cassino (FR) – Italy
{buzzi, d.saturnino}@unicas.it

[2] School of Engineering and Applied Science
Princeton University
Princeton, NJ 08544, USA
poor@princeton.edu

**Summary.** In this paper the issue of energy efficiency in CDMA wireless data networks is addressed through a game theoretic approach. Building on a recent paper by the first two authors, wherein a non-cooperative game for spreading-code optimization, power control, and receiver design has been proposed to maximize the ratio of data throughput to transmit power for each active user, a stochastic algorithm is here described to perform adaptive implementation of the said non-cooperative game. The proposed solution is based on a combination of RLS-type and LMS-type adaptations, and makes use of readily available measurements. Simulation results show that its performance approaches with satisfactory accuracy that of the non-adaptive game, which requires a much larger amount of prior information.

## 1 Introduction

Game theory [1] is a branch of mathematics that has been applied primarily in economics and other social sciences to study the interactions among several autonomous subjects with contrasting interests. More recently, it has been discovered that it can also be used for the design and analysis of communication systems, mostly with application to resource allocation algorithms [2], and, in particular, to power control [3]. As examples, the reader is referred to [4–6]. Here, for a multiple access wireless data network, non-cooperative and cooperative games are introduced, wherein each user chooses its transmit power in order to maximize its own utility, defined as the ratio of the throughput to transmit power. While the above papers consider the issue of power control assuming that a conventional matched filter is available at the receiver, Meshkati et al. [7] consider the problem of joint linear receiver design and power control so as to maximize the utility of each user. It is shown here that the inclusion of receiver design in the considered game brings remarkable advantages, and, also, results based on the powerful large-system analysis are presented. More recently, the results of [7] have been extended in [8] to the case in which also each user's spreading code is included in the tunable parameters for utility maximization. The study [8] thus shows that significant performance gains can be obtained through the joint optimization of the spreading code, the transmit power and the receiver filter for each user.

On the other hand, the solutions proposed in [7] and [8], while providing a general framework for cross-layer resource optimization through a game theoretic approach, describe solutions based on a perfect knowledge of a number of parameters such as the spreading codes, the transmit powers, the propagation channels and the receive filters for all the users. Otherwise stated, the optimization procedure for each user requires a vast amount of prior information not only for the user of interest, but also for all the remaining active users. In this paper, instead, we consider the more practical and challenging situation in which each user performs utility maximization based on the knowledge of its parameters only, i.e. assuming total ignorance of the interference background. This may be the usual scenario in the downlink of a wireless data network, as well as in the uplink of a multicell wireless network, wherein each access point (AP) is disturbed by the interference originating from users served by surrounding AP's. An adaptively learning algorithm capable of approaching with good accuracy the performance of the non-adaptive game is thus presented, based on a combination of the recursive-least-squares (RLS) and least-mean-squares (LMS) adaptation rules. The proposed algorithm assumes no prior knowledge on the interference background and makes use of readily available measurements.

The rest of this paper is organized as follows. Section 2 contains some preliminaries and the system model of interest. Section 3 contains a brief review of the non-adaptive game considered in [8], while the stochastic implementation of the resource allocation algorithm is detailed in Sect. 4. Section 5 contains extensive simulation results, while, finally concluding remarks are given in Sect. 6.

## 2 Preliminaries and Problem Statement

Consider the uplink of a $K$-user synchronous, single-cell, direct-sequence code division multiple access (DS/CDMA) network with processing gain $N$ and subject to flat fading. After chip-matched filtering and sampling at the chip-rate, the $N$-dimensional received data vector, say $\boldsymbol{r}$, corresponding to one symbol interval, can be written as

$$\boldsymbol{r} = \sum_{k=1}^{K} \sqrt{p_k} h_k b_k \boldsymbol{s}_k + \boldsymbol{n} , \tag{1}$$

wherein $p_k$ is the transmit power of the $k$-th user[3], $b_k \in \{-1, 1\}$ is the information symbol of the $k$-th user, and $h_k$ is the real[4] channel gain between the $k$-th user's transmitter and the access point (AP); the actual value of $h_k$ depends on both the distance of the $k$-th user's terminal from the AP and the channel fading fluctuations. The $N$-dimensional vector $\boldsymbol{s}_k$ is the spreading code of the $k$-th user; we assume that the entries of $\boldsymbol{s}_k$ are real and that $\boldsymbol{s}_k^T \boldsymbol{s}_k = \|\boldsymbol{s}_k\|^2 = 1$, with $(\cdot)^T$ denoting transpose. Finally, $\boldsymbol{n}$ is the ambient noise vector, which we assume to be a zero-mean white Gaussian random process with covariance matrix $(\mathcal{N}_0/2)\boldsymbol{I}_N$, with $\boldsymbol{I}_N$ the identity matrix of order $N$. An alternative and compact representation of (1) is given by

$$\boldsymbol{r} = \boldsymbol{S}\boldsymbol{P}^{1/2}\boldsymbol{H}\boldsymbol{b} + \boldsymbol{n} , \tag{2}$$

wherein $\boldsymbol{S} = [\boldsymbol{s}_1, \ldots, \boldsymbol{s}_K]$ is the $N \times K$-dimensional spreading code matrix, $\boldsymbol{P}$ and $\boldsymbol{H}$ are $K \times K$-dimensional diagonal matrices, whose diagonals are $[p_1, \ldots, p_K]$ and $[h_1, \ldots, h_K]$, respectively, and, finally, $\boldsymbol{b} = [b_1, \ldots, b_K]^T$ is the $K$-dimensional vector of the data symbols.

Assume now that each mobile terminal sends its data in packets of $M$ bits, and that it is interested both in having its data received with as small as possible error probability at the AP, and in making careful use of the energy stored in its battery. Obviously, these are conflicting goals, since error-free reception may be achieved by increasing the received SNR, i.e. by increasing the transmit power, which of course comes at the expense of battery life[5]. A useful approach to quantify these conflicting goals is to define the utility of the $k$-th user as the ratio of its throughput, defined as the number of information bits that are received with no error in unit time, to its transmit power [4,5], i.e.

$$u_k = \frac{T_k}{p_k} . \tag{3}$$

Note that $u_k$ is measured in bit per Joule, i.e. it represents the number of successful bit transmissions that can be made for each Joule of energy drained from the battery. Denoting by $R$ the common rate of the network (extension to the case in which each user transmits with its own rate $R_k$ is quite simple) and assuming that each packet of $M$ symbols contains $L$ information symbols and $M - L$ overhead symbols, reserved, e.g., for channel estimation and/or parity checks, the throughput $T_k$ can be expressed as

---

[3] To simplify subsequent notation, we assume that the transmitted power $p_k$ subsumes also the gain of the transmit and receive antennas.

[4] We assume here, for simplicity, a real channel model; generalization to practical channels, with I and Q components, is straightforward.

[5] Of course there are many other strategies to lower the data error probability, such as for example the use of error correcting codes, diversity exploitation, and implementation of optimal reception techniques at the receiver. Here, however, we are mainly interested to energy efficient data transmission and power usage, so we consider only the effects of varying the transmit power, the receiver and the spreading code on energy efficiency.

$$T_k = R \frac{L}{M} P_k \ , \tag{4}$$

wherein $P_k$ denotes the probability that a packet from the $k$-th user is received error-free. In the considered DS/CDMA setting, the term $P_k$ depends formally on a number of parameters such as the spreading codes of all the users and the diagonal entries of the matrices $\boldsymbol{P}$ and $\boldsymbol{H}$, as well as on the strength of the used error correcting codes. However, a customary approach is to model the multiple access interference as a Gaussian random process, and assume that $P_k$ is an increasing function of the $k$-th user's Signal-to-Interference plus Noise-Ratio (SINR) $\gamma_k$, which is naturally the case in many practical situations.

Recall that, for the case in which a linear receiver is used to detect the data symbol $b_k$, according, i.e., to the decision rule

$$\widehat{b}_k = \text{sign}\left[\boldsymbol{d}_k^T \boldsymbol{r}\right] , \tag{5}$$

with $\widehat{b}_k$ the estimate of $b_k$ and $\boldsymbol{d}_k$ the $N$-dimensional vector representing the receive filter for the user $k$, it is easily seen that the SINR $\gamma_k$ can be written as

$$\gamma_k = \frac{p_k h_k^2 (\boldsymbol{d}_k^T \boldsymbol{s}_k)^2}{\frac{\mathcal{N}_0}{2} \|\boldsymbol{d}_k\|^2 + \sum\limits_{i \neq k} p_i h_i^2 (\boldsymbol{d}_k^T \boldsymbol{s}_i)^2} \ . \tag{6}$$

Of related interest is also the mean square error (MSE) for the user $k$, which, for a linear receiver, is defined as

$$\text{MSE}_k = E\left\{ \left( b_k - \boldsymbol{d}_k^T \boldsymbol{r} \right) 2 \right\} = 1 + \boldsymbol{d}_k^T \boldsymbol{M} \boldsymbol{d}_k - 2\sqrt{p_k} h_k \boldsymbol{d}_k^T \boldsymbol{s}_k \ , \tag{7}$$

wherein $E\{\cdot\}$ denotes statistical expectation and $\boldsymbol{M} = (\boldsymbol{SHPH}^T\boldsymbol{S}^T + \frac{\mathcal{N}_0}{2}\boldsymbol{I}_N)$ is the covariance matrix of the data.

The exact shape of $P_k(\gamma_k)$ depends on factors such as the modulation and coding type. However, in all cases of relevant interest, it is an increasing function of $\gamma_k$ with a sigmoidal shape, and converges to unity as $\gamma_k \to +\infty$; as an example, for binary phase-shift-keying (BPSK) modulation coupled with no channel coding, it is easily shown that

$$P_k(\gamma_k) = \left[1 - Q(\sqrt{2\gamma_k})\right]^M \ , \tag{8}$$

with $Q(\cdot)$ the complementary cumulative distribution function of a zero-mean random Gaussian variate with unit variance. A plot of (8) is shown in Fig. 1 for the case $M = 100$.

It should be also noted that substituting (8) into (4), and, in turn, into (3), leads to a strong incongruence. Indeed, for $p_k \to 0$, we have $\gamma_k \to 0$, *but* $P_k$ converges to a small but non-zero value (i.e. $2^{-M}$), thus implying that an

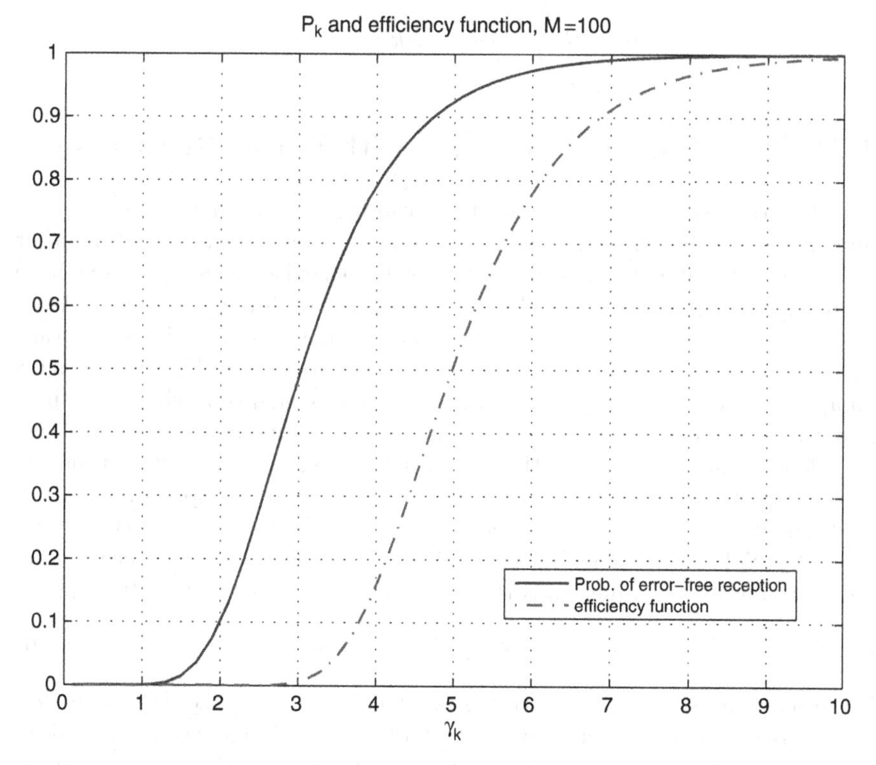


**Fig. 1.** Comparison of probability of error-free packet reception and efficiency function versus receive SINR and for packet size $M = 100$. Note the S-shape of both functions

unboundedly large utility can be achieved by transmitting with zero power, i.e. not transmitting at all and making blind guesses at the receiver on what data were transmitted. To circumvent this problem, a customary approach [5, 7] is to replace $P_k$ with an *efficiency function*, say $f_k(\gamma_k)$, whose behavior should approximate as close as possible that of $P_k$, except that for $\gamma_k \to 0$ it is required that $f_k(\gamma_k) = o(\gamma_k)$. The function $f(\gamma_k) = (1 - e^{-\gamma_k})^M$ is a widely accepted substitute for the true probability of correct packet reception, and in the following we will adopt this model[6]. This efficiency function is increasing and S-shaped, converges to unity as $\gamma_k$ approaches infinity, and has a continuous first order derivative. Note that we have omitted the subscript "$k$," i.e. we have used the notation $f(\gamma_k)$ in place of $f_k(\gamma_k)$ since we assume that the efficiency function is the same for all the users.

Summing up, substituting (4) into (3) and replacing the probability $P_k$ with the above defined efficiency function, we obtain the following expression for the $k$-th user's utility:

[6] See Fig. 1 for a comparison between the Probability $P_k$ and the efficiency function.

$$u_k = R \frac{L}{M} \frac{f(\gamma_k)}{p_k} , \quad \forall k = 1, \dots, K . \tag{9}$$

## 3 The Non-Cooperative Game with Linear Receivers

Based on the utility definition (9), it is natural to wonder how each user may maximize its utility, how this maximization affects utilities achieved by other users, and, also, if a stable equilibrium point does exist. These questions have been answered in recent papers by resorting to the tools of game theory. As an example, non-cooperative scenarios wherein mobile users are allowed to vary their transmit power only have been considered in [4–6]; in [7], instead, such an approach has been extended to the cross layer scenario in which each user may vary its power and its uplink linear receiver, while, more recently, Buzzi and Poor [8] have considered the case in which each user is able to tune the transmit power, the uplink linear receiver and the adopted spreading code.

Formally, the game $\mathcal{G}$ considered in [8] can be described as the triplet $\mathcal{G} = [\mathcal{K}, \{\mathcal{S}_k\}, \{u_k\}]$, wherein $\mathcal{K} = \{1, 2, \dots, K\}$ is the set of active users participating in the game, $u_k$ is the $k$-th user's utility defined in (9), and

$$\mathcal{S}_k = [0, P_{k,\max}] \times \mathcal{R}^N \times \mathcal{R}_1^N \tag{10}$$

is the set of possible actions (strategies) that user $k$ can take. It is seen that $\mathcal{S}_k$ is written as the Cartesian product of three different sets, and indeed $[0, P_{k,\max}]$ is the range of available transmit powers for the $k$-th user (note that $P_{k,\max}$ is the maximum allowed transmit power for user $k$), $\mathcal{R}^N$, with $\mathcal{R}$ the real line, defines the set of all possible linear receive filters, and, finally,

$$\mathcal{R}_1^N = \left\{ \boldsymbol{d} \in \mathcal{R}^N \ : \ \boldsymbol{d}^T \boldsymbol{d} = 1 \right\} ,$$

defines the set of the allowed spreading codes[7] for user $k$.

Summing up, the proposed non-cooperative game can be cast as the following maximization problem

$$\max_{\mathcal{S}_k} u_k = \max_{p_k, \boldsymbol{d}_k, \boldsymbol{s}_k} u_k(p_k, \boldsymbol{d}_k, \boldsymbol{s}_k) , \quad \forall k = 1, \dots, K . \tag{11}$$

Given (9), the above maximization can be also written as

$$\max_{p_k, \boldsymbol{d}_k, \boldsymbol{s}_k} \frac{f\left(\gamma_k(p_k, \boldsymbol{d}_k, \boldsymbol{s}_k)\right)}{p_k} , \quad \forall k = 1, \dots, K . \tag{12}$$

Moreover, since the efficiency function is monotone and non-decreasing, we also have

[7] Here we assume that the spreading codes have real entries; the problem of utility maximization with reasonable complexity for the case of discrete-valued entries is a challenging issue that will be considered in the future.

$$\max_{p_k, \boldsymbol{d}_k, \boldsymbol{s}_k} \frac{f\left(\gamma_k(p_k, \boldsymbol{d}_k, \boldsymbol{s}_k)\right)}{p_k} = \max_{p_k} \frac{f\left(\max\limits_{\boldsymbol{d}_k, \boldsymbol{s}_k} \gamma_k(p_k, \boldsymbol{d}_k, \boldsymbol{s}_k)\right)}{p_k} , \tag{13}$$

i.e. we can first take care of SINR maximization with respect to spreading codes and linear receivers, and then focus on maximization of the resulting utility with respect to transmit power.

Letting $(\cdot)^+$ denoting Moore–Penrose pseudoinverse, the following result is reported in [8].

**Proposition 1.** *The non-cooperative game defined in (11) admits a unique Nash equilibrium point* $(p_k^*, \boldsymbol{d}_k^*, \boldsymbol{s}_k^*)$, *for* $k = 1, \ldots, K$, *wherein*

- $\boldsymbol{s}_k^*$ *and* $\boldsymbol{d}_k^*$ *are the unique fixed stable* $k$*-th user spreading code and receive filter*[8] *resulting from iterations*

$$\begin{aligned} \boldsymbol{d}_i &= \sqrt{p_i} h_i \left(\boldsymbol{SHPH}^T \boldsymbol{S}^T + \tfrac{\mathcal{N}_0}{2} \boldsymbol{I}_N\right)^{-1} \boldsymbol{s}_i \;\; \forall i = 1, \ldots, K , \\ \boldsymbol{s}_i &= \sqrt{p_i} h_i \left(p_i h_i^2 \boldsymbol{D}\boldsymbol{D}^T + \mu_i \boldsymbol{I}_N\right)^+ \boldsymbol{d}_i \qquad \forall i = 1, \ldots, K \end{aligned} \tag{14}$$

  *with* $\mu_i$ *such that* $\|\boldsymbol{s}_i\|^2 = 1$, *and* $\boldsymbol{D} = [\boldsymbol{s}_1, \ldots, \boldsymbol{s}_K]$. *Denote by* $\gamma_k^*$ *the corresponding SINR.*
- $p_k^* = \min\{\bar{p}_k, P_{k,\max}\}$, *with* $\bar{p}_k$ *the* $k$*-th user transmit power such that the* $k$*-th user maximum SINR* $\gamma_k^*$ *equals* $\bar{\gamma}$, *i.e. the unique solution of the equation* $f(\gamma) = \gamma f'(\gamma)$, *with* $f'(\gamma)$ *the derivative of* $f(\gamma)$.

In practice, this result states that the non-cooperative game (11) admits a unique Nash equilibrium, that can be reached as follows. First of all, the equation $f(\gamma) = \gamma f'(\gamma)$ is to be solved in order to determine its unique solution $\bar{\gamma}$; then, an iterative procedure starts wherein the system alternates between these two phases:

a Given the transmit powers, each user adjusts its spreading code and receive filter through iterations (14) until an equilibrium is reached;

b Given the spreading codes and uplink receivers, each user tunes its transmit power so that its own SIR equals $\bar{\gamma}$. Denoting by $\boldsymbol{p} = [p_1, \ldots, p_K]$ the users' power vector, and by $\boldsymbol{I}(\boldsymbol{p})$ the $K$-dimensional vector whose $k$-th entry $\boldsymbol{I}_k(\boldsymbol{p})$ is written as

$$\boldsymbol{I}_k(\boldsymbol{p}) = \frac{\bar{\gamma}}{h_k^2(\boldsymbol{d}_k \boldsymbol{s}_k)^2} \left(\frac{\mathcal{N}_0}{2} \|\boldsymbol{d}_k\|^2 + \sum_{i \neq k} p_i h_i^2 (\boldsymbol{d}_k \boldsymbol{s}_i)^2\right) , \tag{15}$$

the transmit power vector $\boldsymbol{p}$ is the unique fixed stable point of the iteration [3]

[8] Actually the linear receive filter is unique up to a positive scaling factor.

$$p_k = \begin{cases} \boldsymbol{I}_k(\boldsymbol{p}) \ , & \text{for } \boldsymbol{I}_k(\boldsymbol{p}) \leq P_{k,\max} \ , \\ P_{k,\max} \ , & \text{for } \boldsymbol{I}_k(\boldsymbol{p}) > P_{k,\max} \end{cases} \tag{16}$$

for all $k = 1, \ldots, K$.

Steps a and b are to be repeated until convergence is reached. It is crucial to note that computation of the equilibrium transmit power, spreading code and linear receiver for each user needs a lot of prior information. In particular, it is seen from (14) that computation of the $k$-th user receiver requires knowledge of the spreading codes, transmit powers and channel gains for all the active users, while computation of the $k$-th user spreading code requires knowledge of $\boldsymbol{D}$, i.e. the matrix of the uplink receivers for all the users. Likewise, implementation of iterations (16) also requires the same vast amount of prior information. Our next goal is thus to propose a stochastic resource allocation algorithm that alleviates the need for such prior knowledge, and that is amenable to a decentralized implementation, wherein each user may allocate its own resources based only on knowledge that is readily available, and with total ignorance on the interference background.

## 4 Adaptive Energy Efficient Resource Allocation

In order to illustrate the adaptive implementation of the non-cooperative game, a slight change of notation is needed. Indeed, since any adaptive algorithm relies on several data observations in consecutive symbol intervals, we cannot restrict any longer our attention to one symbol interval only, and we thus denote by $\boldsymbol{r}(n)$ the $N$-dimensional received data vector in the $n$-th bit interval, i.e.

$$\boldsymbol{r}(n) = \sum_{k=1}^{K} \sqrt{p_k(n)} h_k b_k(n) \boldsymbol{s}_k(n) + \boldsymbol{w}(n) \ . \tag{17}$$

Equation (17) differs from (1) in that a temporal index has been added to some parameters, to underline their time-varying nature: as an example, $p_k(n)$ and $\boldsymbol{s}_k(n)$ are the transmit power and the spreading code of the $k$-th user in the $n$-th symbol interval. Note also that the channel gain does not depend on $n$, i.e. we are implicitly assuming a slow fading channel, even though generalization to the case of slowly time-varying channels is quite straightforward. In order to obtain an adaptive implementation of the utility maximizing algorithm for the generic $k$-th user, first we focus on iterations (14); as specified in [8, 9], the unique fixed stable point of these iterations achieves the global minimum for the total mean square error (TMSE), which is given by

$$\text{TMSE} = \sum_{k=1}^{K} \text{MSE}_k \ , \tag{18}$$

with $\mathrm{MSE}_k$ defined as in (7). It can be shown, indeed, that minimization of the TMSE leads to a Pareto-optimal solution to the problem of maximizing the SINR for each user. On the other hand, an alternative approach is to consider the case in which each user tries to minimize its own MSE with respect to its spreading code and linear receiver. The $k$-th user MSE can be shown to be written as

$$\mathrm{MSE}_k = 1 + \boldsymbol{d}_k^T \left(p_k h_k^2 \boldsymbol{s}_k \boldsymbol{s}_k^T + \boldsymbol{M}_k\right) \boldsymbol{d}_k - 2\sqrt{p_k} h_k \boldsymbol{d}_k^t \boldsymbol{s}_k \ , \tag{19}$$

with $\boldsymbol{M}_k = \boldsymbol{M} - p_k h_k^2 \boldsymbol{s}_k \boldsymbol{s}_k^T$. Minimization of (19) with respect to $\boldsymbol{d}_k$ and $\boldsymbol{s}_k$, under the constraint $\|\boldsymbol{s}_k\|^2 = 1$, yields the iterations

$$\begin{aligned} \boldsymbol{d}_i &= \sqrt{p_i} h_i \left(\boldsymbol{S}\boldsymbol{H}\boldsymbol{P}\boldsymbol{H}^T\boldsymbol{S}^T + \tfrac{\mathcal{N}_0}{2}\boldsymbol{I}_N\right)^{-1} \boldsymbol{s}_i \ \ \forall i = 1, \ldots, K \\ \boldsymbol{s}_i &= \sqrt{p_i} h_i \left(p_i h_i^2 \boldsymbol{d}_i \boldsymbol{d}_i^T + \mu_i \boldsymbol{I}_N\right)^{-1} \boldsymbol{d}_i \qquad \forall i = 1, \ldots, K \ . \end{aligned} \tag{20}$$

In general, minimization of the TMSE is not equivalent to individual minimization of the MSE of each user; however, in our scenario, i.e. in the case of a single-path fading channel, the two approaches can be shown to be equivalent [10]. As a consequence, we can state that the fixed point of iterations (14) coincides with that of iterations (20). Note however that, despite such equivalence, the spreading code update for each user in (20) depends only on parameters of the user itself, and does not require any knowledge on the interference background. The receiver updates in (14) and (20) are the same, and indeed they coincide with the MMSE receiver. Accordingly, since the utility maximizing linear receiver is the MMSE filter, we start resorting to the well-known recursive-least-squares (RLS) implementation of this receiver. Letting $\boldsymbol{R}(0) = \epsilon \boldsymbol{I}_N$, with $\epsilon$ a small positive constant, letting $\lambda$ be a close-to-unity scalar constant, assuming that the receiver has knowledge of the information symbols $b_k(1), \ldots, b_k(T)$, and denoting by $\boldsymbol{d}_k(n)$ the estimate of the linear receiver filter for the $k$-th user in the $n$-th symbol interval, the following iterations can be considered

$$\begin{aligned} \boldsymbol{k}(n) &= \frac{\boldsymbol{R}^{-1}(n-1)\boldsymbol{r}(n)}{\lambda + \boldsymbol{r}^T(n)\boldsymbol{R}^{-1}(n-1)\boldsymbol{r}(n)}, \\ \boldsymbol{R}^{-1}(n) &= \frac{1}{\lambda}\left[\boldsymbol{R}^{-1}(n-1) - \boldsymbol{k}(n)\boldsymbol{r}^T(n)\boldsymbol{R}^{-1}(n-1)\right], \\ e_k(n) &= \boldsymbol{d}_k^T(n-1)\boldsymbol{r}(n) - b_k(n), \\ \boldsymbol{d}_k(n) &= \boldsymbol{d}_k(n-1) - e_k(n)\boldsymbol{k}(n). \end{aligned} \tag{21}$$

The last line in (21) represents the update equation for the detection vector $\boldsymbol{d}_k(\cdot)$. Note that this equation, in turn, depends on the error vector $e_k(n)$, which, for $n \leq T$, can be built based on the knowledge of the training symbol

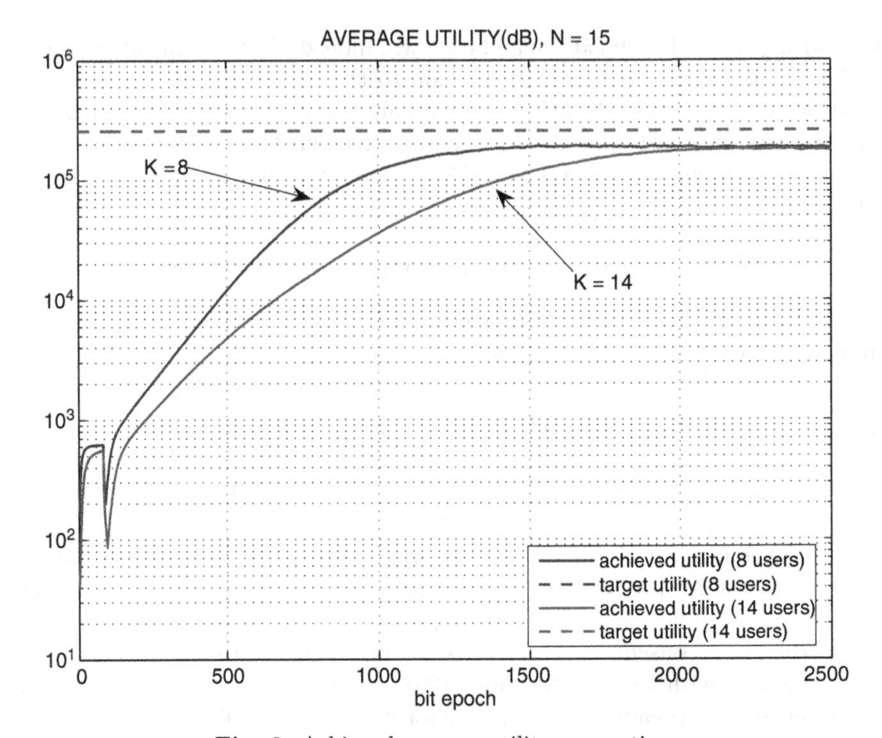


**Fig. 2.** Achieved average utility versus time

$b_k(n)$. Once the training phase is over, real data detection takes place and the error in the third line of (21) is computed according to the equation

$$e_k(n) = \boldsymbol{d}_k^T(n-1)\boldsymbol{r}(n) - \text{sgn}\left[\boldsymbol{d}_k^T(n-1)\boldsymbol{r}(n)\right] . \tag{22}$$

Given its own receive filter $\boldsymbol{d}_k(n)$, the $k$-th user can then modify its spreading code according to the second line of (20), i.e.:

$$\boldsymbol{s}_k(n+1) = \sqrt{p_k(n)}h_k\left(p_k(n)h_k^2\boldsymbol{d}_k(n)\boldsymbol{d}_k^T(n) + \mu_k(n)\boldsymbol{I}_N\right)^{-1}\boldsymbol{d}_k(n) \tag{23}$$

with $\mu_k(n)$ a constant such that $\boldsymbol{s}_k^T(n)\boldsymbol{s}_k(n) = 1$. Note that the update in (23) only requires parameters of the $k$-th user, thus implying that no knowledge on the interference background is needed. Finally, we have to consider the tuning of the transmit power so that each user may achieve its target SINR $\bar{\gamma}$. This is a classical stochastic power control problem that has been treated, for instance, in [11]. A possible solution is to consider a least-mean-squares (LMS) update of the transmit power according to the rule

$$\begin{aligned} p_k(n+1) &= (1-\rho)p_k(n-1) + \rho I_k(n) , \\ p_k(n+1) &= \min(p_k(n+1), P_{k,\max}) . \end{aligned} \tag{24}$$

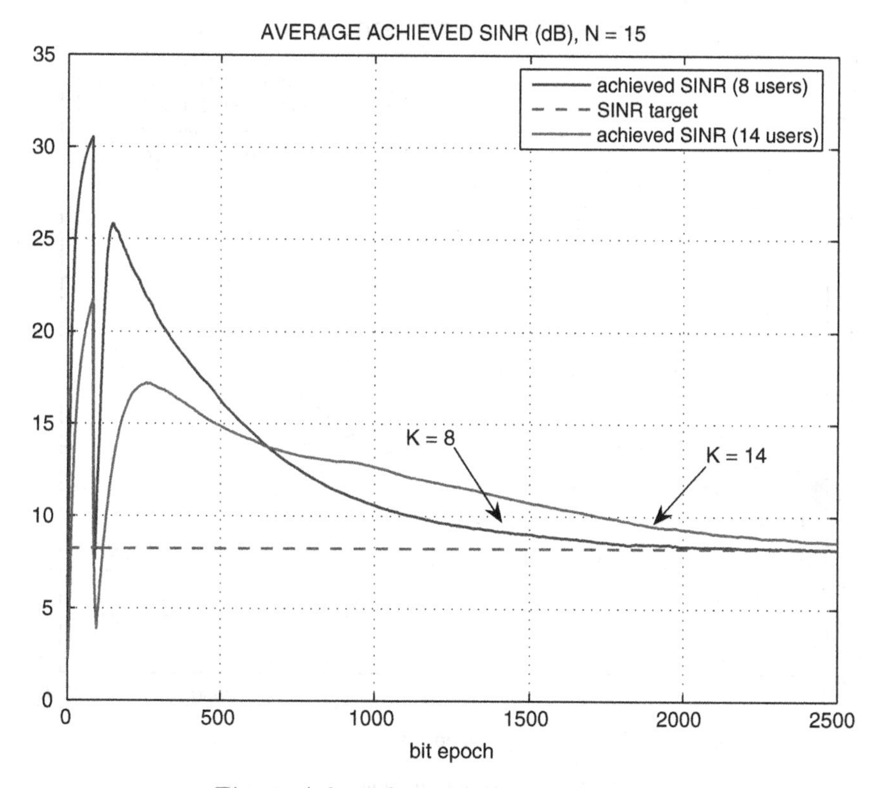


**Fig. 3.** Achieved average SINR versus time

In the above equations, the step-size $\rho$ is a close-to-zero positive constant and $\mathcal{I}_k(n)$ is a stochastic approximation of the $k$-th entry of the vector $\boldsymbol{I}(\boldsymbol{p})$, and is expressed as

$$I_k(n) = \frac{\bar{\gamma}}{h_k^2 \left(\boldsymbol{d}_k^T(n)\boldsymbol{s}_k(n)\right)^2} \left[(\boldsymbol{d}_k^T(n)\boldsymbol{r}(n))^2 - p_k(n)h_k^2(\boldsymbol{d}_k^T(n)\boldsymbol{s}_k(n))^2\right] . \quad (25)$$

Note that also the update (24) does not require any knowledge on the interference. To summarize, the algorithm proceeds as follows: for $n \leq T$, only the RLS update (21) is performed; then, for each $n > T$, the algorithm performs the updates in (21), (23) and, finally, (24). In particular, note that the power update is made in parallel to the spreading code update and receiver update, i.e. without waiting for convergence of the RLS-based adaptive implementation of the MMSE receiver.

Although a theoretical convergence study of this algorithm is definitely worth being undertaken, it is out of the scope of this paper; in Sect. 5, we will discuss the results of extensive computer simulations that will show the excellent behavior of the outlined procedure.

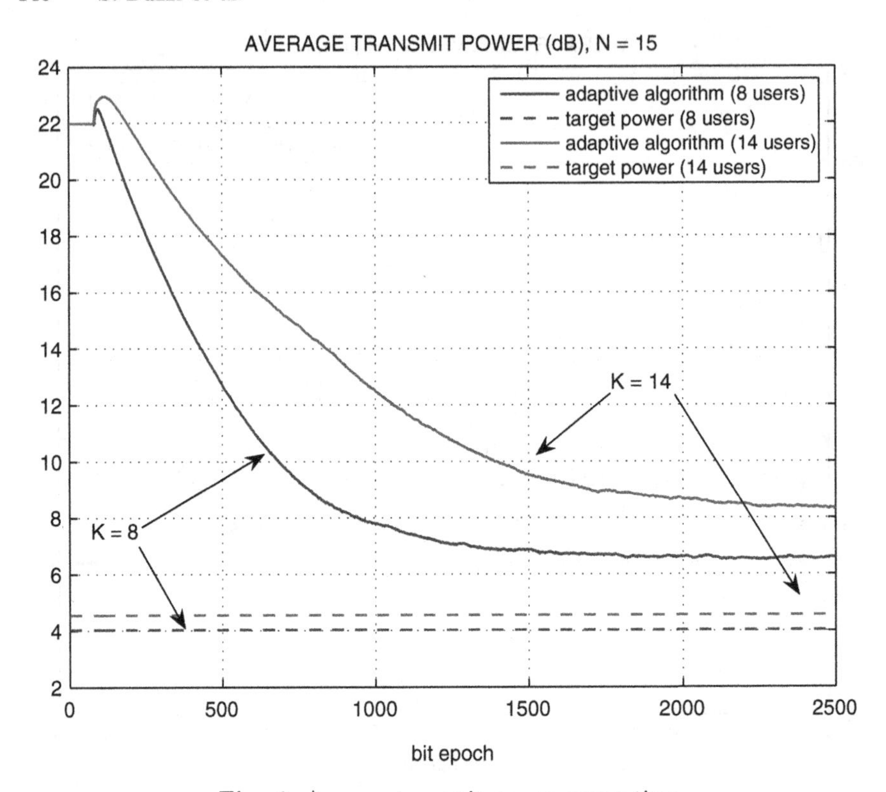


**Fig. 4.** Average transmit power versus time

# 5 Numerical Results

We contrast here the performance of the non-adaptive game discussed in [8] with its adaptive implementation proposed here. We consider an uplink DS/CDMA system with processing gain $N = 15$, and assume that the packet length is $M = 120$, for this value of $M$ the equation $f(\gamma) = \gamma f'(\gamma)$ can be shown to admit the solution $\bar{\gamma} = 6.689 = 8.25\,\text{dB}$. A single-cell system is considered, wherein users may have random positions with a distance from the AP ranging from 10 to 500 m. The channel coefficient $h_k$ for the generic $k$-th user is assumed to be Rayleigh distributed with mean equal to $d_k^{-2}$, with $d_k$ being the distance of user $k$ from the AP[9]. We take the ambient noise level to be $\mathcal{N}_0 = 10^{-5}\,\text{W}\,\text{Hz}^{-1}$, while the maximum allowed power $P_{k,\max}$ is 25 dB. We present the results of averaging over 1,000 independent realizations for the users locations, fading channel coefficients and starting set of spreading codes. More precisely, for each iteration we randomly generate an $N \times K$-

[9] Note that we are here assuming that the power path losses are proportional to the fourth power of the path length, which is reasonable in urban cellular environments.

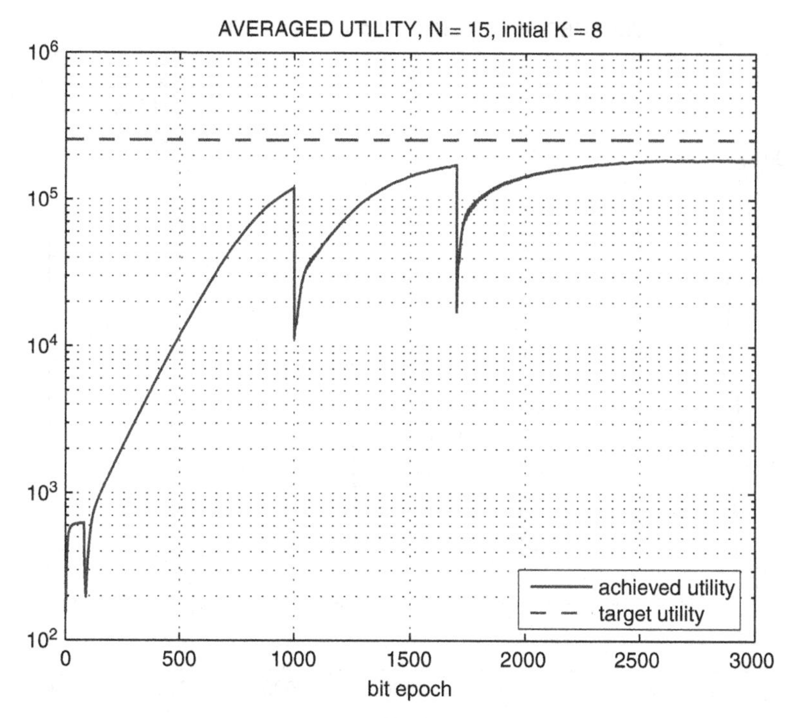


**Fig. 5.** Dynamic environment: average achieved utility versus time

dimensional spreading code matrix with entries in the set $\left\{-1/\sqrt{N}, 1/\sqrt{N}\right\}$; this matrix is then used as the starting point for the considered games. We consider the case in which $T = 80$ training symbols are used, while in (24) the step size $\rho = 0.01$ has been taken. Figures 2–4 report the time-evolution of the achieved average utility (measured in bit per Joule), the average achieved SINR and the average transmit power, respectively, for both the cases in which $K = 8$ and $K = 14$. It is seen that after about one thousand iterations the adaptive algorithm approximate with satisfactory accuracy the benchmark scenario that a non-adaptive game is played as in [8]. In particular, while the target SINR and the achieved utility are quite close to their target values, it is seen from Fig. 4 that the average transmit power is about 3 dB larger than in the non-adaptive case; such a loss is not at all surprising, since it is well-known that adaptive algorithms have a steady-state error, and that their performance may only approach that of their non-adaptive counterparts.

In order to test the tracking properties of the proposed algorithm, we also consider a dynamic scenario with an initial number of users $K = 8$, and with two additional users entering the channel at time epochs $n = 1,000$ and $n = 1,700$. The results are reported in Figs. 5–7. Results clearly show that the algorithm is capable of coping with changes in the interference background.

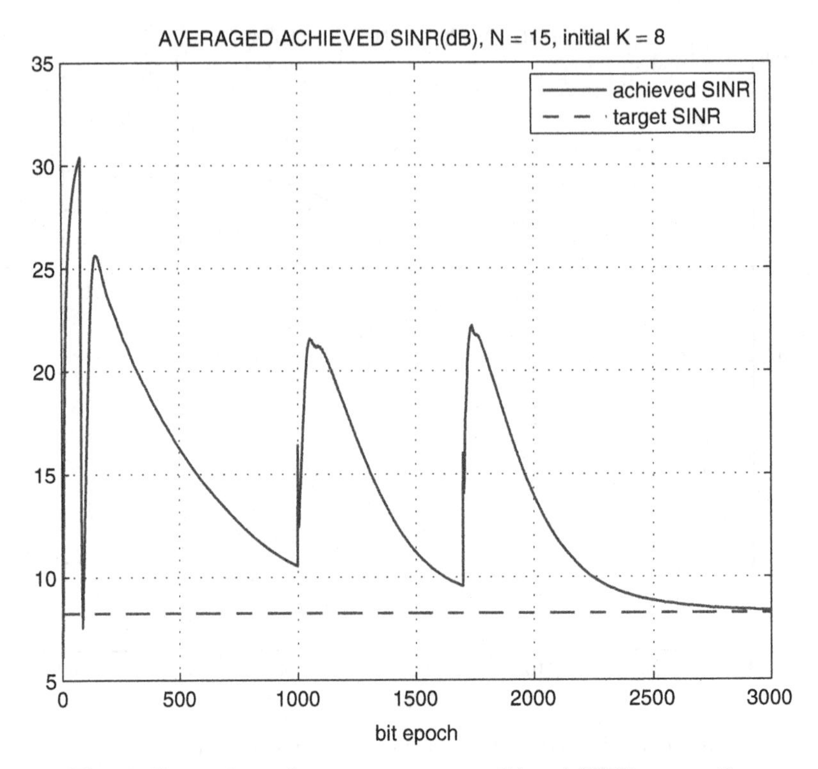


**Fig. 6.** Dynamic environment: average achieved SINR versus time

## 6 Conclusion

In this paper the cross-layer issue of joint stochastic power control, spreading code optimization and receiver design for wireless data networks has been addressed using a game-theoretic framework. Building on [8], wherein a non-cooperative game for resource allocation has been proposed, we have here considered the issue of adaptive implementation of the resource allocation algorithm, based on readily available measurements and assuming no prior knowledge on the interference background. The result is thus a stochastic algorithm that can be realized in a decentralized fashion, wherein each user just needs knowledge of its own parameters. The performance of the proposed scheme has been validated through computer simulations, which showed that the adaptive implementation achieved a performance quite close to that of the non-adaptive benchmark. As interesting research topics worth being investigated we mention the theoretical analysis of the convergence properties of the algorithm, and the development and the analysis of adaptive algorithms able to implement the said game without prior knowledge of the fading channel coefficient of the user of interest.

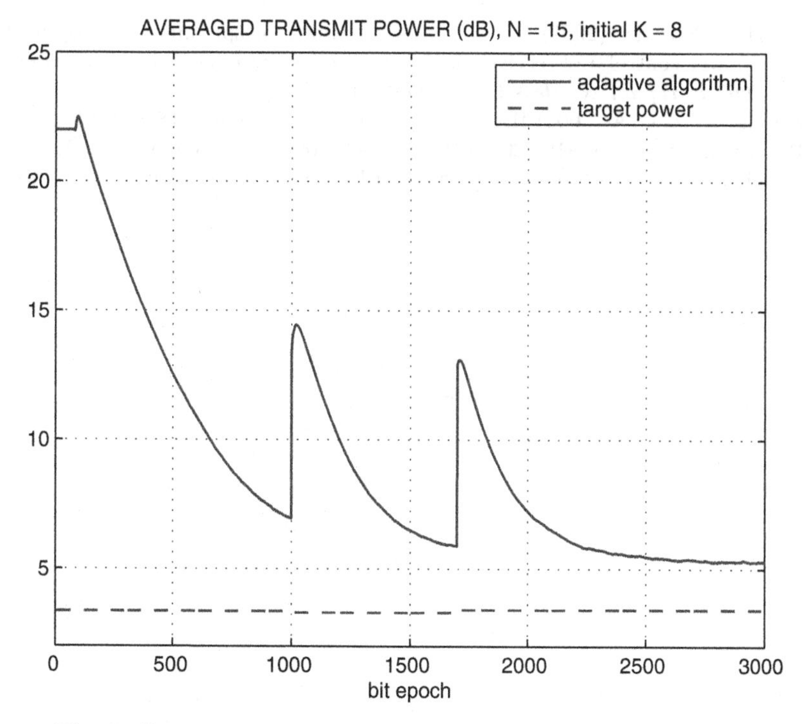


**Fig. 7.** Dynamic environment: average transmit power versus time

## Acknowledgment

This paper was supported by the U.S. Air Force Research Laboratory under Cooperative Agreement No. FA8750-06-1-0252, and by the U.S. Defense Advanced Research Projects Agency under Grant No. HR0011-06-1-0052.

## References

1. Fudenberg D, Tirole J (1991) Game Theory. MIT Press, Cambridge, MA
2. MacKenzie AB, Wicker SB (2001) Game theory in communications: Motivation, explanations, and applications to power control, *Proceedings to the IEEE Global Telecommunications Conference*, San Antonio, TX
3. Yates RD (1995) IEEE J Sel Areas Commun 13:1341–1347
4. Goodman DJ, Mandayam NB (2000) IEEE Pers Commun 7:48–54
5. Saraydar CU, Mandayam NB, Goodman DJ (2002) IEEE Trans Commun 50:291–303
6. Saraydar CU (2001) Pricing and power control in wireless data networks. Ph.D. Dissertation, Rutgers University, Piscataway, NJ, USA
7. Meshkati F, Poor HV, Schwartz SC, Mandayam NB (2005) IEEE Trans Commun 53:1885–1894

8. Buzzi S, Poor HV (2007) Non-cooperative games for spreading code optimization, power control and receiver design in wireless data networks *Proceedings of the 2007 European Wireless Conference*, Paris
9. Ulukus S, Yener A (2004) IEEE Trans Wireless Commun 3:1879–1884
10. Rajappan GS, Honig ML (2002) IEEE J Sel Areas Commun 20:384–395
11. Luo J, Ulukus S, Ephremides A (2005) IEEE Trans Inf Theory 51:2612–2624

# Session 3

## Security and Applications in Wireless Networks

# 12

# Security Overheads for Signaling in Beyond-3G Networks

Dario S. Tonesi, Alessandro Tortelli, and Luca Salgarelli

Università degli Studi di Brescia, Brescia, Italy
darios.tonesi@ing.unibs.it, alessandro.tortelli@ing.unibs.it,
luca.salgarelli@ing.unibs.it

**Summary.** The Third Generation Partnership Project (3GPP) Release 6 is an important step towards all-IP, next generation mobile networks. Compared to previous 3GPP networks, Release 6 defines the IP Multimedia Subsystem, a core network architecture completely based on IETF protocols. Among these protocols, the Session Initiation Protocol (SIP) is used for initiating, managing and terminating media sessions. Even though simple and flexible, SIP has not been defined with security in mind. Therefore, 3GPP technical specifications require SIP signaling to be protected by means of IPSec security associations. The use of SIP together with IPSec can result in a significant increase of both the amount of traffic exchanged and the computational load due to signaling inside the core network. This paper analyzes the computational overheads due to security for signaling in 3GPP Release 6 core networks.

## 1 Introduction

The Third Generation Partnership Project (3GPP) is responsible for the evolution of technical specifications for third generation (3G) cellular networks. It also leads the way to beyond third generation (B3G) networks. Starting with Release 5, and subsequently with Release 6, 3GPP introduces the so called IP Multimedia Subsystem (IMS) [1]. IMS is a framework completely based on IETF protocols, such as IP, UDP, TCP, SIP [2], Diameter [3], etc. It has been introduced mainly because of its expected low cost of operations and maintenance, its flexibility and seamless integration with other IP networks.

However, the introduction of IMS comes at a price, from a security point of view. Legacy 2G cellular networks such as GSM implement clear security boundaries between their core networks and the external world. At the other end of the scale, B3G networks are meant to be completely integrated with the Internet. This makes these new types of networks very vulnerable to many kinds of security attacks. To properly protect signaling flows, 3GPP mandates a series of technical specifications [4] which require new IMS network elements for implementing IPSec security associations. The use of IPSec inside the core network implies several overheads. From a spatial (byte-wise) perspective, IPSec augments the size of every single SIP message. In addition, from a computational-load viewpoint, IPSec requires extra time to manage and execute cryptographic procedures such as encryption, decryption, hash computation and verification.

This paper analyzes the overheads introduced by the use of IPSec for the protection of signaling in B3G core networks. We focus specifically on the most significant part of such overheads, i.e., the computational one. By combining a numeric analysis of the cryptographic overheads with experimental measurements, we show how the implementation of IPSec in 3GPP core networks could require a significant upgrade in the processing power of signaling nodes.

The main research contributions of this paper are:

- A numeric analysis of the cryptographic overheads due to IPSec, with corrections to a previous work on the subject [5]. We validate our numeric formulas through laboratory experiments.
- A report on a series of experimental measurements on the computational overheads introduced by IPSec on SIP signaling messages.
- The evaluation, based on the two points above, of the overall increase in computational power required for 3GPP core network signaling nodes to sustain the introduction of IPSec.

The rest of the paper is organized as follows. In Sect. 2 we describe the architecture of 3GPP Release 6 core networks and the security protections required by 3GPP technical specifications. Section 3 introduces our model for characterizing the signaling overheads. Subsequently, in Sect. 4 we define and validate new formulas for the analysis of the parts of the overheads that are due to cryptography. Such formulas are then used in Sect. 5, together with experimental results, to evaluate the computational overheads on SIP signaling due to IPSec. In Sect. 6 we discuss the possible consequences of such overheads on 3G operators. Finally, Sect. 7 concludes the paper.

## 2 IPSec in 3GPP Release 6

The top part of Fig. 1 represents the generic 3GPP Release 6 UMTS network architecture when security protection is not enabled. The main difference from previous UMTS releases is in the architecture of the core network, where the IMS has been introduced. The main goal of IMS is to support multimedia applications more efficiently than previous generation network architectures. IMS is based on IETF protocols and, among all, the Session Initiation Protocol (SIP) is the one used for signaling.

SIP is a text-based, client-server, end-to-end protocol and it has been chosen mainly because of its simplicity and flexibility. It works at the application layer and it is used to manage multimedia sessions among two or more parties.

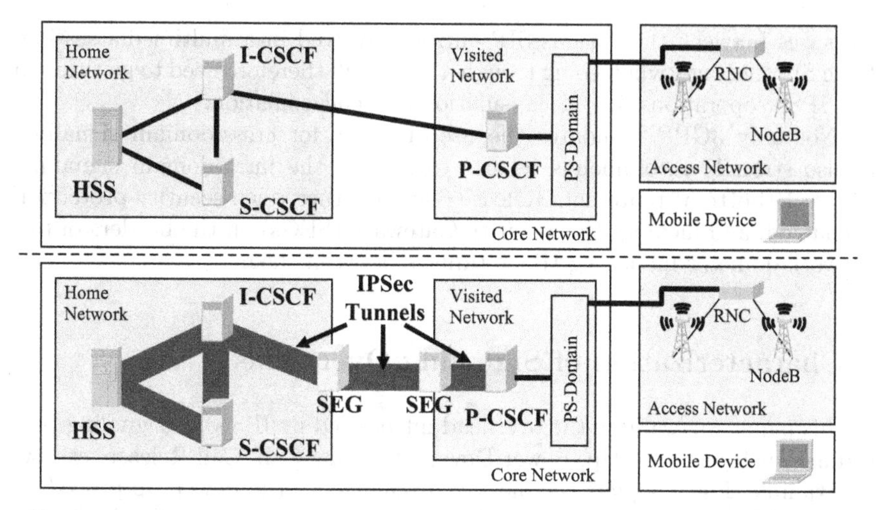


**Fig. 1.** Rel. 6 architecture with (*bottom*) and without (*top*) security protection

Sessions can consist in telephone calls over IP, video conferences or, more in general, any type of multimedia content exchange.

Figure 1 shows the main IMS network elements. The *Proxy-Call Session Control Function* (P-CSCF) is the point of access for a *User Agent* (UA) hosted by a mobile device. The P-CSCF is the SIP server a UA refers to while it is in a visited network. The *Serving-Call Session Control Function* (S-CSCF), is the SIP registrar that maintains the UA information, such as its identity, location, etc., and takes care of session control.The P-CSCF is always located in the UA home network. *Interrogating-Call Session Control Function* (I-CSCF) is the SIP server on the border of the UA's home network: its address is public, so S-CSCFs from other domains or P-CSCFs from visited networks can find it and use it as an entry point for a particular domain. The I-CSCF interrogates the *Home Subscriber Server* (HSS) for obtaining the location of the users and contacting the proper S-CSCF.

3GPP technical specifications require the SIP message exchange inside IMS to be protected by means of IPSec security associations [4]. According to the specifications:

- IPSec must work in tunnel mode
- IPSec must use the Encapsulating Security Payload (ESP) [6]
- As cryptographic protocol either Triple-DES (TDES) [7] or AES-128 [8], both in cipher-block chaining (CBC) mode [9], must be used
- As keyed-hash algorithm, HMAC-SHA1 [10, 11], must be used

Also, 3GPP mandates the introduction of new network elements, the Security Gateways (SEGs), that must establish security associations between different network domains in order to protect SIP signaling messages. By terminating multiple IPSec tunnels, each SEG acts as a forwarder of SIP messages between the various SIP entities involved in a multimedia session. Each SIP message, when going trough a SEG, will therefore need to go through two IPSec operations, i.e., decapsulation and encapsulation.

Not only 3GPP mandates the use of IPSec for cross-domain signaling, it also strongly recommends its use to protect the intra-domain signaling. Figure 1 (bottom) represents Release 6 architecture when security protection is enabled, and includes the Security Gateways (SEGs) on the borders of two network domains managing IPSec tunnels between them.

## 3 Characterization of Signaling Overheads

In this paper we evaluate the overhead introduced by IPSec on signaling procedures such as Power Up, Power Down, Outgoing Call, Call Release, etc. As an example, Fig. 2 depicts the message exchange for the Power Up procedure when the UA is in a visited network and security protection is enabled. The first step for analyzing the overhead introduced by IPSec is to define a model for characterizing the time required to execute a given signaling procedure. We model such time as a sum of different components whose main terms are:

- $t_{Cin}, t_{Cout}$ – *computational time.* The time needed to carry out computations and protocol management on a given IMS node when a SIP message is received ($t_{Cin}$) and the correspondent SIP message is generated ($t_{Cout}$). In general, these two terms can be calculated as simple sums:

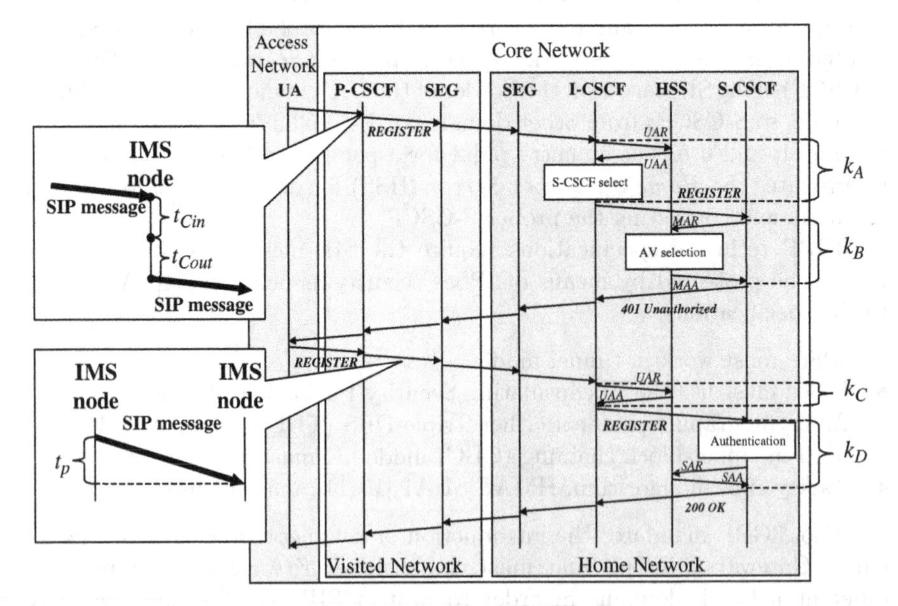


**Fig. 2.** Power Up procedure when UA is in a visited network and security is used

$$t_{Cout} = t_{SIP_E} + t_{UDP_E} + t_{IP_E} + t_{ESP_E} + t_E + t_H, \tag{1}$$

$$t_{Cin} = t_{SIP_D} + t_{UDP_D} + t_{IP_D} + t_{ESP_D} + t_D + t_H. \tag{2}$$

In each expression, the first four terms represent the times needed to encapsulate (decapsulate) data at the application, transport, network and IPSec layer, respectively. The last two terms account for the cryptographic operations that are needed to cipher (decipher) the message and calculate its keyed-hash. The last three terms appear only when IPSec is enabled.

- $t_p$ – *packet delay*, the delay a given SIP message experiences while traveling between two IMS nodes. It depends on the transmission delay and on the delays introduced by intermediate, non signaling-related nodes such as routers and network switches. The transmission delay, in turns, depends on the message size and on the link capacity. Assuming the link capacity is a constant, evaluating how IPSec augments the transmission delay can be as simple as computing the SIP message size overhead introduced by IPSec.
- $K$ – a constant that models delays introduced by protocols other than SIP, such as Diameter, etc. Referring to Fig. 2, $K$ is given by the sum of $k_A, \ldots k_D$ terms. The analysis of $K$ is out of the scope of this paper.

An overall evaluation of security overheads due to IPSec would have to consider, for each call-management procedure, the extra delay added by IPSec to each of the three main components described above. In this paper, we focus on the terms that are going to be the most affected by the introduction of IPSec, i.e., $t_{Cin}$ and $t_{Cout}$.

# 4 Cryptographic Overhead: Numerical Analysis

In order to assess what is the time overhead due to the usage of IPSec in IMS, we evaluated in detail what happens during encryption, decryption and hash computation.

We studied different cryptographic algorithms, included those involved with the types of IPSec security associations required by 3GPP technical specifications. In particular we studied the following cryptographic algorithms: Data Encryption Standard (DES) [7], Triple-DES (TDES) [7] and AES-128 [8], all in CBC mode [9].

Similarly to [5], we decomposed these algorithms in basic operations, such as copy, swap, table look-up, bit permutations, etc. In order to estimate $t_E$ and $t_D$ for these algorithms, we computed the number of basic operations $n_E$ and $n_D$ needed for each algorithm as a function of the size of input messages.

Let us consider DES as an example. DES is a symmetric cipher that works on 64-bit blocks and with a 56-bit key. The algorithm has to perform 19 steps to encrypt or decrypt a block. The first and the last steps are simple

**Table 1.** Basic operations for a single DES 64-bit block

| Step | Operation | Iterations | Total |
|---|---|---|---|
| 1, 19 | 64-bit permutation | 2 | 128 |
| 2,...,17 | 64-bit swap | 16 | 128 |
| | 4-byte xor | 16 | 64 |
| | 48-bit permutation | 16 | 768 |
| | 6-byte xor | 16 | 96 |
| | 8 table look ups | 16 | 384 |
| | 32-bit permutation | 16 | 512 |
| 18 | 8-byte swap | 1 | 8 |
| CBC-mode | 8-byte xor | 1 | 8 |
| Total basic operations | | | 2,096 |

permutations. In step 18 there is a 32-bit swap between the left and the right block. Each of the other 16 steps is repeated 16 times. Table 1 illustrates the number of basic operations for DES. The total number of operations includes eight extra `xor` operations, to account for implementing CBC mode.

Proceeding in a similar way, we obtained the analogous formulas for the other algorithms mentioned above. Such formulas express the number of operations as a function of the number of bytes $B$ to be encrypted or decrypted:

$$n_{E_{DES}}(B) = n_{D_{DES}}(B) = \left\lceil \frac{B}{8} \right\rceil \cdot 2096, \tag{3}$$

$$n_{E_{TDES}}(B) = n_{D_{TDES}}(B) = \left\lceil \frac{B}{8} \right\rceil \cdot 6272, \tag{4}$$

$$n_{E_{AES}}(B) = \left\lceil \frac{B}{16} \right\rceil \cdot 1764, \tag{5}$$

$$n_{D_{AES}}(B) = \left\lceil \frac{B}{16} \right\rceil \cdot 4356. \tag{6}$$

Due to their Feistel structure, DES and TDES encryption and decryption require the same amount of instructions per byte. This is not true for AES where encryption requires more complex operations than decryption. In the same manner, we obtained the analogous formulas for HMAC-MD5 and HMAC-SHA1 [10–12], the keyed-hash message authentication codes required by 3GPP specifications:

$$n_{H_{MD5}}(B) = \left( \left\lceil \frac{B + 4616}{4096} \right\rceil + 2 \right) \cdot 2192 + 332, \tag{7}$$

$$n_{H_{SHA1}}(B) = \left( \left\lceil \frac{B + 4616}{4096} \right\rceil + 2 \right) \cdot 4068 + 860. \tag{8}$$

These formulas allow us to calculate the number of operations required to encrypt, using IPSec, any exchange between IMS nodes, including SIP

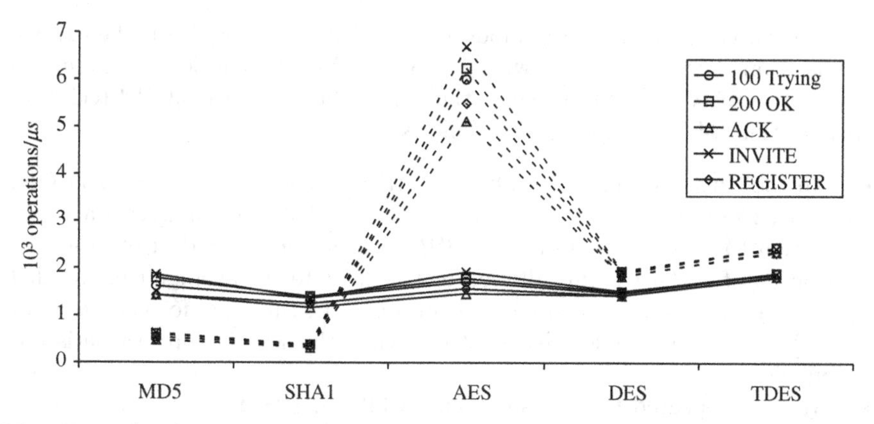


**Fig. 3.** $n_H/t_H$ (for MD5 and SHA1) and $n_E/t_E$ (for AES, DES and TDES): our model (*solid*) vs. [5] (*dashed*)

messages. However, before applying these numeric expressions, we want to validate them through an experimental analysis.

### 4.1 Experimental Validation

In order to validate the formulas reported in the previous section, we measured the time needed to generate actual SIP messages and protect them by using the cryptographic algorithms mandated by 3GPP, such as TDES-CBC, HMAC-MD5, etc. In this way we were able to obtain $t_E$ and $t_H$ and use them for validating our model.

We then divided the number of operations that in theory are needed for a certain algorithm, $n_E$ or $n_H$, by the time we actually measured for that algorithm, $t_E$ or $t_H$.

Figure 3 shows the ratios between the number of calculated operations and the measured execution time for the different algorithms and for five different types of SIP message. The solid lines represent the results of our model, while the dashed lines represent those of the model proposed by Xenakis et al. [5]. The ratios plotted on the graph should be constant and independent of the type of algorithm used or the type and size of the SIP message being ciphered. It seems that the formulas we are proposing approximate reality closer than the ones defined in [5], according to the experimental results we have obtained, especially for AES, MD5 and SHA1.

## 5 Computational Overheads Due to IPSec

In order to estimate the values of computational overheads due to IPSec we evaluated $t_{Cout}$ and $t_{Cin}$ when (a) there is no security protection, (b) when

IPSec is based on TDES-CBC mode with HMAC-SHA1 and (c) when IPSec is based on AES-CBC mode with HMAC-SHA1. We considered these two kinds of security associations since they are required by the 3GPP technical specifications [4]. We proceeded as follows:

- We took into account five different SIP messages: 100 Trying, 200 OK, ACK, INVITE and REGISTER. Using SIPp [13], we configured a workstation to generate thousands of SIP messages per second, up to the saturation of its CPU. This allowed us to calculate the average time needed to generate and send a single SIP message. In this way, for each type of SIP message, we could measure $t_{Cout}$ when the security protection is not enabled.
- We then repeated the measurements while an IPSec security association was established between the source workstation and the target host. These new measured values of $t_{Cout}$ would take into account also the terms of Expression (1) related to IPSec. By repeating the experiments with different IPSec configurations (e.g., null encryption, TDES-CBC encryption, etc.), and operating simple algebraic expressions on the results, we were able to measure not only $t_{Cout}$, but also calculate its main components, $t_{ESP_E}$, $t_H$ and $t_E$.
- As seen above, measuring $t_{Cout}$ was just a matter of having a workstation generate thousands of SIP messages with different IPSec configurations, measure the total time required to *send* such messages, and divide it by the number of messages. However, we could not devise a similar experimental configuration that would allow us to reliably measure $t_{Cin}$: with SIPp it was not possible to accurately measure how many SIP messages a single node can *accept and process in input.* Therefore, we had to estimate the value of $t_{Cin}$ and we proceeded as follows:
  - We assumed that decapsulation (UDP, SIP, etc.) require the same time as encapsulation.
  - We assumed that the hash computation time in $t_{Cin}$ is the same as that in $t_{Cout}$.
  - We used the formulas introduced in Sect. 4 for estimating the decryption times starting from the respective encryption times.

  Thus, by adding up these values we were able to estimate $t_{Cin}$ both in the case when security associations are enabled and when they are not.

### 5.1 Measured Values of $t_{Cout}$

Table 2 reports the measured values of $t_{Cout}$ in the cases of (a) no security protection, (b) TDES with SHA1 and (c) AES with SHA1. The different components of $t_{Cout}$ due to IPSec are also reported: $t_{ESP_E}$, $t_H$ and $t_E$. Notice that, as expected, for each message, $t_E$ due to AES is smaller than $t_E$ due to TDES.

**Table 2.** Measured values of $t_{Cout}$ and its components (in µs)

| SIP Message | Size (bytes) | Plain $t_{Cout}$ | TDES-SHA1 $t_{Cout}$ | AES-SHA1 $t_{Cout}$ | $t_{ESP_E}$ | SHA1 $t_H$ | TDES $t_E$ | AES $t_E$ |
|---|---|---|---|---|---|---|---|---|
| 100 Trying | 580 | 238 | 529 | 325 | 10.4 | 39 | 242 | 38 |
| 200 OK | 695 | 243 | 544 | 330 | 10.9 | 39 | 251 | 37 |
| ACK | 439 | 264 | 494 | 341 | 9.0 | 36 | 185 | 32 |
| INVITE | 845 | 290 | 706 | 401 | 10.6 | 51 | 354 | 49 |
| REGISTER | 435 | 239 | 473 | 316 | 11.3 | 33 | 190 | 33 |

**Table 3.** Estimated values of $t_{Cin}$ and its components (in µs)

| SIP Message | Size (bytes) | Plain $t_{Cin}$ | TDES-SHA1 $t_{Cin}$ | AES-SHA1 $t_{Cin}$ | $t_{ESP_D}$ | SHA1 $t_H$ | TDES $t_D$ | AES $t_D$ |
|---|---|---|---|---|---|---|---|---|
| 100 Trying | 580 | 238 | 529 | 381 | 10.4 | 39 | 242 | 94 |
| 200 OK | 695 | 243 | 544 | 385 | 10.9 | 39 | 251 | 91 |
| ACK | 439 | 264 | 494 | 389 | 9.0 | 36 | 185 | 79 |
| INVITE | 845 | 290 | 706 | 473 | 10.6 | 51 | 354 | 121 |
| REGISTER | 435 | 239 | 473 | 365 | 11.3 | 33 | 190 | 82 |

### 5.2 Estimated Values of $t_{Cin}$

As previously mentioned, we could not devise a way to accurately measure the values of $t_{Cin}$. Therefore we assumed that decapsulation requires the same time as encapsulation. In particular, for the case with no security protection we assumed that $t_{Cin} = t_{Cout}$, and that $t_{ESP_D} = t_{ESP_E}$ when security is enabled. Moreover, we assumed that the verification of a keyed hash is computationally equivalent to its calculation. Therefore the term $t_H$ of $t_{Cin}$ is equal to that of $t_{Cout}$.

Regarding decryption, we calculated $t_D$ by applying the formulas introduced in Sect. 4 to the measured values of $t_E$. In particular, we applied the following proportion:

$$t_D = \frac{n_D(B)}{n_E(B)} t_E. \tag{9}$$

Since we focused on TDES and AES as cryptographic algorithms, we used Expressions (4)–(6), thus obtaining:

$$t_{D_{TDES}} = t_{E_{TDES}}, \tag{10}$$

$$t_{D_{AES}} = 2.47 \cdot t_{E_{AES}}. \tag{11}$$

Table 3 reports the estimated values of $t_{Cin}$ and its components in the cases of no security protection, TDES with SHA1 and AES with SHA1.

**Table 4.** $t_{Cout}$ and $t_{Cin}$ values (in $\mu$s) and the overheads introduced by TDES-SHA1 and AES-SHA1

| SIP | $t_{Cout}$ | | | $t_{Cin}$ | | |
|---|---|---|---|---|---|---|
| Message | Plain | TDES-SHA1 (%) | AES-SHA1 (%) | Plain | TDES-SHA1 (%) | AES-SHA1 (%) |
| 100 Trying | 238 | +122 | +37 | 238 | +122 | +60 |
| 200 OK | 243 | +124 | +36 | 243 | +124 | +58 |
| ACK | 264 | +87 | +29 | 264 | +87 | +47 |
| INVITE | 290 | +143 | +38 | 290 | +143 | +63 |
| REGISTER | 239 | +98 | +32 | 239 | +98 | +52 |
| Average | 255 | +115 | +34 | 255 | +115 | +56 |

### 5.3 Computational Overheads

By combining the measured values of $t_{Cout}$ and the estimated values of $t_{Cin}$ we can now evaluate the average overheads due to IPSec for each type of SIP message. These values are shown in Table 4. As expected, the overhead due to an IPSec tunnel based on TDES-CBC mode with HMAC-SHA1 is higher than the equivalent due to an IPSec tunnel based on AES-CBC mode with HMAC-SHA1.

Using TDES-CBC with HMAC-SHA1 means, on average, more than doubling the computational times (both $t_{Cin}$ and $t_{Cout}$). Instead, using AES-CBC with HMAC-SHA1 makes, on average, $t_{Cout}$ and $t_{Cin}$ increment by 34% and 56%, respectively.

## 6 Discussion

Signaling plays a critical role in any mobile networks, specifically when the Quality of Service (QoS) perceived by the user is significantly related to the way the network elements process signaling information. Many QoS parameters are related to the signaling and traffic loads inside the operator's network: for example, [14] recommends that the "Call release" delay should be kept at around 0.4 s. Also, the "Probability of blocking" should be kept at around $5 \times 10^{-3}$.

The computational overheads introduced by IPSec on the call management signaling used in 3GPP networks, i.e., SIP, can play a significant role in defining the QoS parameters, as indicated by the numerical results described in the previous sections.

We do not know for sure what is the typical CPU load on IMS nodes in 3G core networks today: not many 3G operators are willing to release such information, even for research purposes. However, starting from reasonable assumptions, we can describe as an example the following scenarios:

*Light load.* Assuming that the nodes of a given operator's network use 20% of their computational power, protecting SIP signaling messages with IPSec would not change much:

- With AES-SHA1 the power usage would go from 20 to 27% for $t_{Cout}$) and to 31% (for $t_{Cin}$)
- With TDES-SHA1 the usage would go to 43%

The introduction of IPSec in this case would not require any changes to the existing hardware.

*Normal load.* If we assume that the CPUs of IMS nodes are loaded at 50% capacity, things would get a little worse:

- With AES-SHA1, CPU load would go to 68% and 78% for $t_{Cout}$ and $t_{Cin}$, respectively
- With TDES-SHA1, SIP signaling would saturate CPUs. This would most likely exponentially increase the blocking probability, if no additional hardware were deployed

*Heavy load.* Last, if we assume an initial CPU load of more than 64%, even with AES-SHA1 the nodes would reach 100% of their computational power, thus worsening the overall performance of the network.

Besides the simple example scenarios reported above, in general when 3G operators will need to upgrade their core networks to the latest security specifications (deploying SEGs and protecting with IPSec all SIP signaling), they will have the following options:

- Increment the nodes' computational power, for example adding hardware-accelerated cards for the support of IPSec, therefore maintaining the QoS perceived by the end user and the same number of sessions managed per unit of time. However, this would also imply additional monetary investments for the network provider.
- Maintain the same computational power and the same number of sessions per unit of time per node (no extra expenses). This would however reduce the QoS perceived by the end user, since there would be a significant increase in the call-management delays, according to the numbers and formulas we have introduced.
- Maintain the same computational power and reduce the number of sessions handled by each signaling node, therefore potentially generating less revenues.

## 7 Conclusions and Future Work

3GPP Release 6 core networks rely on IMS, a framework completely based on TCP/IP and SIP. The use of SIP has mandated the implementation of IPSec to protect its security, which in turns can cause significant overheads. This paper proposes an improved set of numerical formulas derived from [5] for evaluating the number of operations needed by a cryptographic algorithm in function of the message size. We validated this model and we used it, together with extensive laboratory experiments, to compute the overheads introduced by IPSec on Release 6 core network signaling. We then discussed at a high level how this overhead can affect the operations and management of Release 6 core networks.

Our work is continuing in the direction of developing a comprehensive, experimentally based model to describe the security overheads on both signaling and user traffic in 3G networks. Besides extending this work to the access network and to user traffic, the final goal is to create a model which can be used to balance the level of security of 3G networks with economical and technical requirements.

## References


1. "Universal Mobile Telecommunications Systems (UMTS); IP Multimedia Subsystem (IMS); Stage 2," TS 123.228 Version 6.12.0 (2005-12), ETSI, December 2005
2. J. Rosenberg, H. Schulzrinne, G. Camarillo, A. Johnston, J. Peterson, R. Sparks, M. Handley, and E. Schooler, "SIP: Session Initiation Protocol," RFC 3261, IETF, June 2002
3. P. Calhoun, J. Loughney, E. Guttman, G. Zorn, and J. Arkko, "Diameter Base Protocol," RFC 3588, IETF, September 2003
4. "3G Security, Network Domain Security (NDS), IP Network Layer Security," TS 133.210, ETSI, 2005
5. C. Xenakis, N. Laoutaris, L. Merakos, and I. Stavrakakis, "A generic characterization of the overheads imposed by IPSec and associated cryptographic algorithms," *Computer Networks*, vol. 50, pp. 3225–3241, December 2006
6. S. Kent, "IP Encapsulating Security Payload (ESP)," RFC 4303, IETF, December 2005
7. "Data Encryption Standard (DES)," FIPS PUB 46-3, U.S. Federal Information Processing Standards Publication, 1999
8. "Advanced Encryption Standard (AES)," FIPS PUB 197, U.S. Federal Information Processing Standards Publication, 2001
9. A. J. Menezes, P. C. van Oorschot, and S. A. Vanstone, *Handbook of Applied Cryptography*. CRC Press, October 1996. ISBN 0-8493-8523-7
10. "Secure Hash Standard," FIPS PUB 180-1, U.S. Federal Information Processing Standards Publication, April 1995
11. H. Krawczyk, M. Bellare, and R. Canetti, "HMAC: Keyed-Hashing for Message Authentication," RFC 2104, IETF, February 1997
12. R. Rivest, "The MD5 Message-Digest Algorithm," RFC 1992, IETF, April 1992
13. http://sipp.sourceforge.net
14. "ITU-T Standard; Network Grade of Service Parameters and Target Values for Circuit-Switched Services in the Evolving ISDN," ITU-T Recommendation E.721, ITU, May 1999

# 13

# Mobility and Key Management in SAE/LTE

Anand R. Prasad[1], Julien Laganier[1], Alf Zugenmaier[1], Mortaza S. Bargh[2], Bob Hulsebosch[2], Henk Eertink[2], Geert Heijenk[3], and Jeroen Idserda[3]

[1] DoCoMo Communications Labs Europe GmbH, Germany
[2] Telematica Instituut, The Netherlands
[3] Twente University, The Netherlands

**Summary.** Often in wireless communications the cryptographic algorithm is considered as 'the security solution' but actually it is only the nucleus. The means for using the cryptographic algorithm is the 'key' used by the algorithm. Thus management of keys and security there-of is an important issue. The security of the key management solution should not impede mobility of devices by adding undue delays. Thus, secure and fast key management during mobility is an important issue for the third generation partnership project (3GPP) activity on system architecture evolution/long-term evolution (SAE/LTE). In this paper we review mobility and security issues with the focus of key management in SAE/LTE and present possible existing solutions together with their analysis.

## 1 Introduction

At times, it is said that security is the holy grail of any communications system. True or false, practice shows that solutions developed without security in mind from the beginning leads to solutions that have severe issues in due course. To avoid such situations it was necessary that the third generation partnership project (3GPP) activity on their system architecture evolution and long term evolution (SAE/LTE) includes security from the very beginning. The first step for SAE/LTE activity was taken in 2004 and it is expected that the specifications will be available around September 2007. Details of 3GPP time-plan and specification can be found in [1].

The SAE part of the 3GPP activity focuses on the core network (CN) of a mobile network and the LTE part focuses on the radio access network (RAN). SAE assumes the core network will be migrated to IP as the basis communication protocol. SAE allows integration of radio access networks based on different radio access technologies into the network, e.g., UMTS, LTE, wireless LAN and WiMAX. LTE specifies a radio access technology (RAT) that is aimed at peak data-rates of 100 Mbps downlink and 50 Mbps uplink. The goal of both SAE and LTE is also to decrease the overall complexity and cost for both operators and end-users. The security goal of the SAE/LTE activity was to provide security that is at least of the level of UMTS today. Of course, the security measures should not impede mobility support that is the essence of a mobile operator's business.

Looking at the security aspect, of the three common security goals confidentiality, integrity and availability, the first two are achieved using cryptography, which in turn requires keys to function. Key management includes key establishment and key distribution besides key generation and key management policies. The usage of the system defines the requirements for the keys. An insecure key management solution can lead to leakage of keys that can cause attack on the system or network. In such a situation the strength of the cryptographic algorithm is irrelevant.

It is of utmost importance that the mobile network is able to provide fast handovers such that there is no impact on perceived service quality by the user. These handovers need to consider security too. It should not happen that a mobile user hooks up to a rogue base station or there is hijack of the session by an intruder. This means that mobility related security requirements should be fulfilled, which includes re-keying when the user moves. Re-keying is also part of the key management and should fulfil the related requirements.

In this paper we will first discuss LTE and SAE development by 3 GPP and their goals in Sects. 2 and 3 respectively. Possible mobility and key management related solutions are discussed in Sects. 4 and 5 with an analysis and comparison of the solutions in Sect. 6 leading to the conclusions in Sect. 7.

# 2 Long Term Evolution

LTE or the Evolved UMTS Terrestrial Radio Access (E-UTRA) and Evolved UMTS Terrestrial Radio Access Network (E-UTRAN) aims at developing standards that will ensure competitiveness of 3GPP in long-term (10 years or more) [5]. There are several scenarios for LTE deployment, but at a high-level one could expect two scenarios [2]. The first one is the standalone deployment and the second scenario is integration and handover with UTRAN and/or GERAN. Further it is expected [1] that in LTE there will be support for (1) shared networks during mobility and initial access, (2) various cell sizes and planned or ad-hoc deployments, and (3) efficient mobility with an intra-LTE handover interruption time of 30 ms. An overview of LTE is given in this section [2–4].

## 2.1 Requirements

Some of the major requirements for LTE are given in Table 1 [2, 10].

Although the concern of cost for operators is addressed in the requirements there is no mention about security. Further the requirements set for handover with legacy solution does not allow seamless perception of service [6].

**Table 1.** LTE major requirements

| | | |
|---|---|---|
| 1 | Bandwidth (MHz) | Scaleable bandwidths of 1.25, 2.5, 5, 10, 15, 20 |
| 2 | Data rate (Mbps) | Peak of 100 Mbps for downlink (5 bps/$Hz^{-1}$) 50 Mbps for uplink (2.5 bps/Hz) at 20 MHz, with 2 Rx antennas and 1 Tx antenna at the terminal |
| 3 | Latency (ms) | C-plane 100 ms for camped to active state and 50 ms between active and dormant state. Transit time between IP layers of UE and RAN less than 5 ms |
| 4 | Capacity (users per cell) | C-plane 200 users per cell in active state for 5 MHz and at least 400 users for higher spectrum allocation<br>Much higher for dormant and camped state |
| 5 | Throughput | Compared to Rel 6 average user throughput per MHz: downlink 3–4 times and uplink is 2–3 times |
| 6 | Mobility | Optimized for 0–15 kmph. High performance for 15–120 kmph. Support upto 350 kmph or 500 kmph<br>Rel 6 voice and real-time CS services provided over PS in LTE with interruption time less than or equal to CS domain handovers in GERAN |
| 7 | QoS | End-to-end QoS shall be supported<br>VoIP with at least as good radio and backhaul efficiency and latency as voice traffic over the UMTS CS |

### 2.2 Physical Layer Parameters

Details of current work on physical layer can be found in [7]. In brief, the downlink (DL) part of LTE uses orthogonal frequency division multiplexing OFDM in which the data is multiplexed onto a number of subcarriers. This number scales with bandwidth. There is frequency selective scheduling in DL (i.e. OFDMA) and adaptive modulation and coding (up to 64-QAM). In the uplink SC-FDMA (Single Carrier-Frequency Division Multiple Access) is used with fast Fourier transform-based transmission scheme like OFDM. The total bandwidth is divided into a small number of frequency blocks to be assigned to the UEs (e.g., 15 blocks for a 5 MHz bandwidth). Multiple antenna are used (two at eNodeB and two receive antennas at UE) for beam-forming and multiple input–multiple output (MIMO).

### 2.3 Architecture

The LTE architecture interconnects the network side termination points of the wireless link (called eNodeBs, eNBs) with each other, the interface for this is called X2 interface [2]. The eNBs are also connected by means of the

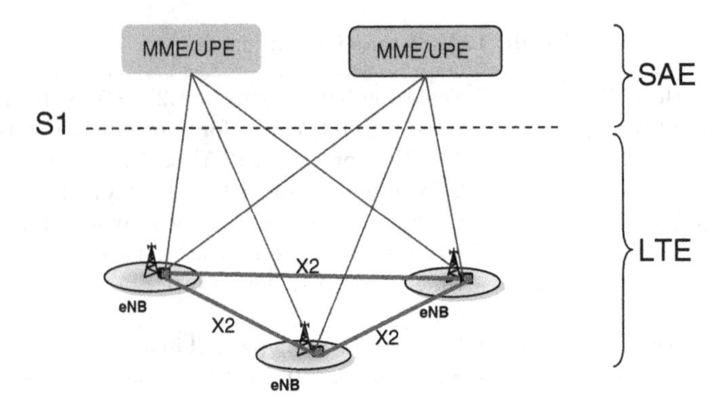


**Fig. 1.** LTE architecture

S1 interface to the core network called Evolved Packet Core (EPC). This EPC includes Mobility Management Entities (MME) and User Plane Entities (UPE) together also known as access gateway (aGW). The LTE architecture is illustrated in Fig. 1. The LTE architecture differentiates between user plane U-plane (carrying the user's applications generated traffic, e.g., voice, mail, web, etc.) and control plane C-plane (carrying the terminal's signalling protocols traffic, e.g., paging, call set-up, etc.). The U-plane and C-plane protocol stack are shown in Fig. 2.

The eNB hosts the radio resource management unit that includes radio bearer control, radio admission control, connection mobility control, and dynamic resource allocation (scheduling) functions. The S1-C (control plane) interface supports, among others, intra- and inter-system mobility of UE; and the S1-U (user plane) interface supports the tunneling of end user packets between the eNB and the UPE as a means to minimize packet losses due to e.g. mobility. The X2–C interface supports UE mobility between eNBs. The X2–U interface supports the tunneling of end user packets between the eNBs as a means to minimize packet losses due to e.g., mobility.

There are several handover scenarios for LTE, dependent on the state of the mobile device, C-plane or U-plane handover, and whether the MME/UPE is involved. In terms of security the RAN and security group specifications [8,9] discuss the termination point that is naturally dependent on the end-point of a given protocol. The Non Access Stratum (NAS) signalling requires confidentiality and integrity protection. U-plane must be confidentiality protected (between UE and eNB), but it is still under study whether or not its integrity shall be protected. For Access Stratum (AS) signalling, MAC security and requirement for confidentiality protection of RRC signalling is yet to be studied, while RRC signalling integrity protection is required.

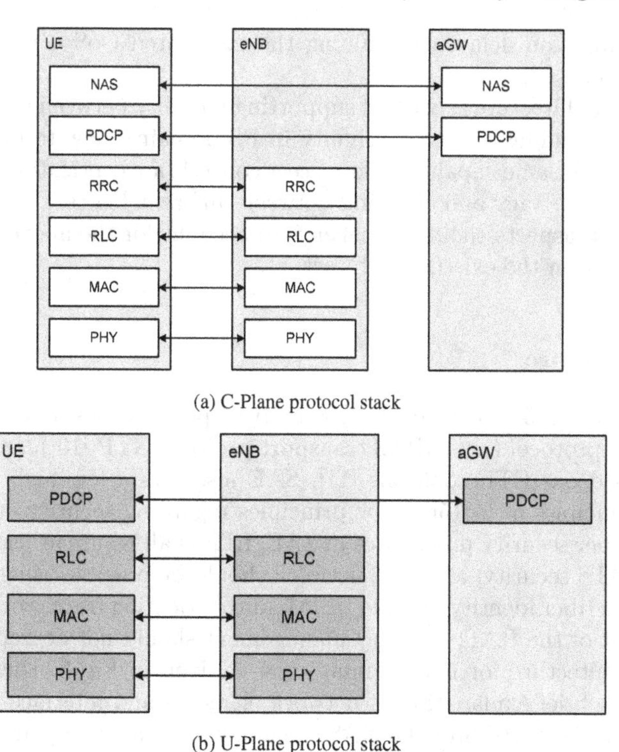


(a) C-Plane protocol stack

(b) U-Plane protocol stack

**Fig. 2.** C-plane and U-plane protocol stacks

# 3 System Architecture Evolution

System architecture evolution SAE focuses on enhancing the capability of the 3GPP system's core network to cope with the rapid growth in IP data traffic. This 3GPP system enhancement includes reduced latency, higher user data rates, improved system capacity and coverage, and reduced overall cost for the operator. IP based 3GPP services will be provided through various access technologies together with mechanisms to support seamless mobility between heterogeneous access networks. In this section the current work of 3GPP regarding SAE are presented from [10–14].

## 3.1 Requirements

The main objectives to address are [14]:

1. The architectural developments should take into account the LTE targets for the evolution of the radio-interface. It should address efficient support of services especially from the PS domain (e.g. VoIP).
2. Overall architecture impacts from support of different RAN/RATs and access selection based on combinations of operator policies, user preferences and RAN conditions; improving the basic system performance e.g.

communication delay; maintaining the negotiated QoS across the whole system; etc. [12].

3. Overall architecture aspects of supporting mobility between heterogeneous RANs (including service continuity in PS domain); how to maintain and support the same capabilities of access control (authentication, authorization); and privacy and charging between different RATs.
4. Migration aspects should be taken into account for the above, i.e. how to migrate from the existing architecture.

### 3.2 Architecture

It was decided in 3GPP to proceed with two specifications; one that utilizes the existing protocol (i.e., GPRS transport protocol GTP [10]), and the other that is based on IETF solutions [11]. SAE also sets a few high level architectural principles in [4,15] A few principles regarding security and mobility are: subscriber security procedures in SAE/LTE shall assure at least the same level of UMTS security; access to network should be possible using Release 99 UMTS subscriber identity module USIM; authentication framework should be independent of the RAT; mobility management should not degrade security.

The architecture for non-roaming case is given in Fig. 3. Due to lack of space only a brief explanation of network elements and interfaces is given in this section: The MME provides NAS signalling and its security, inter CN node

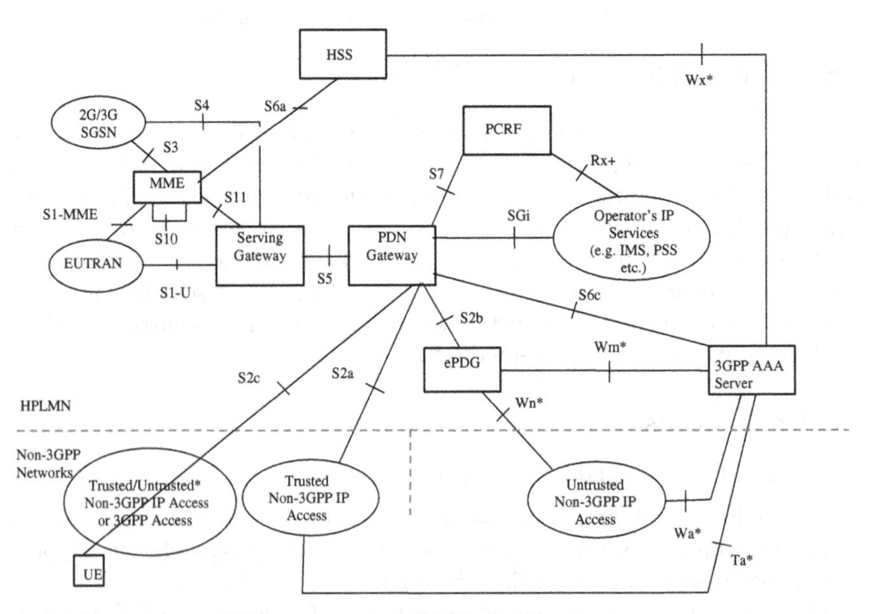


**Fig. 3.** Non-roaming architecture for SAE

signalling for mobility between 3GPP access networks, etc. The Serving GW is the gateway which terminates the interface towards E-UTRAN. For each UE, at a given point of time, there is a single Serving GW with the function of Local Mobility Anchor point for inter-eNB handover, mobility anchoring for inter-3GPP mobility, lawful Interception, packet routing and forwarding. The PDN GW functions include policy enforcement, per-user packet filtering, charging support, lawful interception and UE IP address allocation.

There are several different mobility management concepts in SAE which are dependent not only on the access technology and network layer protocols but also on the state of the UE. Other mobility issues that SAE has to cater for are inter-RAT mobility, dependence of paging/tracking area, context information availability, power saving, etc.

# 4 Mobility Solutions

There are several network layer mobility protocols that could be utilized in SAE/LTE to support mobility within LTE and to/from other RANs. We now know that 3GPP has made a choice of protocols. Anyhow in this section we present the different choices that were possible together with their differences and similarities. In later section we give an analysis.

In traditional IP networks, the IP address of a node is usually bound to its topological location within the network to permit route aggregation. For a mobile node, that means that moving and changing its location implies that it changes its IP address. In the traditional TCP/IP communications paradigm, IP addresses of a node were expected to remain stable. It was thus possible to: (a) reach an IP node knowing only its IP address, (b) bind upper layer communications (e.g. TCP conations) to IP addresses of communication endpoints. With the advent of mobile nodes, this change of IP address role has thus the following implications:

1. It is no longer possible to reach a mobile node knowing only its IP address since it changes when the mobile node moves.
2. Upper layer communications will break with movement of one of the communication endpoint since they are bound to IP addresses that will change.

Because of that, network layer mobility protocols have been designed to restore the two basic properties that were broken by apparition of mobile nodes. In addition, these protocols might, depending on their architecture and mechanisms, offer additional mobility-related functionalities such as:

- Route optimisation between a mobile node and its correspondent nodes.
- Reduction of communication disruption latency upon movement via proactive configuration of care-of address before movement, buffering of packet received at old access router (AR) and tunnelling to new AR, and/or local anchoring.

- Reduction of packet loss upon movement via pro-active configuration of care-of address before movement, buffering of packet received at old access router (AR) and tunnelling to new AR, and/or local anchoring.

Additionally, these protocols may also offer functionalities which are not directly related to mobility such as:

- Network layer multi-homing: Ability to switch between different provider-assigned subnet prefixes to cope with ISP failures. Such prefixes might be assigned on a single interface, or each prefix to a different interface.
- Network layer security: Ability to protect integrity and confidentiality of communications.

With the focus on IPv6 the protocols given below were considered in this paper for which a comparison is given in Table 2.

- Mobile IP version 6 (MIPv6)
- Fast Handover for Mobile IP version 6 (FMIPv6)
- Hierarchical Mobile IP version 6 (HMIPv6)
- Network-based Localized Mobility (NETLMM)
- IKEv2 Mobility and Multi-homing Protocol (MOBIKE)
- Host Identity Protocol (HIP)

# 5 Key Management Solutions

Authentication process is one of the major latency sources that prevents seamless handovers. This latency is mainly due to the signalling overhead that is needed to authenticate a user and for making the association with the new Access Point (AP) secure. Both aspects involve proper key management. Therefore, solutions for fast authentication are dearly needed in order to realize seamless handovers and thereby to improve the user's experience. These solutions boil down to effective and efficient key management schemes that are suitable for intra- and inter-domain handovers as well as for horizontal and vertical handovers.

The Extensible Authentication Protocol (EAP) [16] is a generic framework for network access authentication. The EAP framework allows an authenticator to authenticate a peer (and possibly mutual authentication) and establishes between them two keys, the Master Session Key (MSK) and the Extended MSK (EMSK), which are used to secure communications of EAP lower layers. At the moment only the MSK is used by different lower layers and protocols. The most common usage is in the IEEE 802.11i lower layer to derive the Transient Session Key (TSK) to provide access link security. For instance in 802.11i the first 512 bits of the MSK are used for TSK derivation, 802.11r uses the second 256 bits to derive Pair-wise MKs (PMKs-R1) for fast BSS transition, and 802.16 uses the first 320 bits. The Internet Key

**Table 2.** Differences in the mobility protocols

| | MIPv6 | MIPv6 + FMIPv6 | HMIPv6 | HMIPv6 + FMIPv6 | NETLMM | MOBIKE | HIP |
|---|---|---|---|---|---|---|---|
| Scope of mobility | Global, local | Global, local | Local | Local | Local | Global, local | Global |
| Location of rendezvous point | On routing path to home address | On routing path to home address | On routing path to regional care-of-address | On routing path to regional care-of-address | On routing path to regional care-of-address | On routing path to IPsec inner address | Anywhere |
| Trust model | SA with rendezvous point | SA with rendezvous point | SA with rendezvous point | SA with rendezvous point | SA with access router | SA with rendezvous point | SA with rendezvous point and correspondent node |
| Route optimization | Yes | Yes | No | No | No | No | Required |
| Reduction of communication disruption latency and packet loss | Yes if local anchor | Yes | Yes | Yes | Yes | Yes if local anchor | No |
| Rendezvous point | Home agent (HA) | | Mobility anchor point (MAP) | | Localized mobility anchor (LMA) | Security gateway (SGW) | Rendezvous server (RVS) |
| Routing update | Binding update (BU) | | Local binding update (LBU) | | Routing update (RU) | Update SA address (USA) | Locator update (UPD) |

Exchange protocol (IKEv2) has an authentication mode where one of the IKE peer is authenticated via EAP, thus making use of the MSK as well. IKEv2, however, uses it for entity authentication purposes. This disparate usage of the MSK makes it less suitable for a root key of a key hierarchy that supports fast re-authentication for seamless handovers. For this reason, the IETF HOKEY working group tries to define an EMSK-based key hierarchy for authenticated seamless handovers [17]. Since the EMSK has never been used in any specifications it can be specified in such manner that it is acceptable to all lower layers. A Usage Specific Root Key (USRK) can be derived from the EMSK and used for efficient re-authentication within the EAP framework. In HOKEY terminology this key is called re-authentication Root Key (rRK). The rRK on its turn is used to derive the re-authentication Integrity Key (rIK) and a re-authentication MSKs (rMSK) that is specific to each authenticator that the MN associates with. The rIK is used to prove being a party to the full EAP method-based authentication and is used in a proof of possession exchange between the MN and the AAA-server. Finally, the rMSK is used for deriving the TSK after each re-authentication phase (see Fig. 4).

One of the most important features of the HOKEY key hierarchy is that it doesn't require the MN to interact with the home domain for authentication purposes when roaming within a foreign domain.

Since HOKEY is still work in progress, a number of issues with its usage in 3GPP SAE/LTE haven't been addressed yet. One of them is related to dealing with the heterogeneity of the authentication mechanisms. Different network technologies use different authentication mechanisms. For instance, UMTS networks use the UMTS-AKA authentication mechanism, and EAP-AKA is used in WLANs. Though UMTS AKA and EAP-AKA are almost identical, they differ by the transport method of the AKA protocol: PMM

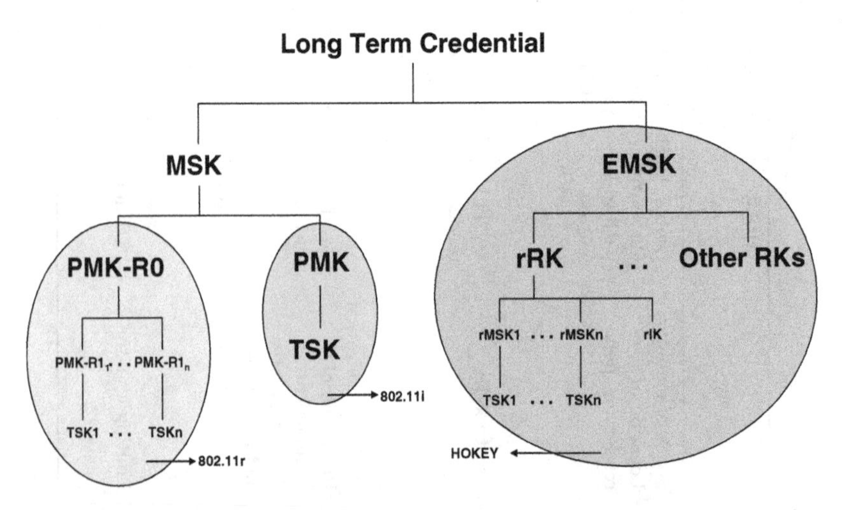


**Fig. 4.** Proposed HOKEY EMSK hierarchy for re-authentication, presented during IETF 66 meeting July 2006

protocol in case of UMTS and EAP protocol in case of WLAN, The former doesn't have a fast re-authentication function, while EAP-AKA [18] does offer such functionality, which makes it better suitable to be used in the EAP-ER framework [19].

Another issue to be solved is the choice of a proper key distribution mechanism. The rMSK must be delivered to the new authenticator following re-authentication. Options for key delivery are either based on a pull or a push model. The push model does not allow randomness contribution by the peer, is not supported by RADIUS, does not scale well, results in keys on target authenticators that the peer may never roam to, target authenticators must store keys, key names, associated nonces, lifetimes, and other attributes for many peers unnecessarily, peer needs to be involved in a re-authentication protocol anyway to receive nonces or other attributes. So there is not much value in the push model. Therefore a peer-initiated, on-demand pull model makes more sense.

For the inter-domain case, key delivery is not straightforward. How does the AAA server know the AP in the foreign domain? How to setup a secure communication channel with the foreign domain? Do the foreign APs communicate with the home rMSK server directly or via their own rMSK server?

IETF PANA [20] is an network access authentication protocol transported over IP, and as such independent of the underlying technology. It authenticates peers with the EAP protocol, and as such is both an EAP transport and an EAP lower layer, like IEEE 802.1x. The Media Independent Pre-Authentication (MPA) approach [21] tries to define a solution for pre-authentication that support both inter-domain and inter-technology handovers. MPA is a mobile-assisted higher-layer authentication, authorization and handover scheme that is performed prior to establishing link layer connectivity to a network where the MN may move in near future. In MPA, the notion of 802.11i pre-authentication is extended to work at higher layer, with additional mechanisms to securely perform early acquisition of an IP-address from the new network as well as pro-active handover to this network while the MN is still attached to the current network. MPA provides a secure and seamless mobility optimization that works for inter-domain heterogeneous handovers.

## 6 Analysis

To evaluate the existing solution we consider a number of principles that serve as guidelines in this paper to evaluate handover solutions. These guiding principles can be related to the architecture, performance and security aspects of the solutions. The architectural guidelines considered are reusability (i.e., to be able to use the solution again to add new functionality with minimum modifications) and modularity (i.e., the solution is composed of components with well defined functionality and interfaces). Requiring a handover solution to be fast, we consider the following performance guiding principles for such a solution: support of different air interface technologies (as mobile devices are equipped with multiple network interfaces nowadays), compatibility of local and global mobility solutions (as mobile devices are going to cross over administrative domain boundaries frequently), and support of multiple air interfaces being active simultaneously (when possible and appropriate). The latter requires the solutions to be energy effective. The security related guidelines include binding L2, L3 and higher layers to the user as identified by its USIM.

For our analysis we consider three categories of protocols or (partial) solutions and evaluate them based on our guiding principles mentioned. These solution categories are: mobility/handover solutions, authentication methods, and authentication transport protocols. Mobility or handover management related solutions that we consider are: MIPv6, HMIPv6, FMIPv6, MOBIKE, NetLMM, MPA, IEEE802.21, IEEE802.16/e and IEEE802.11. For authentication methods we investigate UMTS-AKA, EAP-AKA, EAP-TLS, 802.11i, and EAP-ER. Finally, we consider EAP and combined PANA and IPsec as authentication transport protocols for our analysis. Figure 5 presents a summary

| *State of the art analysis* | architecture | | fast | | | secure | | |
|---|---|---|---|---|---|---|---|---|
| | reuse | modularity | local mobility | global mobility | support multiple air IFs | L2 binding | L3 binding | uses SIM |
| **Mobility** | | | | | | | | |
| MIPv6 | | | ~ | y | y | n | y | |
| HMIPv& | | | y | n | y | n | y | |
| FMIPv6 | | | y | y | y | n | y | |
| MOBIKE | | | | y | y | n | y | |
| NetLMM | | | y | n | n | n | y | |
| MPA | | | | | y | y | y | |
| 802.21 | | | | | y | | | |
| 802.16/e | | | y | n | y | y | n | |
| 802.11 | | | y | n | y | | | |
| **Auth & keying** | | | | | | | | |
| UMTS-AKA | n | n | | | | y | | y |
| EAP-AKA | | | | | y | y | | y |
| EAP-TLS | | | | | y | y | | |
| 802.11i | | | | | y | y | n | |
| EAP-ER | | | | | | | | |
| **Auth transport** | | | | | | | | |
| EAP | | | | | y | y | y | |
| PANA + IPsec | | | | | | | y | |

**Fig. 5.** Comparison of (partial) solutions for supporting mobility in SAE/LTE

of our analysis. One should note that key establishment/distribution aspects that are provisioned in for example IEEE802.11i, IEEE802.11r, IEEE802.16, EAP-ER, HOKEY, IKE and AKA) are already included in one or more categories identified above. A close investigation of the results of Fig. 5 reveals that a complete system architecture is missing to deliver secure and fast handover management. Such architecture must provide integrated security management to deal with threats in all handover phases.

## 7 Conclusions and Future Work

In this paper we have presented an overview of SAE/LTE and IP layer mobility protocols and key management solutions that can be used for SAE/LTE. Based on analysis that utilizes principles coming from SAE/LTE requirements we come to the conclusion that EAP-ER using AKA is the authentication and key agreement solution that should be utilized for SAE/LTE. The study also shows that NetLMM and MIP are the mobility solutions that can be used. The results to some extent are in contradiction to what is currently accepted in 3GPP.

From network layer protocol perspective 3GPP is focusing on NetLMM and MIP but also has accepted GTP. Obviously the acceptance of GTP is due to the fact that existing solutions can be reused. As for key agreement 3GPP has a working assumption of UMTS-AKA. This working assumption certainly works fine for fast mobility between UMTS and LTE but it does not cater for future where there will be integration of other RANs.

This work still leaves us with the need to study the integration of mobility protocol and key management solution in the SAE/LTE architecture. This integration should be done while considering the security and mobility aspects. Another point to study is the key hierarchy required for SAE/LTE. This can easily be concluded by looking at the end-point of different protocols (MAC, RRC, NAS and U-plane) and the confidentiality/integrity requirement. Once all is done a study on remaining threats and performance is also required.

## References

1. 3GPP Gantt Chart, http://www.3gpp.org/ftp/Specs/html-info/GanttChart-Level-2.htm#32085.
2. 3GPP TR 25.913: "Requirements for Evolved UTRA (E-UTRA) and Evolved UTRAN (E-UTRAN)".
3. 3GPP TR 25.912: "Feasibility Study for Evolved UTRA and UTRAN".
4. 3GPP TS 36.300: "Evolved Universal Terrestrial Radio Access (E-UTRA) and Evolved Universal Terrestrial Radio Access (E-UTRAN); Overall description; Stage 2".
5. UTRA-UTRAN Long Term Evolution (LTE) and 3GPP System Architecture Evolution (SAE), http://www.3gpp.org/Highlights/LTE/LTE.htm.

6. mITF: "Mobile IT Forum 4G Mobile System Requirements Document," Ver. 1.1.
7. 3GPP TR 25.814: "Physical Layer Aspects for Evolved Universal Terrestrial Radio Access (UTRA)".
8. 3GPP TR 33.821: "Rationale and Track of Security Decisions in Long Term Evolved RAN/3GPP System Architecture Evolution".
9. 3GPP TS 33.922: "Security Aspects for Inter-Access Mobility Between Non 3GPP and 3GPP Access Network".
10. 3GPP TR 23.401: "General Packet Radio Service (GPRS) Enhancements for Long Term Evolution (LTE) Access".
11. 3GPP TS 23.402: "3GPP System Architecture Evolution (SAE): Architecture Enhancements for Non-3GPP Accesses".
12. 3GPP TR 22.258: "Service Requirements for an All-IP Network (AIPN); Stage 1".
13. 3GPP TR 22.978: "All-IP Network (AIPN) Feasibility Study".
14. 3GPP TR 23.882: "3GPP System Architecture Evolution: Report on Technical Options and Conclusions".
15. 3GPP TR 21.902: "Evolution of 3GPP System".
16. Aboba, B., Blunk, L., Vollbrecht, J., Carlson, J., and H. Levkowetz, "Extensible Authentication Protocol (EAP)", RFC 3748, June 2004.
17. IETF Handover Keying (HOKEY) working group http://www.ietf.org/html.charters/hokey-charter.html.
18. J. Arkko and H. Haverinen, EAP AKA Authentication, Internet Draft draft-arkko-pppext-eap-aka-13, October 2004.
19. V. Narayanan and L. Dondeti, EAP Extensions for Efficient Re-authentication, Internet Draft, draft-vidya-eap-er-02, expires July 23, 2007.
20. IETF Protocol for carrying Authentication for Network Access working group http://www.ietf.org/html.charters/pana-charter.html.
21. Ashutosh Dutta, Tao Zhang, Yopshihiro Ohba, Kenichi Taniuchi, Henning Schulzrinne, "MPA assisted Optimized Proactive Handoff Scheme," mobiquitous, pp. 155–165, The Second Annual International Conference on Mobile and Ubiquitous Systems: Networking and Services, 2005.

14

# Enhanced Operation Modes in IEEE 802.16 and Integration with Optical MANs

Isabella Cerutti[1], Luca Valcarenghi[1], Dania Marabissi[2], Filippo Meucci[2], Laura Pierucci[2], Luca Simone Ronga[3], Piero Castoldi[2], and Enrico Del Re[2]

[1] Scuola Superiore Sant'Anna, Pisa, Italy
isabella.cerutti@sssup.it, luca.valcarenghi@sssup.it, piero.castoldi@sssup.it
[2] Università degli Studi di Firenze, Italy
marabissi@lenst.det.unifi.it, filippo.meucci@lenst.det.unifi.it, pierucci@lenst.det.unifi.it, delre@lenst.det.unifi.it
[3] CNIT – Università degli Studi di Firenze, Italy
ronga@lenst.det.unifi.it

**Summary.** Wireless technology evolution permits users to be always connected to IP-based services through IP-based devices. Furthermore, thanks to wireless technology progresses, wirelessly connected users may now exploit a bandwidth comparable to the one provided by copper-based access technologies (e.g., xDSL). WiMAX (i.e., IEEE 802.16) is one of the wireless technologies with such potential.

This paper addresses IEEE 802.16 aspects, from the enhanced modes of operations to the wireless/wired MAN integration. In particular, the paper focuses on advanced physical layer technologies for wireless transmission such as Multiple Input Multiple Output (MIMO) antennas and Adaptive Modulation and Coding (AMC), the optional IEEE 802.16 Mesh mode of operation, and the integration of wireless and wired/optical Metropolitan Area Networks (MANs). Current status and issues are presented and solutions are proposed.

## 1 Introduction

The explosive growth of communications is driven by two complementary technologies: wireless communications and optical transport. Optical fiber offers the massive bandwidth potential that has fueled the rise in Internet traffic, whilst wireless techniques confer mobility and ubiquitous access through bandwidth-constrained and impairment-prone wireless channels.

To overcome the bandwidth limitations and impairment susceptibility of wireless communications, a wireless technology for broadband wireless access has been recently standardized: IEEE 802.16, also known as WiMAX [13]. IEEE 802.16 provides a wireless alternative to wired Metropolitan Area

Network (MAN) access protocols and technologies, with a potential capacity of up to 70 $\mathrm{Mb\,s^{-1}}$.

So far, wireless and wired MANs have been evolving independently. Advanced techniques that provide high capacity transmissions, such as Adaptive Modulation and Coding (AMC) and Multiple Input Multiple Output (MIMO) systems, and multi-hop transmissions are emerging in wireless MANs (WMANs). On the other hand, wired MANs based on optical fiber and exploiting the Generalized MultiProtocol Label Switching (GMPLS) protocol suite, appear as the solution to support the increasing requests in terms of bandwidth and Quality of Service (QoS) coming from the wireless users. This scenario raises the issue of an efficient integration between the WMAN and the wired MAN to provide high-speed and flexible multimedia services to wireless (possibly mobile) terminals. Thus, cross-domain, i.e. wireless/wired, traffic engineering schemes must be implemented to guarantee a seamless QoS to users that communicate while connected to any network segment.

## 2 Open Issues and Solutions

This section addresses specific issues related to the IEEE 802.16 enhanced operation modes and the wireless/wired integration process, showing the solutions investigated in the paper.

### 2.1 Advanced Access Technologies

In wireless links, the overall system performance degrades markedly due to multipath fading, Doppler, and time-dispersive effects introduced by the wireless propagation. In addition, the limitation of the wireless resources (i.e., bandwidth and power) requires that they are efficiently exploited. In order to enhance the spectral efficiency while adhering to a QoS requirement, multiple antenna systems and link adaptation methods can be considered.

Multiple antennas at the transmitter and receiver (i.e., MIMO systems) achieve diversity gain in a fading environment. By employing multiple antennas, multiple spatial channels are created, and it is unlikely that all the channel will fade simultaneously. If statistical decorrelation among antenna elements is provided, multiple transmit and receive antennas can create independent parallel channels. Decorrelation condition can be satisfied by using antennas well separated (by more than $\lambda/2$) or with different polarization. For MIMO systems, spatial multiplexing is often used.

With link adaptation methods, the signal transmitted to and by a particular station can be modified to account for the signal quality variation. This allows to improve system capacity, peak data rate, and coverage reliability. Traditionally, wireless systems use fast power control as the preferred method for link adaptation. Recently, AMC is offering an alternative link adaptation method that promises to increase the overall system capacity. In a system with AMC, the power of the transmitted signal is held constant but the modulation and coding formats are changed to match the current received signal quality. AMC provides the flexibility to match the modulation-coding scheme to the average channel conditions of each station. Users close to the base station (BS) are typically assigned higher-order modulations and high code rates but the modulation-order and/or the code rate will decrease as the distance from the BS increases.

IEEE 802.16 standard supports different MIMO and AMC schemes to improve the link performance. However, the standard lacks to specify the receiver structure and the adaptation algorithms to be used for MIMO and AMC strategies. The study of these solutions is still an open issue. In addition, different and enhanced versions of MIMO and AMC strategies should be investigated as they may help to further increase the system capacity and flexibility.

For instance, the choice of implementing MIMO strategies based on spatial multiplexing or based on space time coding (STC) depends on the specific service requirements: high capacity or high reliability. The possibility to exploit both techniques, alternatively, may help to achieve both objectives.

From this point of view, algorithms that are able to switch between the different diversity schemes, as the one shown in Fig. 1, are necessary. These algorithms can operate by using either PHY information, such as channel state information, or MAC layer information, such as service type and queue status. Moreover a cross-layer approach (i.e., the joint design of the MAC and PHY layers) shall be considered for link adaptation. The transmission parameters,

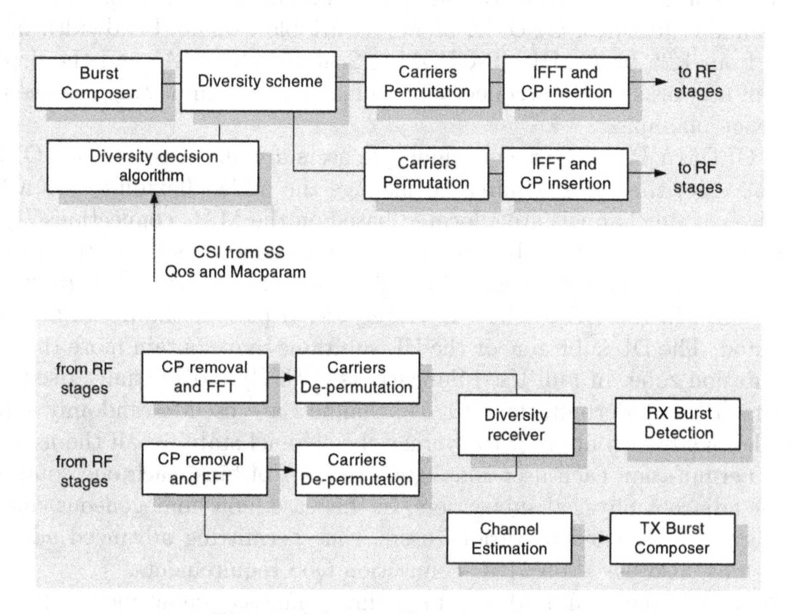


**Fig. 1.** Adaptive diversity scheme

optimized by the adaptive modulation/coding, are selected at the physical layer to match the wireless channel, but they must take into account also the information coming from the higher levels, in particular from the MAC layer.

## IEEE 802.16 MIMO Support

IEEE 802.16d standard [13] introduces multi antenna support. The "Amendment 2 and Corrigendum 1 to IEEE std. 802.16-2004", namely 802.16e version [14], adds several details regarding Adaptive Antenna Systems (AAS) and MIMO.

The standard defines four different PHY layers, each one addressing particular wireless channels. The two single carrier PHY layers (SC and SCa) are designed to operate typically in high capacity links, over 11 GHz Line of Sight (LOS) and under 11 GHz in a non LOS setting, respectively.

OFDM and OFDMA (Orthogonal Frequency Division Multiple Access) PHY layers are designed to be resistant in heavy NLOS environment under 11 GHz.

The main difference between OFDM and OFDM with Multiple Access (OFDMA) consists in the possibility to assign dynamically a subset of the subcarriers to single users. This allows allocation algorithms to shape the traffic load over the available resources in function of the actual channel state. Algorithms have to match the QoS requirements of each MAC connection and the channel state information in the best possible way.

Thanks to the greater flexibility, OFDMA has been selected by the WiMAX Forum as the basic technology for mobile user services.

As added features, OFDMA allows a variable channel bandwidth in the range from 1.25 to 20 MHz (S-OFDMA, Scalable OFDMA) and the deployment of networks with a frequency reuse factor of 1, eliminating the need for frequency planning.

In OFDMA PHY layer, since multiple users are allocated on each OFDM symbol, algorithms are required to manage the added flexibility. As a first step, logical subchannels are allocated based on the MAC connections. Then, before transmitting, the subchannels are permuted and assigned to subcarriers, in order to guarantee some physical related characteristics. Permutation Zones, both in the DL or the UL, define how a part of the frame has been permuted. The DL subframe or the UL subframe may contain more than one permutation zone. In Full Used Subcarriers (FUSC) and Partially Used Subcarriers (PUCS) Permutations the subchannels are pseudo-randomly spread over the bandwidth in order to average the channel state for all the users. In AMC permutation each user is assigned to a set of bins, each one composed of nine adjacent physical subcarriers. In this case, non homogeneous channel states can be assigned to various users, thus permitting advanced adaptive algorithms, also based on MAC-connection QoS requirements.

IEEE 802.16e standard includes three multi-antenna modes (Fig. 2): Alamouti Space Time Transmit Diversity (STTD)[2], Layered Alamouti Space

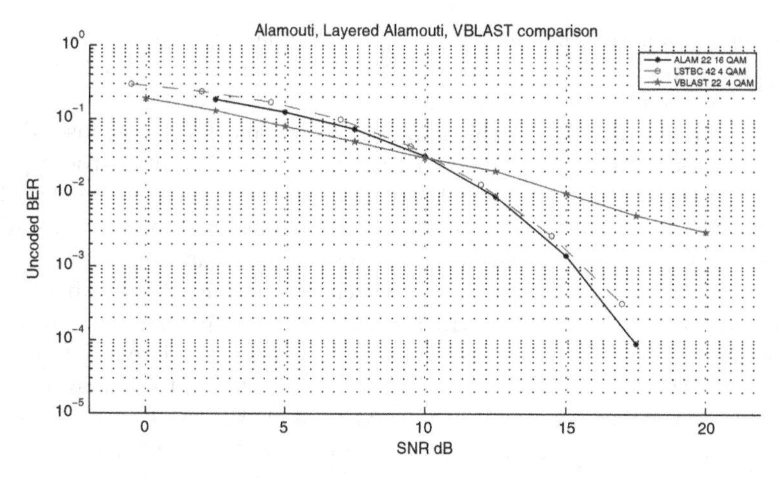


**Fig. 2.** Multi Antenna transmission comparison

Time Block Coding (LSTBC) [20] and VBLAST Spatial Multiplex (SM) [10]. Switching between these modes permits to best adapt to the channel state [11]. Adaptive spatial modulation, jointly with AMC techniques, can offer high flexibility at the physical layer maximizing data throughput and coverage.

Alamouti STTD is standardized for SCa, OFDM and OFDMA PHY Layers. The Alamouti scheme needs 2 transmitting antennas at the BS and provides a transmit diversity of 2. Channel variations that occur during two symbol time interval are the main source of performance degradation. The receiver is a linear combiner [2] where symbols can be reconstructed by using orthogonal properties of space–time coding matrix.

When the BS has three or four antennas, it is not possible to obtain a full-diversity approach, since it has been demonstrated that a full-rate, fully orthogonal Space Time Code only exists for two antennas [12]. When a full rate transmission is desired, Layered (or hybrid) STC schemes have to be used. Data rate is increased at the expense of diversity gain, linearity or orthogonality. In IEEE802.16e, orthogonality is lost but full diversity gain and a linear receiver are preserved. This scheme is a tradeoff between a full-diversity (Alamouti) and a Spatial Multiplex VBLAST approach, which offers maximum capacity gain but no diversity gain.

Layered schemes are subject to interference among symbols transmitted from different antennas. This results in BER degradation with respect to the orthogonal case. The receiver needs a number of receiving antennas $M_r > N_{blocks}$, where $N_{blocks}$ is the number of orthogonal blocks which have been spatially multiplexed (or layered).

Two Alamouti blocks are transmitted at the same time from a subset of the available BS antennas. This solution achieves the same transmit diversity as in the classic $2 \times 1$ Alamouti scheme and a spatial gain which is upper limited to 2. The spatial gain depends on the channel response: the optimum is achieved when the channel matrix is orthogonal. In this case, no detection loss is appreciated with respect to the Alamouti case, since cross-stream interference is null.

The main idea underlying this family of STC comes from multiuser detection strategies. As a matter of fact, IEEE 802.16e standard offers the support for addressing spatially multiplexed data layer to more than a single subscriber station (SS). In downlink, each SS decodes the streams directed to itself on the basis of the map broadcasted at the start of each frame.

As a third option, the system can switch on a Vertical Bell Laboratories Layered space–time (VBLAST) transmission [10]. Independent data streams are spatially multiplexed, i.e., they are transmitted from different antennas in the same OFDM symbol time; each data stream is referred to as a layer. Since the code is operating over a single OFDM symbol, VBLAST is a space-only coding. Layers can be separated at the receiver if the number of antennas is $M_r > N_{streams}$, where $N_{streams}$ is the number of layers. While Alamouti and its hybrid version (two layered Alamouti blocks) do not exploit all the freedom degrees of the MIMO channel, VBLAST can extract the complete spatial gain. On the other hand, VBLAST reaches smaller diversity gain with respect to the Alamouti $2 \times 2$ case.

In [19], optimum decoding order has been demonstrated and is obtained detecting highest SNR layer at each successive cancellation step. At each stage, linear combinatorial nulling is applied to remove interference caused by undetected layers.

### Adaptive Modulation and Coding

The use of modulation and coding allows a wireless system to efficiently exploit the available resources. AMC permits to choose the highest order modulation/coding depending on the channel conditions. In particular, in an Orthogonal Frequency Division Multiplex (OFDM) wireless system, the inherent multi-carrier nature of OFDM allows the use of AMC according to the behavior of the narrow-band channels (subcarriers) and the allocation of different subcarriers to different users to provide a flexible multiuser access scheme, that exploits multiuser diversity. In general, there is a high degree of flexibility for radio resource management in the context of OFDM. Since channel frequency responses are different at different frequencies or for different users, adaptive power allocation, AMC, and subcarrier allocation can significantly improve the performance of OFDM systems. The adaptation algorithms that efficiently assign the resources (power, subcarrier and AMC scheme) must be defined and represent an important open issue. In [6], two adaptive modulation methods for WiMAX systems are proposed. The first, called *Maximum Throughput* (MT) aims to maximize the system throughput without any explicit constraint on target Symbol Error Rate (SER). The second technique is named *Target SER* (TSER) and aims to guarantee a maximum preset target SER, imposed on the basis of a target QoS level. The target value of the SER can be fixed for every Signal to Noise Ratio (SNR) or can vary with it. In the second case, the modulation can be adapted in order to guarantee the theoretical SER of the QPSK modulation, thus obtaining the Minimum SER (MSER) algorithm.

All the techniques use the same system structure. The difference among *Maximum Throughput*, *Target SER* and *Minimum SER* is the threshold calculation, because they are the inputs for the AMC block that controls the adaptive modulation. Each adaptation algorithm is basically characterized by two thresholds, representing the changing events between different modulation schemes. The thresholds are calculated by means of theoretical analysis.

Figure 3 compares the performance in terms of SER for the three static modulation orders and for the three proposed techniques. The *Target SER* technique considers two different values to be guaranteed by the adaptation algorithm, i.e., SER $= 7 \times 10^{-2}$ and $5 \times 10^{-3}$.

In Fig. 4 the performance comparison in terms of throughput is shown. The throughput is expressed as the average useful bits per symbol for different SNR values at the transmitter side. As for QPSK, the best SER performance can be achieved by using either the MSER or the TSER (with SER $= 5 \times 10^{-3}$). However the MSER technique permits to achieve a higher throughput. It is evident that if the Target SER value is higher, the SER performance worsens. The SER curves behavior is reversed in the throughput curves. From the throughput point of view, the best case is represented by the MT method that is always higher than the static modulation throughput. By observing

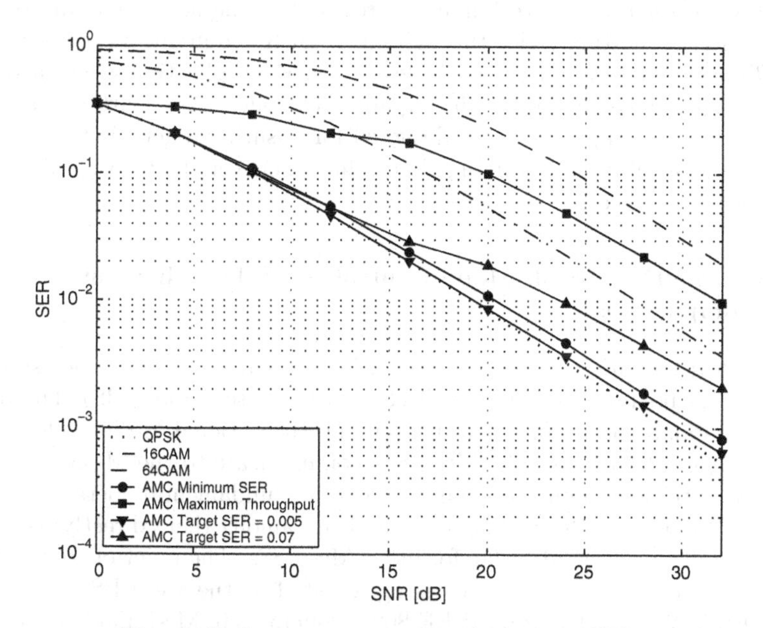


**Fig. 3.** Performance comparison in terms of SER

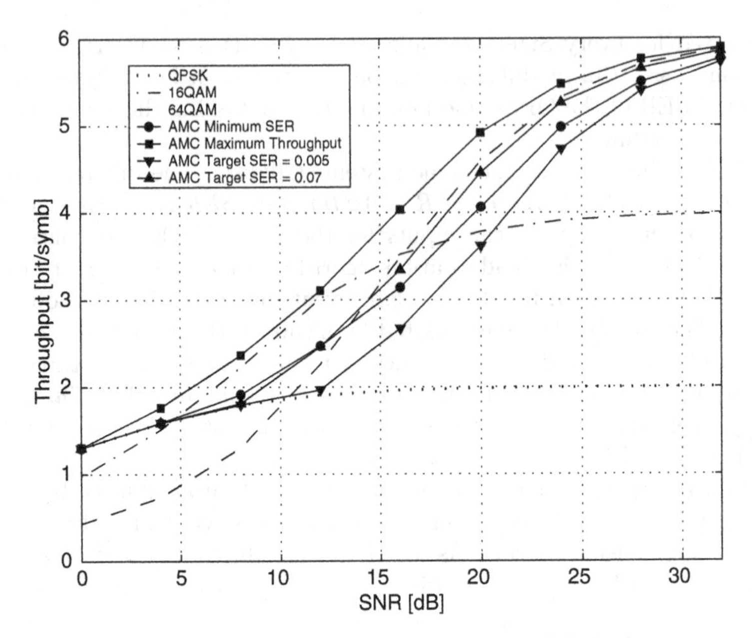


**Fig. 4.** Performance comparison in terms of throughput

Figs. 3 and 4, it is evident that the choice of the best method depends on the specific system requirements. While the *Maximum Throughput* technique and the *Minimum SER* technique represent two methods well suitable for application requesting strict throughput or SER requirements, respectively, the *Target SER* technique is suitable for generic application where a target SER can be adjusted following the requested QoS level. As a matter of fact, the *Target SER* technique has been designed for ensuring a generic target SER, with the possibility of modifying that value whenever the QoS requirements of a specific application change.

## 2.2 IEEE 802.16 Mesh Mode Centralized Scheduling Issues and Comparisons

When WMANs are required to cover a large area, some subscriber stations (SSs) might be unable to connect directly to the base station (BS). The IEEE 802.16 standard comes into help by defining the optional Mesh mode. When operating in Mesh mode, the Mesh SSs communicate to the Mesh BS either directly or through multi-hop. Thus, unlike the fundamental operating mode, i.e., PMP mode, Mesh SSs may communicate among them directly: they exchange traffic or forward traffic for other Mesh SSs. The point of convergence between wireless and wired MAN is represented by the Mesh BS.

When a WMAN based on IEEE 802.16 operates in Mesh mode using centralized scheduling, the routing tree, along which connections are established, and the transmission scheduling must be determined. IEEE 802.16 standard lacks of indications on the strategy to use for designing the routing tree and for planning the scheduling. However, the routing tree selection affects the transmission schedule and the probability of a successful communication from (to) an SS to (from) a BS. It is thus important to optimize the tree selection and the scheduling to improve performance, such as reliability (i.e., the expected probability that SSs can communicate successfully with the BS) and the throughput (i.e., the maximum percentage of bandwidth reserved for any node transmission in the data subframe). Similar to the proposed AMC techniques, strategies that either optimize the reliability or the throughput can be defined for the selection of the routing tree.

A *maximum reliability tree* (MRT) can be optimally found by selecting the shortest path tree rooted at the BS. The shortest paths are calculated on the mesh topology by using $(-\log(1-p_{ij}))$ as link weight, where $p_{ij}$ is the expected probability of an unsuccessful transmission of a data PDU on link $(i,j)$.

A *maximum throughput tree* (MTT) can be found by selecting the minimum hop shortest path tree rooted at the BS. When multiple nodes at minimum hop distance exist, the node with minimum number of two-hop neighbors (i.e., minimum number of interfering neighbors [18]) is selected.

Once the routing tree is designed, the second step consists in optimizing the scheduling. Concurrent transmissions of non-interfering SSs can be allowed [18]. The optimal scheduling is a coloring problem (thus an NP-hard problem) and is solved using a heuristic algorithm.

Performance are compared on a number of randomly generated mesh topologies [1, 7] of 20 nodes and $L$ links to achieve a confidence interval value of 5% or better, at 95% confidence level. One node acts as BS and the others act as SSs and request the same amount of bandwidth for uplink transmissions. It is assumed that concurrent transmissions of SSs that are more than two-hops apart (on the mesh topology) are interference-free. The expected PDU error probability is generated uniformly in $[0, p_{\max}]$.

Figures 5 and 6 show the expected reliability and the expected maximum throughput of the routing tree vs. $p_{\max}$ (i.e., the maximum expected PDU error) for different randomly generated mesh networks with 20 nodes and $L$ links. The plot shows that MRT is more reliable than MTT at the expenses of throughput. Thus the choice of the most suitable tree needs to be carefully evaluated depending on the level of QoS required by the application.

### 2.3 Wireless and Wired MAN Integration

In an integrated wireless and wired MAN infrastructure, an important issue is how to provide a wireless/wired MAN interface. The considered scenario is represented by a wired MAN carrying IP traffic and exploiting GMPLS and an IEEE 802.16 WMAN as depicted in Fig. 7. Different integration approaches can be envisioned and they can be divided in IP-based and sub-IP-based, as explained next.

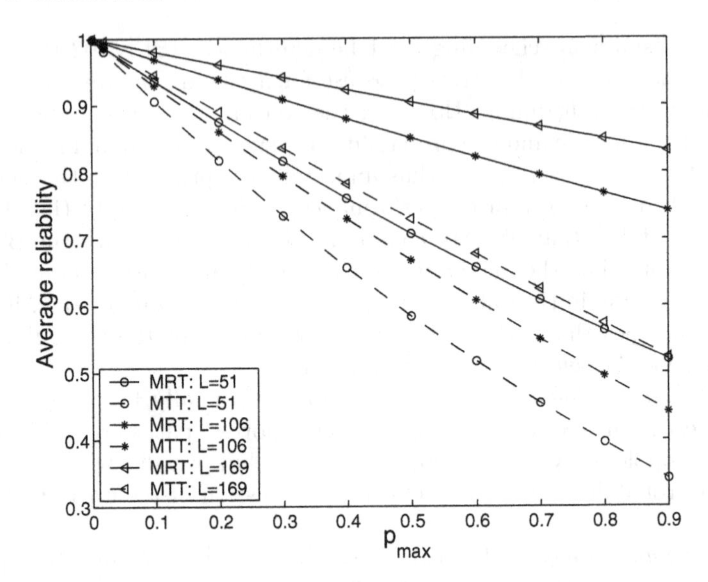


**Fig. 5.** Expected reliability vs. $p_{\max}$

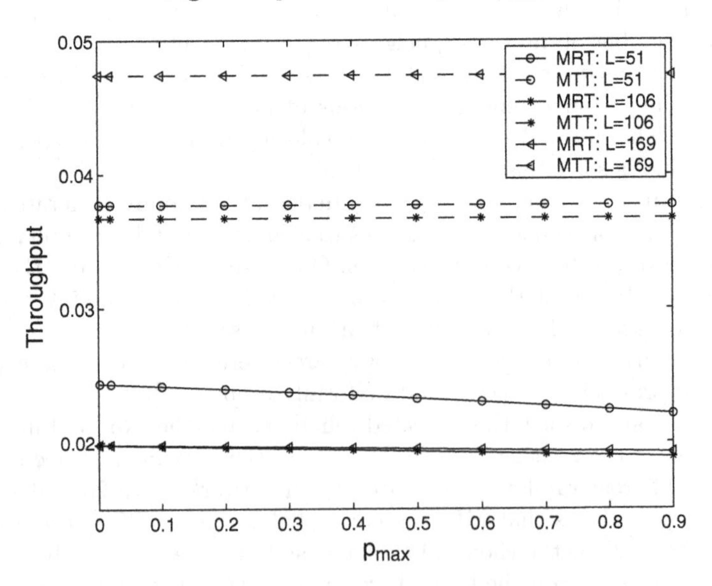


**Fig. 6.** Maximum throughput vs. $p_{\max}$

## IP-Based Integration

The IP-based integration is based on IP paradigm, as proposed in [5]. Both the wireless and the wired nodes should support IP protocol for network control and data transport. As shown in Fig. 8, data and control packets are forwarded at layer 3, according to the IP forwarding tables.

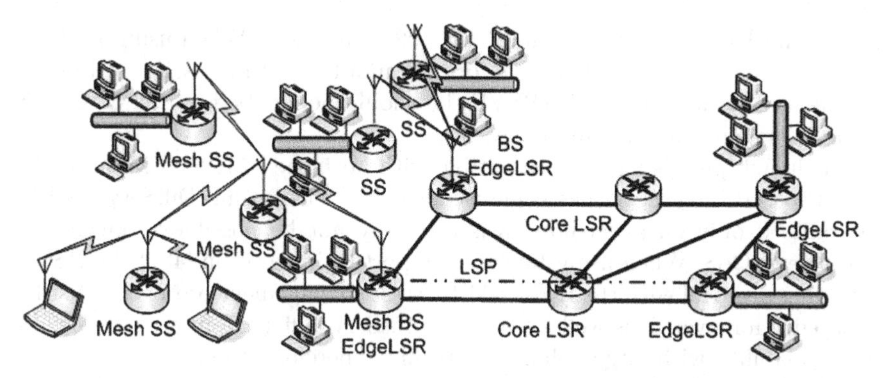


**Fig. 7.** Integrated wireless and wired architecture

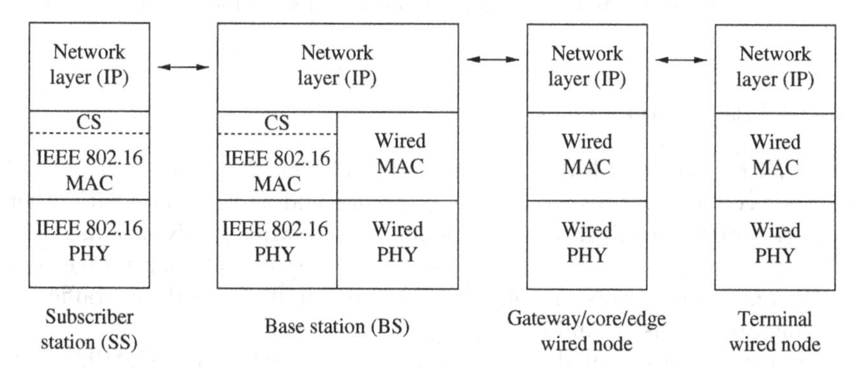


**Fig. 8.** Connectionless integration: integrated wireless and wired network protocol layering

In each wireless node, a service-specific convergence sublayer (CS) for IP packet support is required for interfacing the MAC layer of IEEE 802.16 protocol with the IP layer. Thanks to CS, IP packets can be forwarded to the next hop wireless node as follows. Once the next hop IP address is determined, the IP packet may undergo fragmentation or packing. Each (fragmented or packed) IP packet is then encapsulated with MAC header and transmitted according to PMP or Mesh operation mode of IEEE 802.16 protocol. Similarly, the received MAC packets are decapsulated from the MAC header and passed to the higher layers for the proper handling.

Within IP networks, QoS can be guaranteed using either the integrated service (IntServ) or Differentiated service (DiffServ) approach [17]. In the integrated scenario, the BSs must be provided with additional functionalities for mapping QoS parameters from the IP layer to the MAC layer. When using DiffServ, the Differentiated Service Code Point (DSCP) six bits in the IP header define the per-hop behavior of each packet in both the wired and wireless segment. Data IP packet may be marked at the BS by mapping, for example, the DSCP field into the scheduling services (i.e., UGS, ertPS, rtPS, nrtPS and BE) supported in the IEEE 802.16 network. When using IntServ, the traffic specifications (TSpec) information of the `Path` message used in the resource reservation protocol (RSVP) [3] should be mapped to the MAC QoS parameters, such as rate and delay.

Although promising and compatible with both the PMP and Mesh mode, the IP-based integration has some weaknesses. When using DiffServ, QoS is guaranteed only on a per-hop basis and may not be possible to ensure an end-to-end QoS. When using IntServ, support or integration [4] of the RSVP signaling protocol with the IEEE 802.16 MAC management protocols must be ensured. Moreover, it is well-known that IntServ suffers from the complexity and scalability with large volumes of traffic. Therefore, to ensure end-to-end QoS and to avoid the well-known scalability limitations of IntServ (i.e., the state explosion problem [15]), a sub-IP-based integration is envisioned, as explained next.

### Sub-IP Integration

Except for the Mesh mode operating with distributed scheduling, IEEE 802.16 features a connection-oriented MAC Layer and could well-interface with other connection-oriented sub-IP protocols supported by wired MAN such as (Generalized) Multi-Protocol Label Switched (GMPLS) [9]. The advantage of a sub-IP connection-oriented integration is the possibility to perform traffic engineering and to ensure and control end-to-end QoS.

The connection-oriented nature of both IEEE 802.16 and GMPLS may permit two different integration solutions: wireless (G)MPLS-wired (G)MPLS and wireless MAC-wired (G)MPLS.

#### *Wireless (G)MPLS-Wired (G)MPLS Integration*

Similarly to the IP-based integration approach described in the previous section, the first solution assumes that the wireless network supports (G)MPLS (a similar approach has been proposed for UMTS in [16]) through a well-defined CS at the wireless nodes. (G)MPLS data and management messages should be supported at each wireless node, as shown by the protocol layering in Fig. 9. Notice that IP layer may also be supported on top of (G)MPLS layer. (G)MPLS packets are forwarded along the route of the (G)MPLS connections, referred to as label switched paths (LSPs), according to the shim (G)MPLS labels in the packet header. An extension for traffic engineering of RSVP (RSVP-TE) permits the reservation of the resources, while guaranteeing the required QoS.

At the edge router, either in the wireless or in the wired domain, IP packet DSCP field can be mapped in an EXP-inferred or a label-inferred fashion into LSPs [8]. In the EXP-inferred approach, a single LSP can support IP packets requiring different Per Hop Behaviors (PHB) (i.e., Behavior Aggregates – BAs). In this case, packets with different DSCPs have the same LSP label but

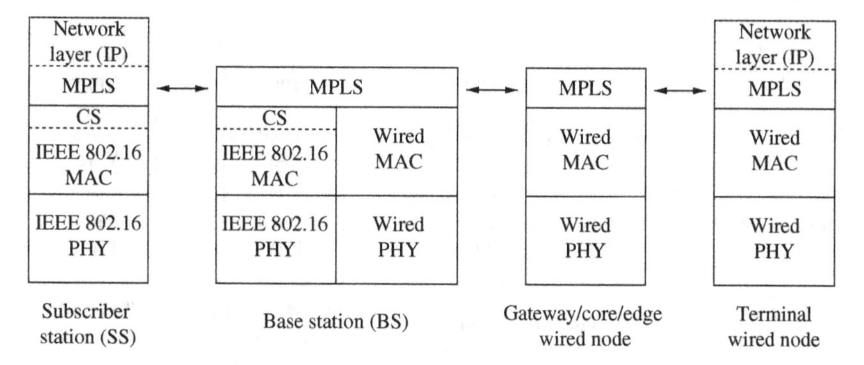


**Fig. 9.** Wireless (G)MPLS-Wired (G)MPLS integration protocol layering

a different EXP field value that is obtained through a DSCP-EXP mapping. With such LSPs, the EXP field of the MPLS Shim Header is used by the Label Switch Router (LSR) to determine which PHB needs to be applied to the packet. In the label-inferred approach, a single LSP carries packets belonging to a single BA. Thus a mapping between Forwarding Equivalence Class (FEC) and BA is necessary. The PHB Scheduling Class (PSC) is explicitly signaled at the time of label establishment, so that after label establishment, the LSR can infer exclusively from the label value the PSC to be applied to a labeled packet.

For an efficient cross-layer integration, RSVP-TE should be interfaced and interlaced with the IEEE 802.16 mechanism to reserve bandwidth for connections (or service flows in PMP mode). In other words, RSVP-TE signaling should be able to trigger signaling or mechanisms for bandwidth allocation as well as to map of QoS parameters, at the BS or at the terminal SS.

*Wireless MAC-Wired (G)MPLS Integration*

Another sub-IP solution for integrating the wireless and the wired networks is sketched in Fig. 10 and is achieved by interlacing the LSPs of the (G)MPLS networks with the IEEE 802.16 connections (e.g., service flows in PMP mode). The (G)MPLS network operates as explained before. The wireless network is based on IEEE 802.16. In the wireless network, end-to-end QoS can be guaranteed at the MAC layer, thus simplifying the cross-layer approach and achieving a better QoS control.

The advantage of this solution is a simplification of the wireless node architecture and protocol stack, at the expense of additional BS complexity. Indeed, when adopting this solution, the BS acts as a protocol translator and thus has the burden to interface RSVP-TE with the MAC signaling for resource allocation in IEEE 802.16, to map QoS parameters, to perform admission control and so on, on behalf also of the SSs.

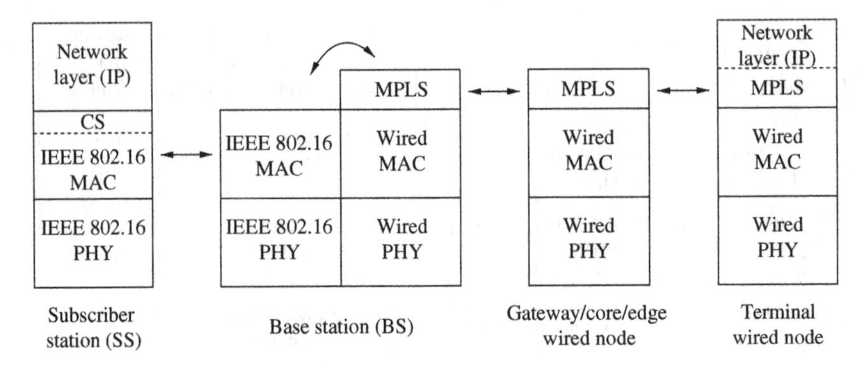


**Fig. 10.** Wireless MAC-Wired (G)MPLS integration protocol layering

## 3 Conclusion

This article focused on IEEE 802.16 standard advanced access technologies such as MIMO antennas, AMC, and the optional IEEE 802.16 Mesh mode of operation. Implementation and integration issues of such advanced technologies and original solutions have been presented and evaluated.

### Acknowledgments

This work has been supported by MIUR under PRIN project "Traffic and terminal self-configuration in integrated mesh optical and broadband wireless networks (TOWN)" (2005095328).

## References

1. Brite: Boston university representative internet topology generator. www.cs.bu.edu/brite/
2. S. M. Alamouti. A simple transmit diversity technique for wireless communications. *IEEE J. Select. Areas Commun.*, 16(8):1451–1458, October 1998
3. R. Braden, L. Zhang, S. Berson, S. Herzog, and S. Jamin. Resource ReSerVation Protocol (RSVP). RFC 2205, Sept. 1997. Standards Track, IETF
4. J. Chen, C. Chi, and Q. Guo. A bandwidth allocation model with high concurrence rate in IEEE802.16 Mesh mode. In *Proc. Asia-Pacific Conference on Communications*, October 2005
5. J. Chen, W. Jiao, and Q. Guo. An integrated QoS control architecture for IEEE 802.16 broadband wireless access systems. In *Proc. IEEE GLOBECOM*, 2005
6. D. Marabissi, D. Tarchi, F. Genovese, and R. Fantacci. A finite state modeling for adaptive modulation in wireless OFDMA systems. Gothenburg, Sweden, April 2007
7. M. Doar and I. Leslie. How bad is naive multicast routing? In *Proc. IEEE INFOCOM*, 1993

8. Ed. F. Le Faucheur. Multi-Protocol Label Switching (MPLS) Support of Differentiated Services. RFC 3270, May 2002. Standards Track, IETF
9. A. Farrel and I. Bryskin. *GMPLS : Architecture and Applications.* Morgan Kaufmann, Los Altos, CA 2005
10. G. J. Foschini. Layered space–time architecture for wireless communication in a fading environment when using multi-element antennas. *Bell Lab. Tech. J.*, 1(2) 41–59, 1996
11. R. W. Heath and A. J. Paulraj. Switching between diversity and multiplexing in mimo systems. *IEEE Trans. Commun.*, 53(6):962–968, June 2005
12. A. Hottinen, O. Tirkkonen, and R. Wichman. *Multi-antenna Transceiver Techniques for 3G and Beyond.* Wiley, New York 2003
13. IEEE. *IEEE Std. 802.16-2004, IEEE Standard for Local and Metropolitan Area Networks, Part 16: Air Interface for Fixed Broadband Wireless Access Systems*, 2004
14. IEEE. *IEEE Std 802.16e-2005 and IEEE Std 802.16-2004/Cor 1-2005: IEEE Standard for Local and metropolitan area networks Part 16: Air Interface for Fixed and Mobile Broadband Wireless Access Systems Amendment 2: Physical and Medium Access Control Layers for Combined Fixed and Mobile Operation in Licensed Bands and Corrigendum 1*, 2005
15. P. Levine, J. Martins, B. Stiller, M. H. Sherif, A. Fumagalli, J. Aracil, and L. Valcarenghi. *Managing IP Networks : Challenges and Opportunities.* IEEE Press, 2003
16. Hung-shih Chueh and K. Wang. An all-MPLS approach for UMTS 3G core networks. In *Proceedings of IEEE Vehicular Technology Conference*, 2003
17. W. Stallings. *High-Speed Networks and Internets: Performance and Quality of Service.* Prentice-Hall, Englewood Cliffs, NJ, USA, second edition, 2005
18. H. -Y. Wei, S. Ganguly, R. Izmailov, and Z. J. Haas. Interference-aware IEEE 802.16 WiMax mesh networks. In *Proceedings of IEEE Vehicular Technology Conference*, June 2005
19. P. W. Wolniansky, G. J. Foschini, G. D. Golden, and R. A. Valenzuela. V-BLAST: an architecture for realizing very high data rates over the rich-scattering wireless channel. In *Signals, Systems, and Electronics, 1998. ISSSE 98. 1998 URSI International Symposium on*, pages 295–300, 29 September–2 October 1998
20. X. Zhuang, F. W. Vook, S. Rouquette-Leveil, and K. Gosse. Transmit diversity and spatial multiplexing in four-transmit-antenna OFDM. In *Proceedings of IEEE ICC*, volume 4, pages 2316–2320, 11–15 May 2003

# 15

# WPAN Applications and System Performance


Yoshinori Nishiguchi, Ryuhei Funada, Yozo Shoji, Hiroshi Harada, and Shuzo Kato

National Institute of Information and Communications Technology, 3–4 Hikarino-oka Yokosuka, Kanagawa 239-0847, Japan
nishiguchi@nict.go.jp, funada@nict.go.jp, shoji@nict.go.jp, harada@nict.go.jp, shu.kato@nict.go.jp



**Summary.** This paper describes recent R&D results on millimeter wave systems and standardization activities in IEEE 802.15.3c for millimeter-wave WPAN (Wireless PAN). The millimeter-wave WPAN system is supposed to transmit 2 Gbps or more as mandatory mode. We propose "Single carrier approach" which will offer easily high data rate and provide "Kiosk file downloading" application more efficiently. Computer simulation results on the proposed systems show excellent performance on "Kiosk file downloading" model.


## 1 Introduction

The IEEE 802.15.3 Task Group 3c was formed in March 2005. TG3c is developing a millimeter-wave-based alternative physical layer (PHY) for the existing 802.15.3 Wireless Personal Area Network (WPAN) Standard 802.15.3-2003 [1]. This mmWave WPAN will operate in the new and clear band including 57–64 GHz unlicensed band in the USA and 59–66 GHz in Japan and Europe. The millimeter-wave WPAN will allow high coexistence (close physical spacing) with all other microwave systems in the 802.15 family of WPANs. In addition, the millimeter-wave WPAN will allow very high data rate over 2 Gbps applications such as high speed internet access, streaming content download (video on demand, HDTV, home theater, etc.), real time streaming and wireless data bus for cable replacement. While 60 GHz millimeter wave communications systems have many good aspects as wide frequency bands which will allow to transmit very high data as several Gbps, they have also drawbacks as high directivity, high path loss and limited device transmission power.

- Proposers shall refer following documents in detail
- TG3c Call for Proposals: 07/586r2
- TG3c Usage Model Document: 06/055r22
- TG3c System Requirements Document: 07/583r1

- TG3c Selection Criteria Document: 05/493r27
- Channel Model Sub-committee Final Report: 07/584r1
- Channel model Matlab code: 07/648r0
- CM Golden set PHY simulation CM13: 07/580r1
- CM Golden set PHY simulation CM23: 07/581r1
- CM Golden set PHY simulation CM31: 07/582r1

To contribute and propose a "single carrier air interface" PHY to the above IEEE standard aiming at commercializing millimeter wave communications systems ASAP, a consortium named as CoMPA (Consortium of Millimeter-wave Practical Applications) was established in July 2006. The CoMPA is composed of 19 Japanese institutes currently and four non-Japanese companies have been in joining process. The CoMPA has created six working groups to cover whole technologies required for millimeter wave communications systems and been very active in contributing to IEEE802.15.3c standardization technically. The CoMPA will be cooperating with other institutes that are also promoting "single carrier approach". Please note The CoMPA is open for all (not individual) as far as they support "Single carrier approach" for air interface/system selection at IEEE 802.15.3c standardization meeting.

## 2 Usage Model

A number of use cases and usage models have been discussed and narrowed down to five major usage models. They are categorized into mandatory usage models (UM1 and UM5) and optional usage models (UM2, 3, and 4). UM1 targets uncompressed high speed video streaming at 1.78 or 3.56 Gbps (MAC-SAP rates) over 5–10 m transmission ranges in LOS (Line of sight) and/or NLOS (Non line of sight) conditions in Residential channel environments. The target error rate is $10^{-9}$ TADS CER (Transition Minimized Differential Signaling Character Error Rate, 24 bit per character) which will be evaluated by $10^{-6}$ BER (Bit error rate) in PHY simulation. The UM1 with images of use cases with some descriptions are given Fig. 1.

UM5 targets "Kiosk file down loading" type file transfer applications at 1.5 or 2.25 Gbps (MAC-SAP rates) over 1 m transmission range in LOS condition in Office channel environments [2]. The target PER (Packet error rate) is 8% with a packet size of 2 K bytes. This application was proposed by CoMPA (Consortium for Millimeter-Wave Practical Applications) to solve "directivity issue" practically by pointing the "Kiosk server" like a remote controller pointing a TV set. This can be used for PC peripherals as well and has been getting a lot of interests. The UM5 with images of use cases with some descriptions are given Fig. 2.

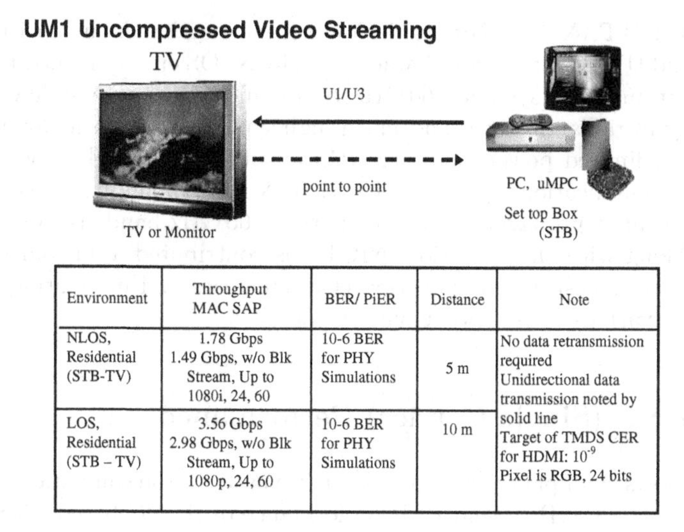


| Environment | Throughput MAC SAP | BER/ PiER | Distance | Note |
|---|---|---|---|---|
| NLOS, Residential (STB-TV) | 1.78 Gbps 1.49 Gbps, w/o Blk Stream, Up to 1080i, 24, 60 | 10-6 BER for PHY Simulations | 5 m | No data retransmission required Unidirectional data transmission noted by solid line Target of TMDS CER for HDMI: $10^{-9}$ Pixel is RGB, 24 bits |
| LOS, Residential (STB – TV) | 3.56 Gbps 2.98 Gbps, w/o Blk Stream, Up to 1080p, 24, 60 | 10-6 BER for PHY Simulations | 10 m | |

**Fig. 1.** Usage models (UM1)

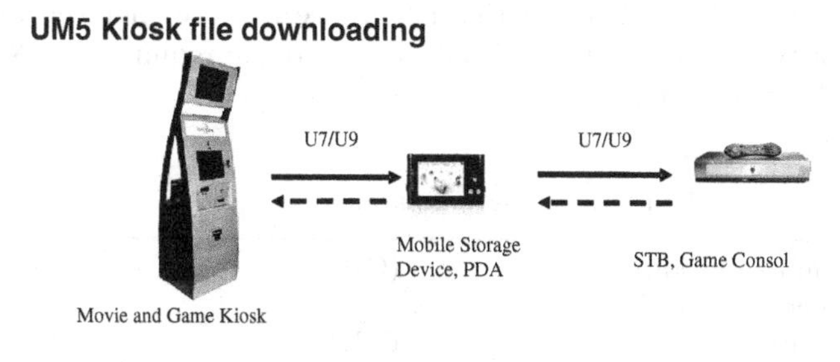


| Environment | Throughput (MAC SAP) | BER/PER | Distance | Note |
|---|---|---|---|---|
| LOS-office LOS-KIOSK (Server-PDA or PDA-STB) | 1.50 Gbps burst (Server-PDA or PDA-STB) | 8% PER before retransmission 2K Byte | 1 m | Asymmetric download/Upload<br><br>Low data rate reverse link |
| LOS-office (Server-PDA or PDA-STB) | 2.25 Gbps burst (Server-PDA or PDA-STB) | 8% PER before retransmission 2K Byte | 1 m | |

**Fig. 2.** Usage models (UM5)

# 3 Channel Model

A number of channel models were studied and proposed aligned with various applications scenarios, then they have been narrowed down to nine channel models. The proposed channel models are categorized into two by having LOS component or not. The latter is a typical Wireless LAN environment and for many other WPAN environments in lower frequency bands, up to 10 GHz as WLAN and UWB systems are characterized by NLOS environments. However, different from these systems, 60 GHz communications systems have strong directivity in radiated waves and much higher path loss than many others in addition to limited power amplifier (PA) output. These lead to deployment of LOS communications systems other than NLOS communications systems. Thus, a peculiar propagation characteristic of 60 GHz bands is LOS but not NLOS. Along with this direction, NICT has contributed a lot and played a major role in measuring propagation characteristics and in creating channel models for various application environments [3–5].

## 4 PHY-SC (Single Carrier) Performance

Table 1 shows example system parameters we used for the simulation analysis. Figure 3 shows the BER and PER performance of the SC-based PHY but in UM5 environment. The channel to used in UM5 environment is basically LOS KIOSK channel but this result is considering the effects of PA non-linearity, and phase noise as well [6, 7]. From this result, we can see that the use of concatenation of two coding schemes is effective and the required $E_b/N_0$ for PER of 8% is about 4.7 dB.

**Table 1.** Simulation parameters

| | |
|---|---|
| System band width | 9 (GHz) |
| Number of channel | 4 |
| Modulation | QPSK |
| Detection | Coherent |
| M-array modulation level | 2 |
| Center frequency | 60 (GHz) |
| Over sampling | 8 |
| Symbol rate | 1.6 (Gsps) |
| Roll off rate (alpha) | 0.35 |
| Pulse-shaping filter | 21 (taps) |
| Effective bandwidth per channel | 2.16 (GHz) |
| Coding scheme | Concatenation of two scheme |
| Outer coding scheme | Reed-solomon code (255, 238) |
| Inner coding scheme | Convolutional code (R = 3/4) (K = 5) |
| Coding rate | 0.7 |

| | |
|---|---|
| PSDU data transmission rate | 2.24 (Gbps) |
| PSDU in one packet | 2048 (Byte) |
| Channel model | CM9 (LOS-KIOSK) |
| Channel realizations | 100 |
| Packets/Ch. Realizations | 100 |
| FDE | NONE |
| synchronization | ideal |

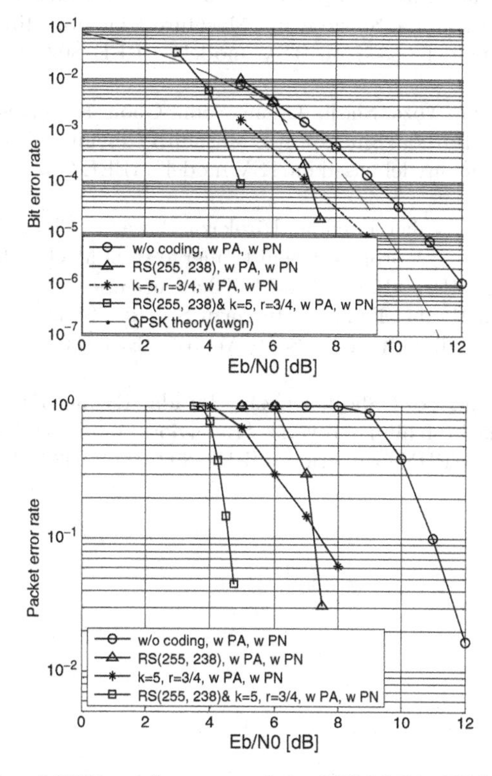


**Fig. 3.** BER and PER performance of the PHY-SC in UM5 environment

## 5 Conclusions

This paper reported current status of standardization activities in IEEE802.15.3c (TG3c), mandatory Usage Models (UM) for the TG3c, and our recent R&D results on the millimeter-wave system. System performance based on our “Single carrier approach” based PHY parameters was studied considering the millimeter-wave WPAN channel environment for UM5. We clarified that the use of concatenation of two coding schemes is effective and 4.7 dB of $E_b/N_0$ will be required for the UM5 environment.

## References

1. http://www.ieee802.org/15/pub/TG3c.html
2. Yozo Shoji, Chang-Soon Choi, Shuzo Kato, Ichihiko Toyoda, Kenichi Kawasaki, Yasuyuki Oishi, Kazuaki Takahashi, Hiroyuki Nakase “Re-summarization of merged Usage Model Definitions parameters”, IEEE 802. 15-06-0379-02-003c, September 2006
3. Hirokazu Sawada, Yozo Shoji, Chang-Soon Choi, Katsuyoshi Sato, Ryuhei Funada, Hiroshi Harada, Shuzo Kato, Masahiro Umehira, Hiroyo Ogawa: “LOS office channel model based on TSV model”, IEEE 802. 15-06-0377-02-003c, September 2006
4. Hirokazu Sawada, Yozo Shoji, Chang-Soon Choi, Katsuyoshi Sato, Ryuhei Funada, Hiroshi Harada, Shuzo Kato, Masahiro Umehira, Hiroyo Ogawa: “LOS residential channel model based on TSV model”, IEEE 802. 15-06-0393-00-003c, September 2006
5. Hiroshi Harada, Ryuhei Funada, Hirokazu Sawada, Chang-soon Choi, Yozo Shoji, Shuzo Kato, “A proposed plan to merge Intel MATLAB code and NICT MATLAB code”, IEEE 15-06-0516-01-003c, December 2006
6. Hiroshi Harada, Ryuhei Funada, Hirokazu Sawada, Shuzo Kato, “CM Golden Set for PHY Simulation of TG3c (CM1.3)” IEEE 802. 15-07-0580-00-003c, January, 2007
7. Chang-Soon Choi, Yozo Shoji, Hiroshi Harada, Ryuhei Funada, Shuzo Kato, Kenichi Maruhashi, Ichihiko Toyoda, “Behavioral model of 60 GHz-band power amplifier for SYS/PHY evaluation”, IEEE 802.15-06-0396-01-003c, September, 2006

# 16

# Improving User Relocatability, Practicality, and Deployment in the Web Stream Customizer System

Jesse Steinberg and Joseph Pasquale

Department of Computer Science and Engineering, University of California, San Diego, USA
jsteinbe@cs.ucsd.edu, pasquale@cs.ucsd.edu

**Summary.** We present improvements to a Web middleware system that supports flexible Web content and stream customizations such as filtering, compression, encryption, remote caching, and remote buffering, especially applicable for wireless Web access. The new extensions provide support for user relocation, and make the system more practical by emphasizing the use of a personal server for running customization modules and for maintaining user configuration and application data, in addition to the selected use of third party servers to bolster performance, increase fault tolerance, and satisfy special security requirements.

## 1 Introduction

A common approach to remote Web customization is the use of HTTP proxies as intermediaries between the client and server. In this model, requests generated by the client are sent to the proxy, which then forwards the request to the Web server. The Web server receives the request, processes it, and returns a response to the proxy. The proxy then has the opportunity to customize the response before it is returned to the client. This approach is transparent to Web servers since they see the proxy as a client, and is also transparent to clients since most popular Web browsers have a proxy mechanism that allows them to automatically forward their requests to a proxy.

In the Web Stream Customizer (WSC) architecture, the role of the proxy as described above is expanded and played by a number of components. First, there is the Customizer Server (CS), which provides an execution environment for running Customizers. A user typically will have multiple Customizers that are active, which depend on their function, e.g., filtering, compression, encryption, caching, etc., and which may apply only to specific sites, e.g., cnn.com, yahoo.com, etc. Figure 1 shows a client using multiple Customizers, each of which are running on a separate CS (including a special one designated as the PCMS, discussed below), along with other parts of the architecture described next.

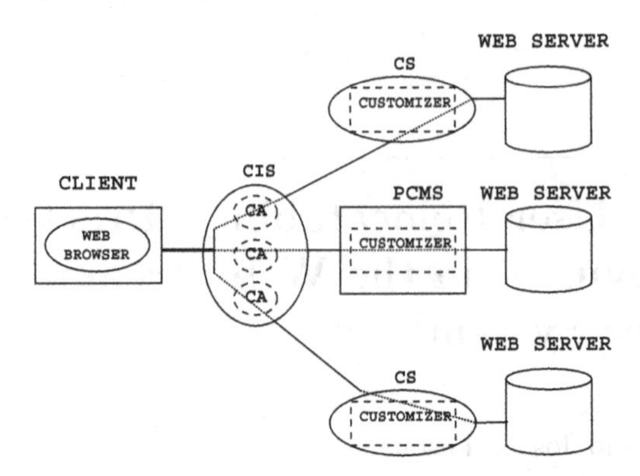


**Fig. 1.** Client using multiple Customizers

To support adaptability, each Customizer has an associated helper module called a Customizer Assistant (CA). The CA runs on a Client Integration Server (CIS), which tends to be located on or near the client device, and acts primarily as an extension of the browser (given that the browser code itself cannot be modified). Serving as a proxy for the browser, all of the browser's requests are sent to the CIS.

Thus, when a Customizer is being used, the request passes from the client to a CA (which runs on the CIS), then to a Customizer (which runs on a CS), and then to the Web server (and vice versa for responses in the opposite direction). Despite the additional stops a request and response must make, we have found the resulting overhead to be insignificant relative to typical end-to-end Web request/response times without Customizers, as reported in [29], and summarized in Sect. 3.

Given its close coupling with the client, the CA is generally responsible for tasks that require knowledge of resource availability and system conditions at or near the client, which may then be communicated to the Customizer (e.g., such as relaying local system or network performance status so that the Customizer can adapt in order to improve performance). In addition, the CA will also reverse data transformations done by the Customizer, such as compression/decompression or encryption/decryption.

Associated with each Customizer is its *domain of applicability* (DA), which is a list of all of the sites for which the Customizer will act. This is how the CIS knows which CA should be given a request from the browser (and consequently to which Customizer the request is passed). When a Customizer is deployed (described below), its DA, along with its corresponding CA, is downloaded to the CIS. The CIS then uses the DAs to perform the matching of requests to CAs and Customizers.

Figure 2 shows how HTTP requests are handled by the CIS. When the CIS gets a request from the browser, it first matches the URL of the request to the DA of all loaded Customizers. If there is a match to a Customizer, the

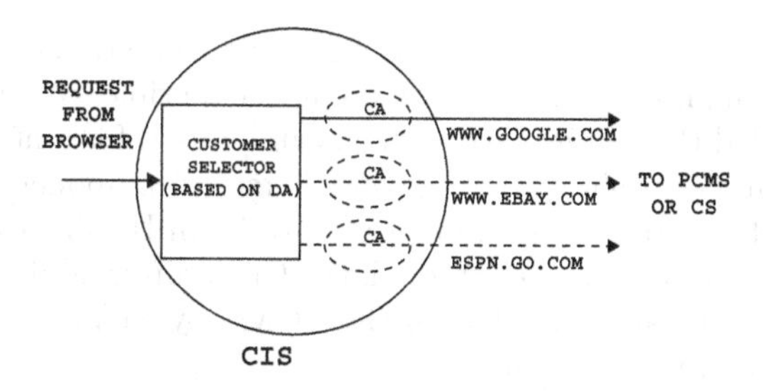


**Fig. 2.** Selecting a Customizer based on the DA

request is passed to the corresponding CA, and then to the location specified in the CR for that Customizer. The Customizer does not need to be explicitly loaded at that location, as described below.

Typically, a user will own a PC at their home or office (or at least have access to a machine at least as powerful on which they have an account), and that has reliable Internet connectivity relative to the wireless PDAs we have been considering as clients. To facilitate user relocation and improve the practicality of our customization system, such a machine can be configured to serve as the user's Personal Customizer Management Server (PCMS). The PCMS is a storage place for the code modules of the user's Customizers, along with their preferences and state (collectively called a *Customizer Package*).

The PCMS can make dynamic decisions about where the Customizer code will run, based on the location of the user and the constraints of the Customizer (such as security restrictions), and it can carry out Customizer reloading without any direct involvement by the low-powered client and its possibly unreliable low-bandwidth wireless connection. Customizer reloading provides a measure of fault tolerance, because a Customizer may also be moved if there is a significant lapse in response time from the current CS on which a Customizer is running.

The PCMS contributes to practicality because it can also be used as a (user-controlled) location for actually running Customizers, which is especially useful if there is no available CS that is willing to host the Customizer.

## 2 Applications

We have experimented with a number of applications that focus on improving performance and reliability.

### 2.1 Remote Filtering

We have implemented an adaptive Image Filter Customizer that can modify the resolution, color-depth, and compression rate of images. The CA measures the approximate throughput at the client, and relays it to the Customizer.

This allows the Customizer to adapt its filtering to maintain consistent transfer times as throughput fluctuates. We are also experimenting with a general Customizer, called the Filter–Saver, which can be used for a number of applications. As the name implies, it consists of two major components, a Filter and a Saver. The Filter reduces the amount of data in an HTTP response coming from the Web Server as it passes through the Customizer, and the Saver takes advantage of the storage available on the PCMS by saving original versions of the objects before they are filtered.

Storing the objects at the PCMS ensures that they will be available even if they are removed from the original source, and allows the objects to be retrieved quickly regardless of network problems between the PCMS and the original source (in fact, the objects are available immediately if the user accesses them from the host on which they are running their PCMS).

In addition to filtering objects based on their data type, content-specific filtering can be done. For example, the Filter could remove the commentary from a product review site and return only the final score; research papers might be filtered so that only the pages containing the abstract, introduction, and bibliography are returned; scene change detection could be used to reduce the frame rates of videos.

### 2.2 Transaction Recording for Reliability

The Transaction Recorder addresses the problem of connection failures during a transaction by storing the results of transactions on the PCMS. The user can easily check on the outcome of the transaction once connectivity is reestablished via the Customizer's configuration page, which is accessible by the user from a Customizer control Web page provided by the CIS.

The Connection Smoother Customizer is an extension of the Transaction Recorder. It stores objects requested by the browser on the PCMS. If there is a short-term lapse in connectivity before the browser receives a stored object, the CA automatically repeats the request for that object, in order to mask the connection failure from the user. Since objects are only stored for short periods of time, this Customizer performs best when it runs on the PCMS or the CS has a cache for the data sent to the PCMS.

### 2.3 Multimedia Streaming

So far we have described applications that customize HTTP transactions. It is also possible to use Customizers for applications that use other protocols. To enable this, Customizers that are trusted by a CS are given privileged access to resources that are persistent across multiple HTTP transactions, including memory for data, hard disk storage, and threads of execution. The Customizer must also have network I/O privileges.

One non-HTTP application that especially benefits from the ability to reload a Customizer near the client is multimedia streaming. A wireless resource-limited device may not have enough memory to adequately buffer streams, whereas a Customizer running near the client can buffer the stream and periodically feed the client. The Customizer can also filter the stream to match the capabilities of the client device.

On the Web, multimedia streams are typically initiated either from a link within a page that contains a URL that causes the browser to launch the multimedia client, or by a page having a link to a metafile that contains information about the stream, and which is downloaded by the browser using HTTP and then passed to the appropriate streaming client. Customizers can be used to intercept multimedia streams for filtering by using the initiating Web page or metafile as a hook.

When a Web server replies to a request with a page linking a stream, or a request for a metafile specifying the location of a multimedia file, the Customizer modifies the location of the video stream. It replaces the identity of the source of the video with itself, so that the client's video application tries to retrieve the video stream from the Customizer. The Customizer then acts as a proxy for the video streaming protocol. The Customizer appears to the video server to be the video client, while the client sees the Customizer as the video server.

For customization that requires local action before being sent to the client, the Customizer can use a CA, and they can act in concert as a pipelined pair of proxies operating on the data. This approach allows Customizers to act on streams between the multimedia client and server if it is able to implement the appropriate protocol.

It is also possible for Customizers to use their own multimedia client, and even stream objects that were not initially set up for streaming by the content provider, by use of Customizer-specific helper-applications preinstalled at the client. This allows the Customizer to control the streaming more directly. In this case, when the Customizer gets a request for a multimedia object, or a metafile representing a multimedia stream, instead of modifying the server directly in the document, the Customizer replaces the requested document in the response with a specialized metafile type associated with the Customizer-specific helper application. The specialized metafile will specify the Customizer as the multimedia server. When the browser receives the specialized metafile, it will launch the Customizer-specific helper application, which will then act in concert with the Customizer to stream the data and display it to the user.

This mechanism can be used to create streams from objects that would normally be downloaded fully by the browser. For example, suppose the user clicks on an HTTP request for an mpeg file. Assuming HTTP is being used, the standard browser behavior is to download the entire file and hand it to the appropriate helper application (in this case a video player). However, if the Customizer responds with a specialized metafile that it generates on the fly instead of the mpeg clip, the browser will open the Customizer-specific helper application associated with the metafile's extension (or the HTTP content type header field sent with the metafile). The Customizer-specific helper application then contacts the Customizer to stream the video, which the Customizer retrieves from the Web server specified in the URL of the initial browser request. If the user relocates, the Customizer can be reloaded on a nearby CS, to provide the best possible playback of the stream.

# 3 Performance

We present an experiment that shows the benefits of using a Customizer for multimedia streaming, that prefetches and buffers data to reduce both delay and jitter. As a motivating application, we model a video player for a wireless client device, designed to play videos that are retrieved using HTTP from a nonstreaming server. Due to the limited memory of the device, the application is not able to buffer an entire video file before playing it. Therefore, it must download the video in small pieces, progressively playing each of the pieces while requesting subsequent ones, the amount being limited to the amount of extra buffer space available. This is accomplished by designing the player to make repeated HTTP 1.1 requests using the "Range" header, whereby a range of bytes is specified, and the Web server sends only those bytes. The problem with this approach is that if the latency of retrieving each piece is high and variable, as is generally the case when communicating with distant servers, the video cannot be played smoothly, if at all, if there is not ample buffering at the client.

A Smoothing Customizer is used to improve the performance of this application. This Customizer runs on a machine near the client such as a PCMS, to ensure low latency (in terms of both average and variance), and where it can be assumed there is plenty of memory to support a large buffer on behalf of the player. When the application makes the initial request for the first frame or portion of the video file, the Customizer not only relays this request, but also initiates an asynchronous prefetch of the rest of the video file. All future requests will then be handled by the Customizer, without having to contact the origin Web Server. This has the effect of reducing response times (after the initial wait) and jitter, resulting in smoother playback.

This is implemented by using the specialized metafile technique discussed above, which causes the browser to launch a Customizer-specific version of the application when the video file is first clicked. If the application allows an HTTP proxy to be specified, it can use Customizers directly to handle its HTTP requests. (If the CIS has been started and the Customizer has been loaded, the application's proxy could just be set to be the CIS).

We ran experiments to measure the performance benefits of the Smoothing Customizer. The Web server (`http://www.w3.org`) is located at MIT, and the client is at UC San Diego (thus separated by a distance of roughly 3,000 miles). We used a test program emulating a video application that retrieves and plays a 2.21 MB file, making requests in 100 KB pieces at 1-s

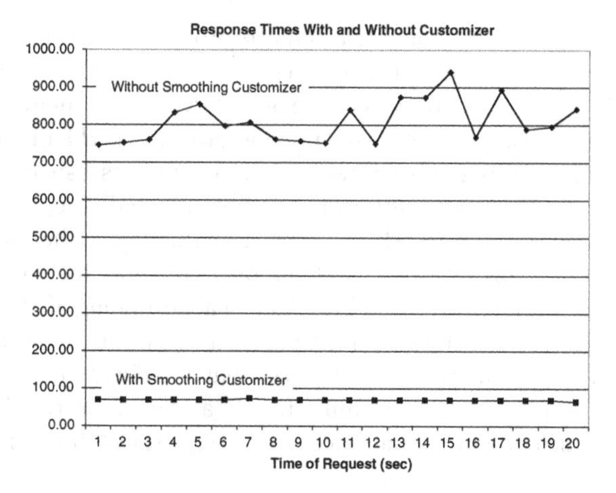


**Fig. 3.** Comparison of response times with and without the Smoothing Customizer

intervals. This simulates the scenario where at most one second's worth of video could be buffered at the player.

In the test environment, the client and Customizer each ran on different machines, both of which were PCs based on 933 MHz Pentium III processors running FreeBSD release 4.62, connected via Ethernet. Rather than running the client on a resource-limited device, we chose a standard PC (and therefore chose to artificially limit the player's buffer capacity) both for convenience, and to not introduce artifacts in the performance measurements due to a resource-poor client device (where it is problematic to reliably make and record accurate measurements without introducing interference), rather than simply the network (and Web server) latencies.

We measured response times (time between making a request for a 100 KB piece and getting the response) with and without the Smoothing Customizer in place. Figure 3 shows a sample run of 20 requests. The response times without the Smoothing Customizer are both significantly higher and more variable than those where the Smoothing Customizer is active: average response time was 810 ms vs. 70 ms, and jitter was 79 ms vs. 0.65 ms. The two-order-of-magnitude reduction in jitter is especially important, as this is the major determinant for buffer size at the client.

# 4 Related Work

There exists a large body of research results verifying the benefits of remote customization of Web data using proxies, mobile code, some combination thereof [2, 7, 9, 27, 29], and on the benefits of multimedia filtering [9, 10, 12, 21, 33]. Our distributed architecture allows these methods to be exploited more effectively.

A number of systems use a single remote proxy for customizing the Web, with communication initiated through the browser's proxy mechanism, including some in which the proxy is a personal server. This includes image and video filtering (sometimes called Multimedia gateways), HTTP request modifications, HTML filtering, user interface improvements especially for small screens, remote caching, and support for disconnected operation and user-selected background retrieval [1, 5, 7, 8, 13, 22, 28]. Other systems have made use of the two-proxy (local and remote) concept, for such customizations as filtering, prefetching, and intelligent cache management at the local proxy [22, 23]. In [11, 19] the server either on its own, or in cooperation with specialized proxies, works to Customize content for clients or to improve the performance of prefetching and caching. In [15] a pair of intermediaries is used to transparently provide fault tolerance, security, and timeliness in distributed object systems.

Research that is closest to ours combines the use of proxies or multimedia gateways with mobile code to support dynamic downloading of filters to a remote host [16, 25, 34]. There are also customization systems that do not use proxies per se, but rather use more general mobile code mechanisms to support remote processing at arbitrary hosts, typically at the servers themselves [14, 18, 20, 26, 27, 30, 31, 32].

A related issue is adaptability, where information is provided to the client application, typically from the operating system, to help it adapt to changes in resource availability and network connectivity [2, 4, 24]. Some of these systems include applications using an adaptable interface, including adaptable protocols.

The Internet Content Adaptation Protocol (ICAP) [17] is a solution developed by an industry coalition for distributing Internet content to edge servers. ICAP is server-centric in that content providers control distribution of functionality. This differs from the WSC architecture, where the client controls the deployment of Customizers. Open Pluggable Edge Services (OPES) [3] is an IETF effort to standardize the tracing and control of proxies for content adaptation. Multiple edge services can be chained together, and callout servers can be used to offload computation. Their services can be server-centric or client-centric. Simple Object Access Protocol (SOAP) [6] is a simple XML-based protocol for exchange of information and RPC for Internet applications. None of the above approaches take advantage of mobile code to dynamically deploy a service, nor do they include the dynamic downloading of a local component on or near the client.

Our work differs from that of others in a number of ways. First, a personal server stores Customizers and their configurations, which facilitates Customizer reloading. We combine the use of both a user-owned personal proxy that provides dependable and trusted resources for the user's Customizers, with third party Customizer servers. Using them together enhances flexibility of the system to benefit performance, satisfy security needs of both the user and the Customizer authors, and provide a measure of fault tolerance.

Second, we have focused on a customization system designed specifically for the Web, allowing us to make a number of simplifying assumptions regarding the programming model, the user model, and the system design and implementation. We use a very restricted and therefore more simplified form of mobile code, rather than providing a generalized mobile code solution that, while more powerful, is less practical and is more complex in terms of usability and security. Other unique features of our system include the use of a CIS that supports dynamic selection of multiple, simultaneously active, Customizers, and the use of CAs running on the CIS to support client-side processing and adaptability.

## 5 Conclusions

We described the design of the Web Stream Customizer architecture that provides the following capabilities:

- Dynamic placement of customization modules that can process and buffer data at intermediate points between a browser client and Web server
- The ability to adapt to network conditions because of its distributed control
- Support for user relocation by allowing efficient movement of customization modules

The architecture is tightly integrated with the existing Web model. Reliance on a user-owned server, the PCMS, for execution of customization modules as well as their storage, increases practicality. The PCMS also plays an important role in providing support for Customizer relocation for execution on third-party servers that may be closer to the user.

As an illustration of the benefits of remote customization, we presented the design of a Smoothing Customizer, which acts as a remote buffer for a limited-memory video player client. When accessing video from a distant server, the performance benefits of using the Smoothing Customizer were dramatic, including a reduction in jitter by two orders of magnitude.

## References

1. E. Amir, S. McCanne, and R. Katz, "An Active Service Framework and its Application to Real-time Multimedia Transcoding," *Proc. SIGCOMM*, pp. 178–189, Aug. 1998.
2. D. Andersen, D. Bansal, D. Curtis, S. Seshan, and H. Balakrishnan, "System Support for Bandwidth Management and Content Adaptation in Internet Applications," *Proc. OSDI*, pp. 213–226, Oct. 2000.
3. A. Barbir, R. Chen, M. Hofmann, H. Orman, and R. Penno, "An Architecture for Open Pluggable Edge Services (OPES)," *Network Working Group Internet Draft.* `http://www.ietf.org/internet-drafts/draft-ietf-opes-architecture-04.txt`.

4. V. Bharghavan and V. Gupta, "A Framework for Application Adaptation in Mobile Computing Environments," *Proc. IEEE Compsac '97*, Aug. 1997.
5. H. Bharadvaj, A. Joshi, and S. Auephanwiriyakul, "An Active Transcoding Proxy to Support Mobile Web Access," *Proc. 17th IEEE Symp. Reliable Distributed Systems*, Oct. 1998.
6. D. Box, D. Ehnebuske, G. Kakivaya, A. Layman, N. Mendelsohn, H. F. Nielsen, S. Thatte, and D. Winer, "Simple Object Access Protocol (SOAP) 1.1," `http://www.w3.org/TR/SOAP/`.
7. C. Brooks, M. S. Mazer, S. Meeks, and J. Miller, "Application-specific Proxy Servers as HTTP Stream Transducers," *Proc. WWW*, pp. 539–548, Dec. 1995.
8. O. Buyukkokten, H. Garcia-Molina, A. Paepcke, and T. Winograd, "Power Browser: Efficient Web Browsing for PDAs," *Proc. CHI*, Apr. 2000.
9. S. Chandra, A. Gehani, C. S. Ellis, and A. Vahdat, "Transcoding Characteristics of Web Images," *Proc. MCN*, Jan. 2001.
10. D. Clark, S. Shenker, and L. Zhang, "Supporting Real-Time Applications in an Integrated Services Packet Network: Architecture and Mechanism," *Proc. SIGCOMM*, pp. 14–26, Aug. 1992.
11. E. Cohen, B. Krishnamurthy, and J. Rexford, "Improving End-to-End Performance of the Web Using Server Volumes and Proxy Filters," *Proc. SIGCOMM*, pp. 241–253, Sept. 1998.
12. S. Deshpande and W. Zeng, "Scalable Streaming of JPEG2000 Images using Hypertext Transfer Protocol," *Proc. ACM Multimedia*, Oct. 2001.
13. A. Fox, S. Gribble, Y. Chawathe and E. A. Brewer, "Adapting to Network and Client Variation Using Active Proxies: Lessons and Perspectives," *IEEE Personal Comm.*, vol. 5, no. 4, Aug. 1998.
14. R. S. Gray, "Agent Tcl: A Transportable Agent Aystem," *Proc. CIKM Workshop on Intelligent Information Agents*, Dec. 1995.
15. J. He, M. A. Hiltunen, M. Rajagopalan, and R. D. Schlichting, "Providing QoS Customization in Distributed Object Systems," *Proc. IFIP/ACM Intl. Conf. Distributed Systems Platforms*, pp. 351–372, Nov. 2001, LNCS 2218
16. A. Hokimoto, T. Nakajima, "An Approach for Constructing Mobile Applications Using Service Proxies," *Proc. ICDCS*, May 1996.
17. Internet Content Adaptation Protocol, www.i-cap.org/home.html.
18. D. Johansen, R. van Renesse, and F. B. Schnieder, "Operating System Support for Mobile Agents," *Proc. Hot Topics in Op. Sys.*, May 1995.
19. B. Krishnamurthy, C. E. Wills, "Improving Web Performance by Client Characterization Driven Server Adaptation," *Proc. WWW.* May 2002.
20. T. Kunz and J. P. Black, "An Architecture for Adaptive Mobile Applications," *Proc. Wireless Comm.*, pp. 27–38, July 1999.
21. J. Li, G. Chen, J. Xu, Y. Wang, H. Zhou, K. Yu, K. T. Ng, and H. Shum, "Bi-level Video: Video Communication at Very Low Bit Rates," *Proc. ACM Multimedia*, Oct. 2001.
22. M. Liljeberg, T. Alanko, M. Kojo, H. Laamanen, and K. Raatikainen, "Optimizing World Wide Web for Weakly-Connected Mobile Workstations: An Indirect Approach," *Proc. SDNE*, pp. 132–139, June 1995.
23. T. S. Loon and V. Bharghavan, "Alleviating the Latency and Bandwidth Problems in WWW Browsing," *Proc. USITS*, Dec. 1997.
24. B. Noble, "System Support for Mobile, Adaptive Applications," *IEEE Personal Computing Systems*, vol. 7, no. 1, pp. 44–49, Feb. 2000.

25. W. T. Ooi, and R. van Renesse, "Distributing Media Transformation Over Multiple Media Gateways," *Proc. ACM Multimedia*, Oct. 2001.
26. H. Peine and T. Stolpmann, "The Architecture of the Ara Platform for Mobile Agents," Rothermel K., Popescu-Zeletin R. (Eds.),*Mobile Agents, Proc. MA '97*, pp. 50–61, Springer Verlag, Apr. 7–8, 1997, LNCS 1219.
27. S. Perret and A. Duda, "Implementation of MAP: A System for Mobile Assistant Programming," *Proc. IEEE Intl. Conf. Parallel and Distributed Systems*, June 1996.
28. H. Rao, Y. Chen, M. Chen, J. Chang, "A Proxy-Based Personal Portal," *Proc. WebNet99*, Oct. 1999.
29. J. Steinberg and J. Pasquale, "A Web Middleware Architecture for Dynamic Customization of Content for Wireless Clients," *Proc. WWW*, May 2002.
30. M. Straßer, J. Baumann, and F. Hohl, "Mole – A Java Based Mobile Agent System," *Proc. ECOOP'96 workshop on Mobile Object Systems*, Linz, Austria, July 1996.
31. A. Vahdat, M. Dahlin, T. Anderson, and A. Aggarwal, "Active Names: Flexible Location and Transport of Wide-Area Resources," *Proc. USITS*, Oct. 1999.
32. Y. Villate, D. Gil, A. Goni, and A. Illarramendi, "Mobile Agents for Providing Mobile Computers with Data Services," *Proc. DSOM*, Oct. 1998.
33. N. Yeadon, A. Mauthe, D. Hutchison, and F. Garcia, "QoS Filters: Addressing the Heterogeneity Gap," *Proc. IDMS '96*, Mar. 1996.
34. Zenel and D. Duchamp, "A General Purpose Proxy Filtering Mechanism Applied to the Mobile Environment," *Proc. 3rd Annual ACM/IEEE Intl. Conf. Mobile Computing and Networking*, pp. 248–259, Sept. 1997.

17

# Cross-Layer Error Recovery Optimization in WiFi Networks

Dzmitry Kliazovich, Nadhir Ben Halima, and Fabrizio Granelli

DIT, University of Trento, Via Sommarive 14, I-38050 Trento, Italy
klezovicdit.unitn.it, nadhirdit.unitn.it, granelli@dit.unitn.it

**Summary.** This paper presents a novel approach for cross-layer error control optimization in WiFi networks. The focus is on the reduction of the overhead deriving from the duplicate ARQ strategies employed at the link and transport layers. The proposed solution, called ARQ proxy, substitutes the transmission of a transport layer acknowledgement with a short request sent at the link layer. Specifically, TCP ACKs are generated based on in-transit traffic analysis and stored at the Access Point. Such TCP ACKs are released towards TCP sender upon a request from the mobile node, encapsulated into link layer acknowledgement frame. TCP ACK identification is computed at the Access Point as well as at the mobile node in a distributed way. ARQ proxy improves TCP throughput in the range of 25–100% depending on the TCP/IP datagram size used by the connection. Additional performance improvement is obtained due to RTT reduction and higher tolerance to wireless link errors.[1,2]

## 1 Introduction

Wireless Local Area Networks (WLAN) represented by IEEE 802.11 standard, often referred to as WiFi, provide mobile access to networks and services – omitting the requirement for a cable (and fixed) infrastructure, thus enabling fast and cost-effective network organization, deployment and maintenance.

As a drawback, the capacity offered by wireless links is relatively low as compared to wired networks. Such capacity limitations derive from the very physical nature of the wireless medium, characterized by limited bandwidth, time-varying behavior, interference, etc.

In particular, Bit Error Rate (BER) on wireless links ranges from $10^{-3}$ to $10^{-1}$ as opposed to $10^{-6}$ to $10^{-8}$ in wired links [1]. This difference of several

[1] This work is supported by the Italian Ministry for University and Research (MIUR) under grant "Wireless 802.16 Multi-antenna Mesh Networks (WOMEN)".

[2] The ARQ Proxy approach is currently patent-pending under EP 07425087.9 "Cross-Layer Error Recovery Optimization for 3G LTE Systems".

orders of magnitude results in poor performance of Transmission Control Protocol (TCP) [2] which accounts for over 95% of Internet traffic [3]. The reason for that is in TCP congestion control mechanism which treats all packet losses as congestion related and halves the outgoing rate for every loss detected.

In order to counteract such variation of error rates, IEEE 802.11 standard employs an Automatic Repeat Request (ARQ) at the link layer. Following a "stop-and-wait" approach, it does not allow the sender proceeding with next frame transmission until positive acknowledgement is received for the previous frame. Lack of positive acknowledgement triggers frame retransmission until a maximum number of retransmissions is exceeded.

However, the link layer is not the only layer which acknowledges packet delivery: TCP reliability is obtained through the utilization of a positive acknowledgement scheme, which specifies TCP receiver to acknowledge data successfully received from the sender. TCP header reserves special fields for enabling it to carry acknowledgement information. As a result, the TCP receiver can produce a TCP acknowledgment (TCP ACK) as standalone packet or, in case of bi-directional data exchange, encapsulate it into outgoing TCP segments.

Considering data transmission over an IEEE 802.11 link using the TCP/IP protocol stack (Fig. 1), whenever a TCP segment is transmitted over the wireless link, the sender first receives an acknowledgement at the link layer. Then, TCP entity at the receiver generates an acknowledgement at the transport layer. This acknowledgement represents ordinary payload data for the link layer, which should be acknowledged by the link layer protocol of the sender node. As a result, a single application data block is acknowledged three times: one at the transport level and two times at the link layer.

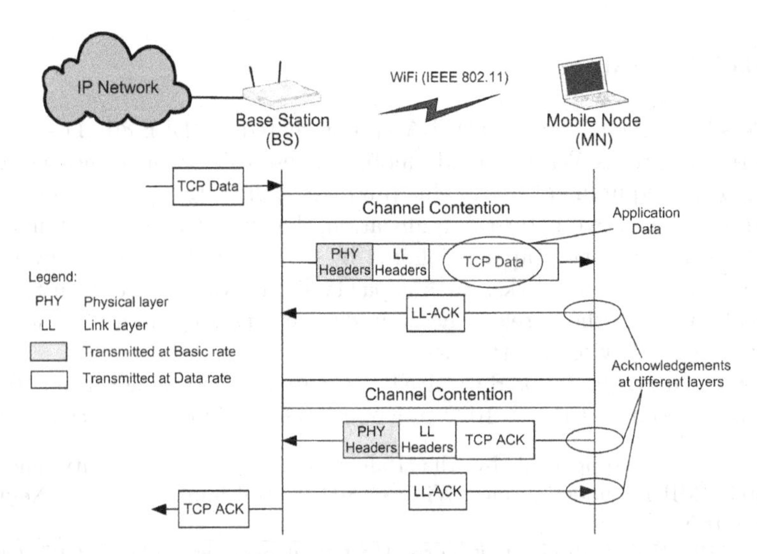


**Fig. 1.** TCP data packet delivery over IEEE 802.11 wireless link

In this paper, we propose a joint optimization of ARQ schemes operating at the transport and link layers using a cross-layer approach called ARQ proxy. The main idea behind ARQ proxy is to substitute the transmission of TCP ACK packet (including the associated physical and link layers overheads) with a small link layer request which is encapsulated into the link layer acknowledgement frame – which does not require any additional bandwidth resources. As a result, ARQ proxy releases network resources associated with TCP ACK transmission over the shared link, thus allowing the corresponding resources to be utilized for a concurrent data transmission originated at any station located within the cell.

The rest of the paper is organized as follows: Sect. 2 provides design and implementation details of ARQ proxy approach focusing on the infrastructure network scenario; Sect. 3 provides ARQ proxy performance evaluation in terms of TCP throughput and delay performance with the respect to TCP/IP datagram size and wireless link error rate; Sect. 4 concludes the paper with summary, conclusions, as well as directions for future work on the topic.

## 2 ARQ Proxy

ARQ proxy design is primarily focused on infrastructure network scenario – the most widely deployed WLAN scenario nowadays [7]. Implementation details in single-hop and multi-hop scenarios are discussed afterwards, as they have significant similarities.

The main idea of the proposed approach is to avoid the transmission of standalone TCP ACK packets over the radio channel on the link between the Base Station (BS) and Mobile Node (MN). In order to support this functionality, no changes are needed to the TCP protocol, but new software entities need to be introduced: the ARQ proxy and ARQ client (see Fig. 2).

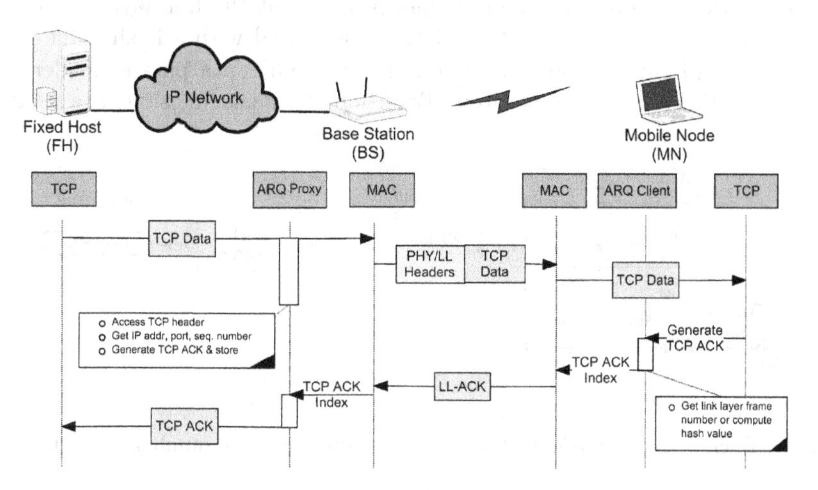


**Fig. 2.** ARQ proxy and ARQ client functionality

*ARQ proxy* is a software module located in the protocol stack of the wireless Base Station (BS) or Access Point (AP). Having access to TCP and IP headers of the in-transit traffic, ARQ proxy generates TCP ACK for every TCP data packet destined to MN which confirms successful data reception up to the flow segment carried in this TCP data packet. ARQ proxy does not require any flow-related state information or TCP layer implementation in a conventional sense. Indeed, TCP ACK is generated using a simple memory copy operation applied to the fields (IP addresses, port numbers, and flow sequence numbers) of the received TCP data packet into a previously generated template of TCP ACK.

The fact that no TCP flow state-related information is used in TCP ACK generation process implies the assumption that all the segments of a given TCP flow are successfully received at the destination node. Since this assumption is not always true, TCP ACKs generated by ARQ proxy module are not released to the Fixed Host (FH) immediately, but stored in BS memory until requested by the ARQ client.

*Packet Identification.* TCP ACKs generated by ARQ proxy should be easily identifiable by ARQ client without direct communication between these parties. There are two alternative approaches that satisfy this property: frame sequence numbers and hash values (see Fig. 3).

The IEEE 802.11 standard specifies that every sender needs to mark outgoing frames with continuously incremented, 12-bit long sequence numbers at the link layer. The reader should note that in case of TCP/IP datagram fragmentation at the link layer frame sequence number remains the same for all the fragments. As a result, ARQ client located at the MN can indirectly identify TCP ACK generated by ARQ proxy, by referring to the frame sequence number added by the BS at the link layer to the TCP data packet used in TCP ACK generation.

An alternate approach for packet identification that can be used in wireless network with no sequence numbers provided at the link layer is the use of hash values. In this way, TCP ACK is associated with a hash value computed by applying a proper hash function to TCP data packet headers for which the TCP ACK is generated. Traditionally, hash functions are used in

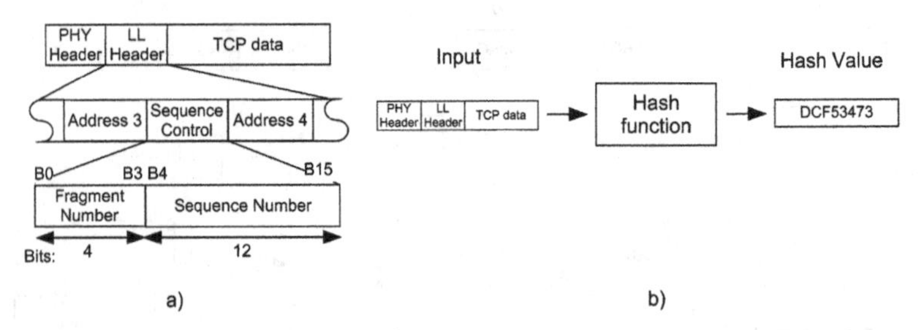


**Fig. 3.** Packet identification techniques: (**a**) frame sequence numbers and (**b**) hash values

cryptography, data storage and search applications. In networking, the use of hash functions is mostly limited to integrity check, error detection and error correction techniques – commonly performed using Cyclic Redundancy Check (CRC) or MD5 algorithms.

In this paper we limit our choice to frame sequence numbers due to simplicity, while for further details on hash functions the reader is directed to [4].

*ARQ Client* is a software module which logic position is between the link and transport layers of the MN protocol stack. It suppresses all outgoing standalone TCP ACK packets and replaces them with MAC layer requests for the appropriate TCP ACK transmission initiated at ARQ proxy.

In order to do so, whenever a standalone TCP ACK is produced at MN transport layer, a TCP ACK suppression request is scheduled for the transmission at the link layer immediately, while the original TCP ACK packet travels down the protocol stack which involves corresponding processing at each layer, output queuing delay, shared medium access and other procedures.

Whichever comes first to the physical layer (the TCP ACK or the corresponding suppression request) will be transmitted, while the other one cancelled.

TCP ACK suppression request includes identification associated with TCP ACK generated by ARQ proxy. This identification depends on the chosen packet identification technique: a frame sequence number or a hash value. At the link layer, TCP ACK identification is inserted into the next outgoing link layer acknowledgement (LL-ACK) frame. In particular, it is inserted into the reserved portion of the "duration" field of LL-ACK frame (see Fig. 4), which does not require modification of the frame structure specified by IEEE 802.11 standard.

The use of the reserved portion of LL-ACK frame favors incremental deployment of the proposed technique enabling operation in the mixed network environment where nodes which implement ARQ proxy co-exist with those not implementing the proposed approach.

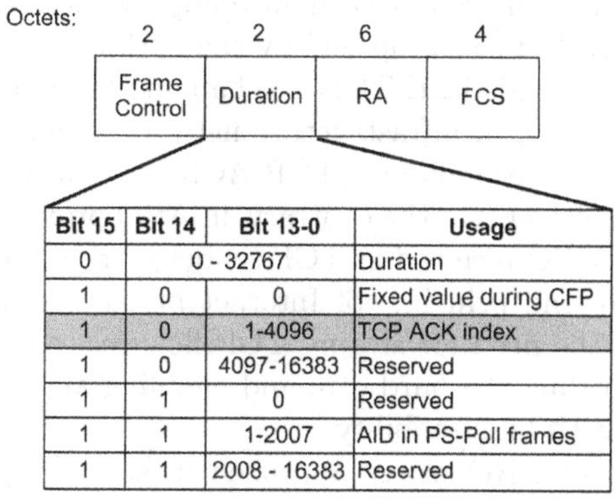

| Octets: | 2 | 2 | 6 | 4 |
|---|---|---|---|---|
| | Frame Control | Duration | RA | FCS |

| Bit 15 | Bit 14 | Bit 13-0 | Usage |
|---|---|---|---|
| 0 | 0 - 32767 | | Duration |
| 1 | 0 | 0 | Fixed value during CFP |
| 1 | 0 | 1-4096 | TCP ACK index |
| 1 | 0 | 4097-16383 | Reserved |
| 1 | 1 | 0 | Reserved |
| 1 | 1 | 1-2007 | AID in PS-Poll frames |
| 1 | 1 | 2008 - 16383 | Reserved |

**Fig. 4.** IEEE 802.11 ACK frame with extension capable to carry TCP ACK index

The lack of TCP flow related information at the BS allows ARQ proxy to generate TCP ACK which acknowledges only in-sequence segment delivery. For that reason, in order to maintain TCP error recovery procedure, ARQ client does not request TCP ACK generated at ARQ proxy in the following cases:

- During TCP connection establishment and connection termination phases, which are explicitly marked by SIN and FIN flags in the packet headers. These packets carry initial sequence numbers, maximum window sizes, and other parameters required by connection setup, and cannot be substituted.
- TCP ACK encapsulated into outgoing TCP data packet. In case of bidirectional data transfer and delayed-ACK option enabled, TCP receiver delays TCP ACKs assuming to have outgoing data segment in order to encapsulate TCP ACK using ACK bit and ACK sequence number fields into the packet header. Consequently, with such encapsulation, TCP ACKs do not create any additional overhead, and thus are not a subject for ARQ proxy optimization.
- Duplicate TCP ACKs. Upon out-of-order segment reception, TCP receiver must generate duplicate ACK for the last successfully received in-sequence segment. Due to the lack of TCP state related information at the BS, duplicate ACKs can not be generated by ARQ proxy and should be transmitted by MN.
- TCP ACK advertising exhausted buffer resources at the receiver (reported in *rwnd* field of TCP header). This ensures the receiver is not running out of the buffer space in case not being able to process traffic at the incoming link rate.

TCP ACKs generated at the BS are associated with a lifetime timer at the moment of generation. Upon expiration of this timer, which recommended value should be equal to or greater than TCP timeout, TCP ACK is silently dropped from the buffer. This lifetime technique is designed to clean up resources from TCP ACKs not requested by ARQ client module.

By the level of the achieved performance improvement the proposed technique is similar to LLE-TCP, proposed by the authors in [6]. However, ARQ proxy conceptually extends LLE-TCP by changing the point triggering TCP ACK generation. In fact, in infrastructure network scenario, LLE-TCP base station is completely responsible for TCP ACK generation with no feedback available from the receiver. On the contrary, in ARQ proxy approach, the generation of all TCP ACKs received at TCP sender is triggered by the receiver – following the end-to-end principle of Internet protocol design. Additionally, ARQ proxy avoids the need for storing TCP flow related information at the base station – unloading the hardware and enabling application of the technique in scenarios with high mobility.

Implementation of ARQ proxy approach in single hop ad hoc network scenario is the same as in infrastructure network scenario presented above. The only difference is that ARQ proxy module is located at the mobile sender node and not at the BS, and TCP ACKs it produces are not routed through the network but immediately directed to the transport layer following ARQ client request sent by wireless receiver at the link layer.

In a multi-hop ad hoc network scenario, ARQ proxy technique can be applied at the last hop of multi-hop connection.

## 3 Performance Evaluation

In order to analyze the performance of the proposed scheme, the corresponding modules of the ns-2 network simulator (version 2.31) [5] are added supporting ARQ proxy and ARQ client functionality. ARQ proxy module is attached to the BS, while ARQ client is located in MN protocol stack. The configuration of the wireless link between BS and MN follows IEEE 802.11b specification parameters with 11 $\mathrm{Mb\,s^{-1}}$ physical data rate. Parameters of the wired link (100 $\mathrm{Mb\,s^{-1}}$ 15 ms) model the situation when a mobile user is connecting to an Internet server physically located within the same metropolitan area. The BS ingress buffer is limited to 700 packets, and RTC/CTS exchange is turned off at the MAC layer as the most appropriate configuration widely used in infrastructure network scenario. Obtained results are averaged over 10 runs with different seeds used for random generator initialization.

TCP NewReno is chosen for performance evaluation as the most widespread TCP version in Internet nowadays. However, it is important to underline that ARQ proxy approach is not constrained to any specific TCP implementation.

Connection throughput and Round Trip Time (RTT) are chosen as main performance metrics of TCP flow evaluated against variable TCP/IP datagram size as well as Packet Error Rate (PER) on the wireless link.

Figure 5 shows the throughput level achieved by TCP NewReno for different TCP/IP datagram sizes. The throughput and level of performance

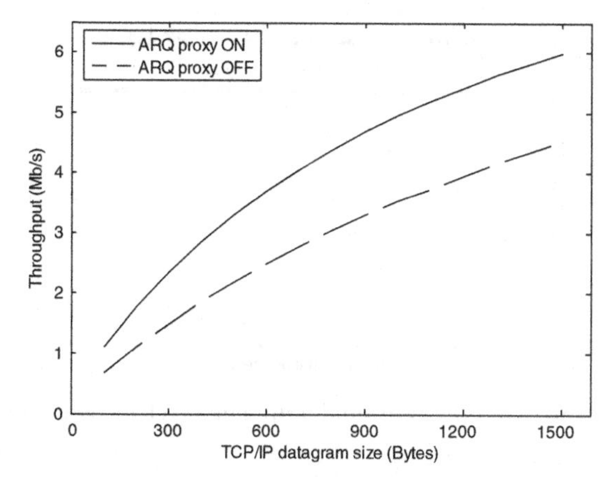


**Fig. 5.** ARQ proxy TCP throughput comparison

improvement of ARQ proxy approach is reversely proportional to the size of TCP data packet. Indeed, the smaller the packet the larger the resources released from TCP ACK substitution. For the maximum considered TCP/IP datagram size of 1,500 bytes (which corresponds to the Ethernet MTU), ARQ proxy performance improvement is only 25–30%. For small packets (40–200 bytes), it is in the range of 60–70%. However, the general rule is that for TCP data packets which tend to be similar in size to TCP ACK packets the throughput improvement can reach 100%.

Along with throughput performance improvement, ARQ proxy reduces RTT of TCP connection. TCP ACKs generated at the BS by ARQ proxy agent avoid transmission, propagation and queuing delays experienced at the wireless link. As it can be observed in Fig. 6, this delay is typically in the order of several milliseconds for IEEE 802.11b.

RTT reduction leads to TCP flow performance increase due to faster window evolution and faster reaction to packet drops performed by Additive Increase Multiplicative Decrease (AIMD) flow control mechanism [8].

Figure 7 illustrates TCP throughput with variable link error rate and TCP/IP datagram size equal to 1,500 bytes. While the throughput level is linearly decreasing, ARQ proxy performance improvement remains constant and corresponds to around 30% for PERs of up to 0.25. Additionally, by enabling ARQ proxy, TCP NewReno is able to sustain higher PERs (see Fig. 7 for PER > 0.25). This is motivated by the fact that in such scenario no wireless link errors propagate into TCP ACKs generated at the base station.

Summarizing, performance evaluation results validate the proposed approach and confirm ARQ proxy design initiatives. In details, ARQ Proxy provides:

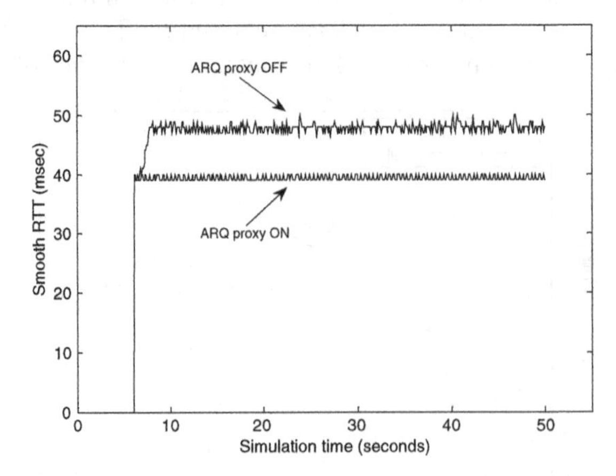


**Fig. 6.** ARQ proxy Round Trip Time (RTT) reduction

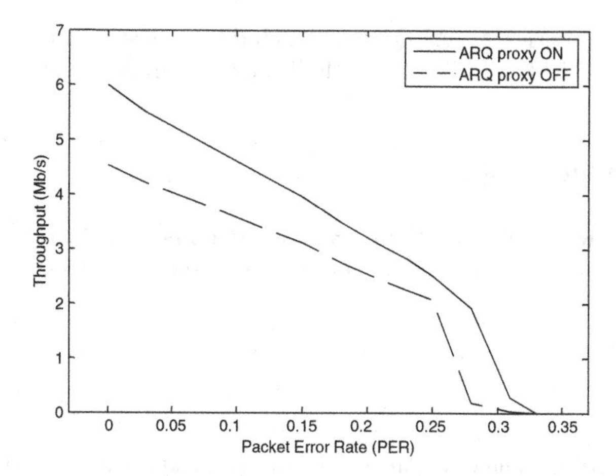


**Fig. 7.** TCP throughput against wireless link errors

- TCP throughput improvement from 25% to 100%, depending on TCP/IP datagram size
- RTT reduction for TCP ACK delivery over the wireless link (typically several milliseconds for IEEE 802.11b standard)
- Higher tolerance to link errors

# 4 Conclusions

The paper introduced a novel approach aiming at TCP performance improvement in WLAN networks. The core of the work lies in the substitution of the transmission of TCP ACK packets with a short link layer request sent over the radio link. Specifically, TCP ACKs are generated by an ARQ proxy located at the base station based on in-transit traffic analysis and stored in the buffer until requested by the ARQ client (located at the mobile node). All TCP ACK transmissions are triggered by the mobile end-node, which maintain end-to-end TCP semantics.

Two packet identification methods are considered for the proposed scheme: using frame sequence numbers available from the link layer (link layer dependant) and using hash values (link layer independent). In the latter case, hash values can be obtained by applying a predefined hash function onto the raw packet headers' data.

No TCP flow related information is stored at the BS, enabling applicability of the technique in a scenario with high mobility as well as incremental deployment in already operational networks.

Performance evaluation results demonstrate that ARQ proxy brings TCP throughput improvement in the range of 25–100% (depending on payload size), reduction of RTT of the connection for several milliseconds, as well as higher tolerance to errors on the wireless link.

Ongoing activities on ARQ proxy include application of the presented technique in 3G LTE and WiMAX (IEEE 802.16) environments.

## Acknowledgement

The authors would like to thank Simone Redana and Nicola Riato from Siemens (Italy) for their contribution and valuable comments on the ARQ proxy approach.

## References

1. W.C.Y. Lee, "Mobile Communications Design Fundamentals," 2nd Ed., Wiley, 1993.
2. J.B. Postel, "Transmission Control Protocol," RFC 793, September 1981.
3. C. Fraleigh, S. Moon, B. Lyles, C. Cotton, M. Khan, D. Moll, R. Rockell, T. Seely and S.C. Diot, "Packet-level traffic measurements from the Sprint IP backbone," IEEE Network, vol. 17(6), pp. 6–16, Nov.-Dec. 2003.
4. A. Partow, "General Purpose Hash Function Algorithms," http://www.partow.net/programming/hashfunctions/
5. NS-2 simulator tool home page, http://www.isi.edu/nsnam/ns/, 2000.
6. D. Kliazovich and F. Granelli, "A Cross-layer scheme for the TCP Performance Improvement in Wireless LANs," IEEE Global Communications Conference, GLOBECOM'04, Dallas, U.S.A., December 2004.
7. "Fixed Wireless, WiMax, and WiFi Market Opportunities, Strategies, and Forecasts, 2005 to 2010," WinterGreen Research, Inc., May 2004.
8. F. Kelly, "Mathematical Modelling of the Internet," B. Engquist and W. Schmid (eds.), "Mathematics Unlimited – 2001 and Beyond", Springer, Berlin Heidelberg Newyork, pp. 685–702, 2001.

# Session 4

## Qos and Efficiency in Multimedia Heterogeneous Wireless Networks

18

# Technology-Independent Service Access Point for QoS Interworking

Mario Marchese[1], Maurizio Mongelli[1], Vincenzo Gesmundo[2], and Annamaria Raviola[2]

[1] DIST, Department of Communication Computer and System Science, University of Genoa, Via Opera Pia 13, 16145 Genoa, Italy
Mario.Marchese@unige.it, Maurizio.Mongelli@unige.it

[2] Selex Communications S.p.A., via Pieragostini 80, 16151 Genoa, Italy
Vincenzo.Gesmundo@selex-comms.com, Annamaria.Raviola@selex-comms.com

**Summary.** The paper deals with protocol architectures to support mapping of *Quality of Service* (QoS) between protocol layers of telecommunication networks. A *Technology-Independent Service Access Point* (TI-SAP) is studied to this aim. Inherent TI-SAP mapping operations introduce the generalization of the regular concept of *equivalent bandwidth* (EqB). Some performance evaluation is proposed to highlight EqB dimensioning at the TI-SAP interface.

## 1 Introduction

Protocol stacks in telecommunication networks are composed of functional layers. *Quality of Service* (QoS) provision depends on the performance achieved at each layer and is based on functions performed at layer interfaces. Having in mind the OSI paradigm, QoS derives from reliable physical and link layers that can offer specific transport services to the upper network layers. The flows generated by the network layers (or bundles of them) are forwarded down to a physical interface that transports the information along a channel.

Even if network layer implements efficient QoS mechanisms (IP IntServ, IP DiffServ, MPLS), it is topical that layers below can assure connection to the channel with specific degrees of performance. Otherwise the implementation of complex QoS mechanism is useless. As a consequence, the QoS requirements must flow vertically and need to be received and satisfied by all the layers of the protocol stack. More specifically, the link layer must implement appropriate mechanisms to support the *Service Level Agreement* (SLA) defined at the network layer. In some cases, in particular in wireless environments, the link layer acts in cooperation with the physical layer through the application of specific cross-layer design solutions. The interaction between the layers in this context is called here "QoS Mapping". It leads to some technological problems that nowadays constitute open areas of standardization and research.

The paper proposes an insight into QoS Mapping issues. A detailed analysis is reported for the protocol architectures necessary to support QoS Mapping. A *Technology-Independent Service Access Point* (TI-SAP) is introduced to this aim.

The remainder of the paper is organized as follows. The next two sections introduces the technological elements of the TI-SAP architecture at both user and control plane levels. Section 4 specifies the abstraction methodology used in [1] to hide the local implementation of the QoS within a given layer 2 technology. The concept of QoS mapping is introduced in Sect. 5. Performance analysis is proposed in Sect. 6 to highlight how TI-SAP bandwidth dimensioning is a hard task in the presence of QoS mapping operations. Conclusions are summarized in Sect. 7 together with possible directions of future research.

# 2 The Concept of Technology-Independent Service Access Point

The protocol stack considered here is partially derived from analysis of the ETSI BSM (*Broadband Satellite Multimedia*) architecture [1]. It is related to satellite communication, but it may have a wider application. Generalizing the approach of [1], dedicated to satellite access points, the protocol stack studied in the following separates the layers between *Technology Dependent* (TD) and *Technology Independent* (TI). The interface between TI and TD is defined as *Technology-Independent Service Access Points* (TI-SAPs). QoS requirements must flow through TI-SAPs and be implemented at TD layers. An appropriate set of primitives (called *Technology Independent Adaptation Functions*, TIAF, and *Technology Dependent Adaptation Functions*, TDAF) may be defined (following the guidelines of [1]) to support resource reservation invocation at the different levels of the protocol stack. It allows decoupling responsibilities of resource control for each independent component of the system. A description of the approach is reported in the following.

Some issues are topical when traffic is forwarded from TI to TD. On one hand, there is the need that TD layers provide a service to the TI layers, but, on the other hand, it should be done with the minimum information TD–TI bi-directional exchange. Ideally, the exchange should be limited to the performance requirements and its matching or not to simplify the structure of TIAF and TDAF primitives. The problem is therefore connected to automatic bandwidth adaptation. TD layer needs to compute the exact bandwidth to be assigned at TD buffer (i.e., the service rate) so that the performance requirements fixed by TI layers can be satisfied. In the next three sections, the key features of the architecture are detailed.

# 3 TI-SAP Protocol Architecture

## 3.1 User Plane

The case of a portion of a network traffic conveyed along a specific TD technology (for example, a specific MAC layer such as ATM, DVB or WiMAX MAC) is considered. The external network traffic is seen at the TI-SAP as "*technology independent*" (TI). The QoS paradigm on the TI portion of the TI-SAP stack is supposed here to be IP DiffServ. The presence of IntServ is implicitly considered since DiffServ flows may be the result of some IntServ over DiffServ aggregations performed before entering the TI-SAP. The application of IntServ directly at the TI-SAP can be taken into account, too. DiffServ classes are classified using DSCP field at TI level. DiffServ management at TI-SAP means that DSCP are mapped to TD queues. Since TD classes are system dependent, there is the need to abstract from the lower layers and to provide TI layer with a common and agnostic interface. This concept leads to the definition of *Queue Identifiers* (QIDs), firstly introduced in [1].

QID represents an abstract queue at the SAP level. Each QID is formally a relationship between TI (IP) queues and TD queues. Each QID defines a class of service for transfer of IP packets into the TD core. The application of QID principle allows hiding to TI layers the local implementation of the QoS within the TD system. A QID is "virtual" in that it represents how a specific subset of IP queues are mapped over a specific subset of TD queues. For example, all *Assured Forwarding* (AF) queues may be aggregated together in a single TD queue or traffic shaping may be applied to IP flows before being forwarded to the TD queues. The application of shaping or other policing depends on the SLA between the two providers, the first one related to the TI portion, the second one, managing the TD core. On the other hand, the QID is "real" because it characterizes the real properties of the TD queues, for example, in terms of available TD (buffer and bandwidth) resources.

Figure 1 depicts the relationship between traditional DiffServ IP queue management, which is performed above the TI-SAP, and the lower layers queues (TD queues and QIDs). The relationship is shown for user plane on the left side of Fig. 1 and for control plane on the other side. IP-to-QID mapping means the definition of which outgoing flow from IP queues is forwarded down to a specific subset of QID queues. The same definition applies to QID-to-TD mapping.

## 3.2 Control Plane

The control plane consists of one module above the TI-SAP (the IP resource manager), one module below (the TD resource manager), and one interfacing module (the *QID Resource Manager*, QRM), which, through the set of TIAF and TDAF primitives:

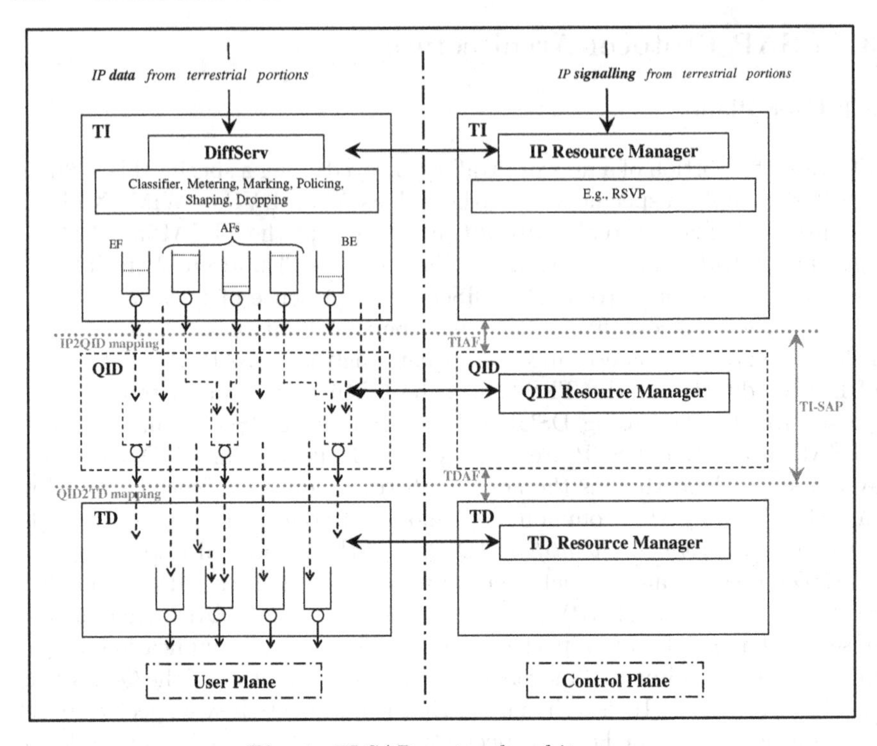


**Fig. 1.** TI-SAP protocol architecture

(1) Receives requests from the IP module to allocate, to release or to modify resources
(2) Translates these requests into QID allocation/release/modification actions
(3) Checks with the TD module how the upper layer requests can be mapped to real resources

The control plane may be statically or dynamically configured.

### Static Configuration

No specification to dynamically reserve resources or to receive indications of network resource availability may be implemented on the terrestrial portion of the network. In this case, user traffic entering the TI-SAP is distinguished through DSCP and served through static trunks. The term *static* denotes the manual management of network resources over large time scales. For instance, the pre-allocation of a bandwidth trunk for a specific traffic class. Neither traffic prediction nor real time reaction to congestion, including *Call Admission Control* (CAC), is provided. In this case, also the TD resources may be statically configured. The QIDs present to higher layers static resource reservations. Obviously, the TI static trunks need to be mapped on the private TD system in dependence of the specific IP-to-QID and QID-to-TD mappings implemented.

### Dynamic Configuration

Another approach consists of designing a dynamic control plane. It means triggering resource reservation on the basis of time varying TI traffic and TD channel conditions. Thus, the control plane is based on some signalling scheme. RSVP may be used on the TI side. Usually, RSVP is applied in conjunction with IntServ. In case aggregation operations like IntServ over DiffServ are applied directly at the TI-SAP, RSVP may be explicitly interfaced with DiffServ. So, TI-SAP can react to RSVP requests by allocating resources based on aggregated TI flows using regular RSVP agents as in common IntServ routers. The IP resource manager in the ST is therefore responsible for:

- Translating RSVP requests into QID allocation requests
- Forwarding the QID allocation requests to the QRM
- Receiving the reply from the QRM and configure the IP queuing module in the user-plane
- Sending appropriate RSVP signalling

In turn, the QRM, which received a QID allocation request, performs the following operations:

- Check for available TD resources to accommodate the request
- If resources are available, make the required changes in the IP-to-QID or QID-to-TD mappings, and notify the IP resource manager
- If TD resources are not available, request allocation of new resources to the TD resource manager, wait for response, make the required changes in the IP-to-QID or QID-to-TD mappings, and notify the IP manager

It is worth noting that comparable operations may be performed if other signalling schemes are applied outside the TI-SAP. Possible examples may be: RSVP-TE, if the TI-SAP interfaces MPLS networks, or extensions made on RSVP to support CAC information in DiffServ environments (called *RSVP – Pre Congestion Notification*) or priority information of the calls (*Emergency RSVP*). The *Common Open Policy Service* (COPS) may be also used as interfacing protocol to define policy-based SLAs. Future evolutions of the *Next Steps In Signalling* (NSIS) IETF working group may be of interest for the TI-SAP standardization, too.

## 4 QID Management

Besides the possible typologies of dynamic service invocations, the topical point is how the invocations are mapped over the TI-SAP control plane. The key mapping operation is QID management. QID management is the function that identifies the TD queues in play, the TD resource reservation, modifies the properties of the abstract queue that is associated with those queues, and makes the association with TI queues. In brief, QIDs interface the IP queues to a set of available TD queues. TD layers are responsible for assigning capacity to the TD queues, and thus to the abstract queues. These operations occur below the TI-SAP, so they are hidden to the TI layers. Also the TD capacity control may be static or dynamic.

### 4.1 Static Resource Allocation

According to the SLA and a pre-configured set of DiffServ queues, a consistent IP-to-QID mapping is created and statically maintained. No signalling is required across the TI-SAP. The QID-to-TD also has to be static. The way in which QIDs are (statically) associated to TD queues is left to the implementers and to the specifics of each TD technology.

### 4.2 Dynamic Resource Allocation

QID modification is supported by two possible cases:

#### IP Manager Action

Requests about allocating or releasing resources may come from TI layers through explicit signalling or by monitoring the IP queue occupancy. The requests may introduce changes of the IP-to-QID mapping or of the QID QoS characteristics (called QIDSPEC). The request is forwarded down to the QRM, which, if accepted, adapts the QIDs to the new needs of the TI layers and, in turn, sends feedback to the TD manager if needed. The set of DiffServ queues remains unchanged.

### 4.3 QID Resource Manager Action

Also without explicit messages coming from TI layer, QIDs can adapt to the time varying incoming IP traffic by means of feed-forward information received on TI queues state.

Also the QID-to-TD mapping can be either static or dynamic. It is static in the case the number of QIDs remains fixed, but other QID characteristics may change, namely, either the IP-to-QID mapping or the QIDSPEC. In any case, the TD resource manager must allocate/deallocate reservations accordingly. The QID-to-TD mapping or the related reservations are dynamic in the case the number of QIDs and/or of TD queues changes. For example, in the presence of time varying channel degradation conditions. Note that QID-to-TD mapping does not need to be signalled across the TI-SAP because QIDs hides TD state to the upper layers.

## 5 QoS Mapping

Dynamic QID management introduces a complex resource control problem for the TD manager. On the basis of the problem solution, the TD manager may update the TD queues reservation state. It may also trigger feedback up to the QRM. For example, in case the TD manager is not able to support the desired level of performance. In turn, the QRM may perform its specific reallocation decisions and/or forward feedback up to the IP manager.

Considering the bandwidth dimensioning viewpoint, QID abstraction hides the following "mapping" operations:

### 5.1 Change of Information Unit

It is the consequence of IP traffic transport over a TD portion that implements a specific technology, e.g., ATM as often done in industrial systems, or DVB. The TD overhead has an impact on bandwidth dimensioning (see [2] and references therein). Moreover, the adoption of specific fragmentation and packing procedures (e.g., [3] p. 125), implemented below the TI level, make the bandwidth dimensioning problem analytically intractable.

### 5.2 Heterogeneous Traffic Aggregation

The due association of IP QoS classes to TD transfer capabilities is limited by hardware implementation constraints. As outlined in [4], "*it is accepted in the BSM industry that at the IP level between 4 and 16 queues are manageable for different IP classes. These classes can further be mapped below into the satellite dependent priorities which can be from 2 to 4 generally*". The problem is how much bandwidth must be assigned to each TD queue so that the TI IP-based SLA is guaranteed.

### 5.3 Fading Counteraction

Finally yet importantly, many transmission environments, as well as satellite or wireless links, need to tackle time varying channel conditions due to fade.

## 6 Performance Evaluation

In this perspective, optimal TD bandwidth dimensioning faces the mentioned technological problems (change of information unit, heterogeneous traffic aggregation, fading counteraction). A possible solution was firstly introduced in [2]. The algorithm of [2] is used here to give an example of bandwidth shift arising at the TI-SAP interface due to the mapping operations reported above.

The presence of two traffic buffers at TI layer is considered without fading degradation at TD physical level. The first buffer offers a VoIP service. Each VoIP source is modeled as an exponentially modulated on–off process, with mean on and off times (as for the ITU P.59 recommendation) equal to 1.008 s and 1.587 s, respectively. All VoIP connections are modeled as 16.0 kbps flows voice over RTP/UDP/IP. The IP packet size is 80 bytes. The required end-to-end performance objective of a VoIP flow for ITU P.59 is end-to-end loss below 2%. The SLA here is $PLP^*_{VoIP} = 0.02$. The second one is dedicated to video service. Different video traces, taken from the web site referenced in [5], are used. Video data are H.263 encoded and have an average bit rate of 64 kbps as well as a peak bit rate of 340 kbps. TI rate allocation (240 kbps) for video (fixed off line through simulations analysis) assures $PLP^*_{video} = 10^{-3}$, which composes the video SLA (TI video buffer size is 75,000 bytes). Both the outputs of the TI buffers are conveyed towards a single queue at the TD layer. DVB encapsulation (header 4 bytes, payload 184 bytes) of the IP packets through the LLC/SNAP (overhead 8 bytes) is implemented in this case. TD buffer is of 500 DVB cells.

Figure 2 depicts the increase in percentage (with respect to the TI layer) of the bandwidth provision at the TD layer necessary to guarantee the SLAs outlined. Different video traces are applied (again taken from [5]). Video data are H.263 encoded and have an average bit rate of 260 kbps as well as a peak bit rate ranging from 1.3 to 1.5 Mbps, depending on the specific trace. The TI rate allocation for video is around 350 kbps (depending on the video trace in play) to assure the mentioned QoS thresholds.

The exact TD rate allocation is obtained through the algorithm in [2] and guarantees the maintenance of the required SLA for VoIP and video services

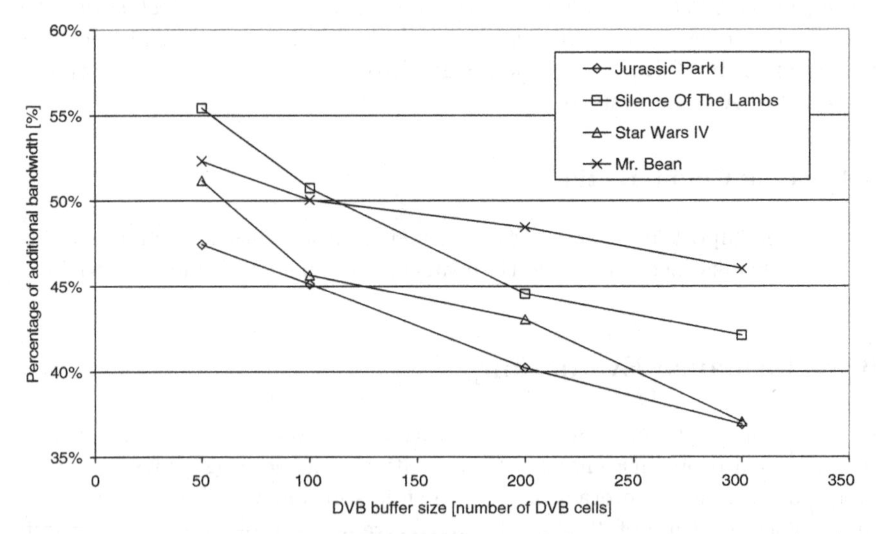


**Fig. 2.** Bandwidth increase at TD layer measured by algorithm in [2]

within the DVB portion, by tuning the bandwidth properly. As mentioned in first part of the paper, the QoS guaranteed over network portions external to the TD core should be guaranteed at TD level. For example, the SLA between providers may consist of the guarantee of the assurance of the same levels of QoS achieved in each portions, despite technologies changes (in this case from IP to DVB). Actually, the QoS support must be implemented in each portion as if technology changes were transparent to end users. In this perspective, RCBC reveals to be a suitable instrument for network planning, too. From the operative viewpoint, if a TD static trunk must be provisioned for the aggregation case presented, the bandwidth provision of the traffic entering the TD queue must be increased up to 50% in order to assure the TI QoS. It is worth noting that the 50% of bandwidth increase holds true in the worst case, namely, with 50 DVB cells in the TD queue. For other buffer allocations, the increase is lower and depends on the specific trace in play. The procedure may help the network operator to measure resource reservation sensitivity as a function of video statistical behaviour. The results may be used to deploy a static trunk serving the aggregation of VoIP and video flows. Overprovisioning may be applied by considering the worst statistical behaviour, thus simplifying resource reservation over a large time scale.

## 7 Conclusions and Future Work

The paper has presented the key elements of a TI-SAP protocol architecture used to map the QoS between layers of a communication protocol stack. Some results are proposed to give an example of bandwidth dimensioning at the TI-SAP. Details concerning the protocol architecture and the application of the control algorithm of [2] in more complicated scenarios can be found in [6].

Directions for future research may rely on (1) study of the control scheme in [2] applied to elastic traffic (e.g., TCP/IP) encapsulated in DVB frame, (2) application of the QID queues abstraction principle for cross-layer design in specific wireless environments (such as WiMAX).

## References

1. ETSI. Satellite Earth Stations and Systems. Broadband Satellite Multimedia Services and Architectures: QoS Functional Architecture. Technical Specification, Draft ETSI TS 102 462 V0.4.2, Jan. 2006.
2. M. Marchese, M. Mongelli, "On-line Bandwidth Control for Quality of Service Mapping over Satellite Independent Service Access Points," *Computer Networks*, vol. 50, no. 12, Aug. 2006, pp. 1885–2126.
3. IEEE. Standard for Local and Metropolitan Area Networks – Part 16: Air Interface for Fixed Broadband Wireless Access Systems, Revision of IEEE Sts 802.16-2001, 1 Oct. 2004.

4. ETSI. Satellite Earth Stations and Systems (SES). Broadband Satellite Multimedia. Services and Architectures; BSM Traffic Classes. ETSI Technical Specification, TS 102 295 V1.1.1, Feb. 2004.
5. `http://www-tkn.ee.tu-berlin.de/research/trace/trace.html`.
6. M. Marchese, *QoS over Heterogeneous Networks*, John Wiley & Sons Inc., Hoboken, NJ, Apr. 2007.

## 19

# A Rate-Controlled VoIP System Based on Wireless Mesh Network Infrastructure: Design Issues and Performance Analysis

Francesco Licandro, Carla Panarello, and Giovanni Schembra

DIIT, University of Catania, Catania, Italy
flicandro@diit.unict.it, cpana@diit.unict.it, schembra@diit.unict.it

**Summary.** Wireless Mesh Networks (WMNs) are increasingly proposed as an economic and effective solution to easily delivering a large variety of services anytime anywhere. However, wireless communications suffer from some shortcomings which limit the use of real-time applications, especially VoIP applications. In fact, the effective available bandwidth can be, sometimes, too low to accept new calls, in addition to the ongoing ones, without the call quality becoming unacceptable. Moreover large delay variations and packet loss dramatically affect speech quality and speakers interactivity. This paper refers to a wireless mesh network scenario used to provide VoIP services to a group of buildings.

Application of Automatic Repeat-reQuest (ARQ) to reduce losses due to the wireless link however causes high variability on the output transmission rate of the buffer in the multiplexer located in each building. So, to avoid buffer congestion, in this paper we propose an architecture where VoIP sources use Adaptive MultiRate encoder (AMR), in such a way that they can be controlled according to the state of the buffer queue in the multiplexer. To this end a new rate controller scheme is defined, and its performance is evaluated in order to achieve some hint to design its parameters.

## 1 Introduction

In the last few years, Wireless Mesh Networks (WMNs) are increasingly proposed as an economic and effective solution to broadband networking [1]. Through gateway and bridging functions of mesh routers, WMNs provide network access for both mesh and conventional clients and connectivity to other networks such as the Internet, WiMAX, Wi-Fi, sensor and cellular networks. Wireless personal, local and metropolitan area networks can be quickly and economically deployed even in underdeveloped regions or where the cost and the complexity of deploying wired networks are prohibitive. In fact, unlike standard wired networks, WMNs are easy and less expensive to deploy by simply placing wireless mesh routers in desired locations. Devices such as laptops, desktops, phones, PDAs, equipped with wireless network interface cards can connect directly to wireless mesh routers. Otherwise, users without wireless network interface cards can access to the WMNs by connecting to mesh routers through wired connections.

In this paper we will refer to the VoIP system shown in Fig. 1. It is constituted by several VoIP sources located into the same building, and connected to the Voice Multiplexer device through high-speed links. The Voice Multiplexer Device is equipped with a WiFi card to access an inter-building WMN. Through this network VoIP sources are connected to the Voice Call Manager, constituting the interface through the telephone network and the Internet. Let us notice that the Voice Call Manager can reside in the Gateway of the Wireless Mesh Network, or alternatively anywhere on the Internet. In such a way, a large variety of services can be delivered anytime anywhere.

Although WMNs can be adopted today as a valid solution for delivering network services, many research challenges and open issues remain in all protocol layers. Wireless communications, for example, suffer from some shortcomings which limit the use of real-time applications, especially VoIP applications. In fact, the effective available bandwidth can be, sometimes, too low to accept new calls, in addition to the ongoing ones, without the call quality becoming unacceptable. Moreover large delay variations and packet loss dramatically affect speech quality and speakers interactivity. More specifically, in order to reduce the errors introduced by the wireless channel, Automatic Repeat reQuest (ARQ) channel technique is often applied [3] to the transmitted flow. According to this technique, each packet is retransmitted several times until either the packet is received from the next node or a maximum number of retransmission is reached. Clearly, this mechanism improves the reliability of the wireless link but it introduces a variable delay before transmitting the packet. This makes the queue length in the transmission buffer grow faster when the number of retransmissions per each packet is high. As soon as the buffer overflows, packet losses occur and the call quality may

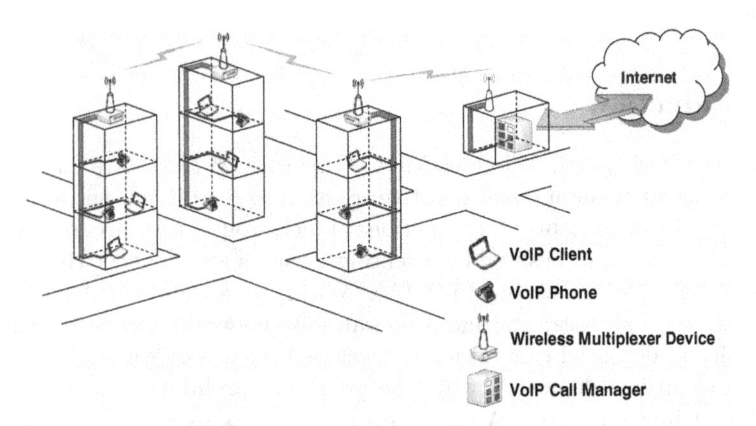


**Fig. 1.** VoIP system architecture

significantly decreases. To reduce losses and avoid buffer congestion, in this paper, we propose to implement a "queue-length-based" Rate Controller combined with the Adaptive MultiRate (AMR) encoder of VoIP sources. The Rate Controller monitors the queue length in the buffer and sends a feedback signal to the VoIP sources which change their emission bit rate accordingly.

The remainder of the paper is organized as follow. In Sect. 2, we describe a method to control the rate of VoIP sources. In Sect. 3, we show some preliminary results of our simulations and discuss the relationship between the speech quality and the number of active calls the system can manage. Finally, Sect. 4 concludes the paper.

## 2 The Rate Controller

The Voice Multiplexer wireless output link constitutes the bottleneck of the system in Fig. 1, given that the buffer is loaded by a high number of VoIP sources, and the service rate sometimes can be strongly reduced by the retransmission number increasing due to losses in the wireless channel. In order to avoid congestion even when channel conditions worse, VoIP sources are adaptive according to the state of the Voice Multiplexer buffer queue, $s_Q \in [0, K]$. More specifically, a Rate Controller is located in the Voice Multiplexer. It implements a feedback law that, according to the value of $s_Q$, at each reaction period $\Delta_S$, in the following defined with a duration equal to the voice source frame length, randomly chooses one source among the $N$ active sources, and sends it a "decrease-rate" or "increase rate" message with a probability dependent on the queue state.

More specifically, as depicted in Fig. 2, the queue range $[0, K]$ is divided into three intervals:

- $[0, q_{MIN} - 1]$: when the queue is in this interval, the Rate Controller produces an "increase-rate" message
- $[q_{MIN}, q_{MAX}]$: when the queue is in this interval, the Rate Controller produces no message
- $[q_{MAX} + 1, K]$: when the queue is in this interval, the Rate Controller produces a "decrease-rate" message

where $q_{MIN}$ and $q_{MAX}$ are defined in such a way that $K/2 - q_{MIN} = q_{MAX} - K/2$.

The message is sent to one source randomly chosen among the $N$ active sources with a probability $\Phi_P(s_Q)$ defined as shown in Fig. 2, that is:

$$\Phi_p(s_Q) = \begin{cases} -\frac{s_Q}{q_{MIN}} + 1 & \text{if} \quad 0 \leq s_Q \leq q_{MIN} \\ 0 & \text{if} \quad q_{MIN} \leq s_Q \leq q_{MAX} \\ \frac{s_Q - q_{MAX}}{K - q_{MAX}} & \text{if} \quad q_{MAX} \leq s_Q \leq K. \end{cases} \tag{2.1}$$

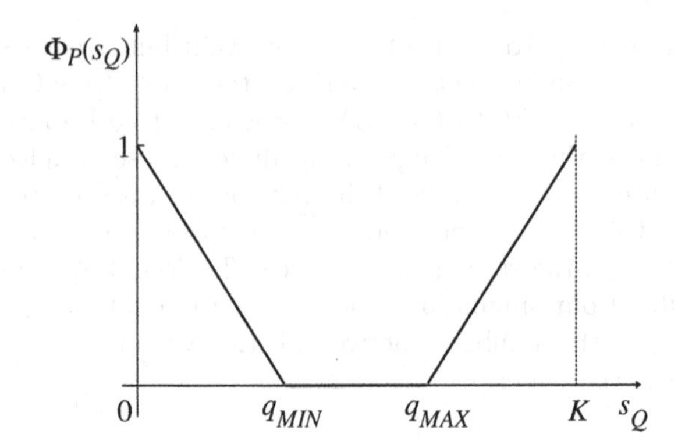


**Fig. 2.** Rate controller mask

Let $\underline{\Psi}^{(S)}$ be the encoding rate array, containing all the emission bit rate levels of the VoIP source, $\underline{\Omega}^{(S)}$ be the quality array containing the quality levels associated to each encoding mode, in terms of Mean Opinion Score (MOS) values, and $G$ the number of encoding levels, that is, the cardinality of the sets $\Psi^{(S)}$ and $\Omega^{(S)}$. In the following we will consider the Adaptive MultiRate (AMR) encoding standard [4], which is one of the most widely adopted multi rate encoding technique for wireless channels. However let us note that the proposed architecture is independent of the particular encoder. In the considered case the set of possible emission bit rates, associated to each encoding mode, is: $\underline{\Psi}^{(S)} = [4.75, 5.15, 5.90, 6.70, 7.40, 7.95, 10.2, 12.2]\,\mathrm{kb\,s^{-1}}$; the set of the relative quality levels [5] is $\underline{\Omega}^{(S)} = [3.50, 3.50, 3.72, 3.77, 3.83, 3.91, 4.01, 4.06]$. Therefore the number of encoding levels is $G = 8$.

Each VoIP source, at each $\Delta_S$ interval (typically $\Delta_S = 20\,\mathrm{ms}$), encodes voice at the imposed rate, and sends an UDP and consequently an IP packet to the Voice Multiplexer. Therefore, the Voice Multiplexer receives $N$ IP packets at each $\Delta_S$ seconds. In order to reduce packetization delay, as usual for voice transmission (see for example [2]), the Voice Multiplexer extracts data units from packets coming from each source, and creates a single UDP packet, to be sent to the Voice Call Manager in an IP packet. The size of this new packet, of course, is the sum of the header of the RTP, the UDP and the IP protocols, plus the sum of the sizes of information units coming from the $N$ sources. A number of bits can be added within the payload to identify the VoIP source for each data unit. Let $p$ be the size of this IP packet.

In order to reduce errors introduced by the wireless channel, Automatic Repeat reQuest (ARQ) channel encoding is applied [3] to the transmitted flow. According to this technique, each packet is subdivided in blocks of $h$ bits. Let $C$ be the transmission rate on the wireless link. Thus the time needed to transmit one block is $\Delta_{ARQ} = h/C$. Each block is retransmitted up to $s_{MAX}^{(R)}$ times. If after $s_{MAX}^{(R)}$ times the block arrives corrupted, it is discarded.

## 3 Numerical Results

In this section we show that the Rate Controller behaves correctly, being able to control VoIP sources in order to avoid too high oscillations of the Voice Multiplexer buffer queue.

For this purpose we present a set of performance tests achieved by simulation. We considered $N$ VoIP sources located into the same building, and connected to the *Voice Multiplexer* device through high-speed links. Every VoIP source uses an Adaptive MultiRate (AMR) encoder with $G = 8$ encoding modes. The AMR Coding Rate, the relative quality levels and the frame size for each coding mode are presented in Table 1. The Voice Multiplexer Device is equipped with a WiFi card to access to the Wireless Mesh network. We assume the capacity of the wireless link is 500 kb s$^{-1}$. The reaction period $\Delta_S$ is defined as the transmission time of an ARQ packet. We assume the ARQ packet size is 60 bytes. Consequently the reaction period $\Delta_S$ is 0.96 ms. The Voice Multiplexer buffer size $K$ is set to 20 IP packets.

We will compare results obtained with different shapes of $\Phi_P(s_Q)$, given by different values of $q_{MIN}$ between 0 and $K/2$. In order to make results independent from the buffer size, let us define a shape coefficient $A$ as $A = (K - 2q_{MIN})/K$. Values of $A$ range between 0 and 1, determining $q_{MIN}$ ranges between 0 and $K/2$. The value $q_{MAX}$ is chosen as the symmetric of $q_{MIN}$ in respect of $K/2$, that is, $q_{MAX} = K - q_{MIN}$.

When the value of $A$ is 0 means $q_{MIN} = q_{MAX} = K/2$, that is, the system reaction is maximum. On the contrary, when $A = 1$, we have $q_{MIN} = 0$ and $q_{MAX} = K$, and therefore no rate control is applied. In this case sources transmit with a constant bit rate, according to the decided encoded mode. For comparison analysis purposes in the following we have considered the performance achieved with all the possible eight encoding modes.

Figure 3 shows the average queue length of the Voice Multiplexer buffer normalized in respect to the buffer size, versus the number of VoIP sources, for different values of $A$. Figure 4 shows the congestion loss probability, calculated as the ratio between the number of sent packets and the number of packets discarded in the VoIP Multiplexer Buffer. The curves relating to the case

**Table 1.** AMR parameters

| Coding modes | AMR coding rate (kb s$^{-1}$) | MOS | Frame size (bit) |
|---|---|---|---|
| 1 | 12.2 | 4.06 | 244 |
| 2 | 10.2 | 4.01 | 204 |
| 3 | 7.95 | 3.91 | 159 |
| 4 | 7.40 | 3.83 | 148 |
| 5 | 6.70 | 3.77 | 134 |
| 6 | 5.90 | 3.72 | 118 |
| 7 | 5.15 | 3.50 | 103 |
| 8 | 4.75 | 3.50 | 95 |

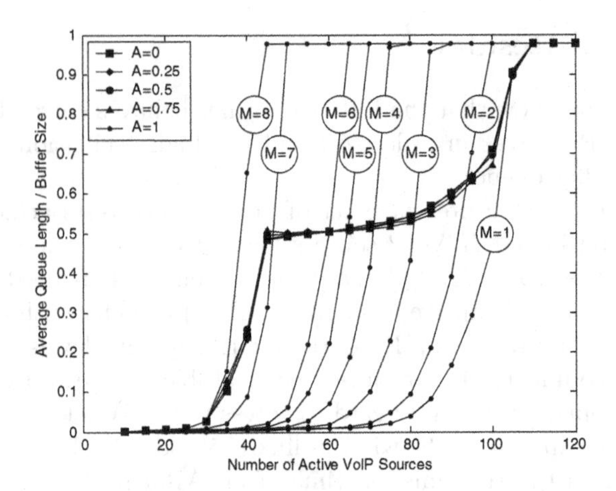


**Fig. 3.** Normalized average queue length

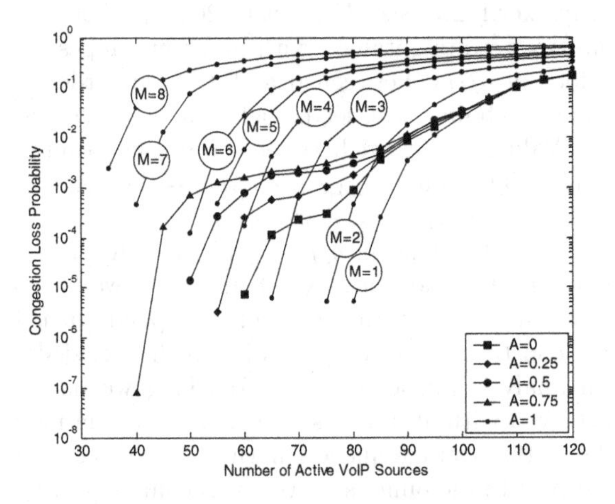


**Fig. 4.** Congestion loss probability

$A = 1$ are marked with the letter $M$ followed by the used encoding mode, which remains constant because no rate control is applied in this case.

In the above figures we can observe that, in the cases of $A = 1$, for high values of the output bit rate, high values of the queue length (Fig. 3) and, consequently, drastic levels of congestion (Fig. 4) are reached even with a small number of sources. On the contrary, if the encoding bit rate is lower, the bandwidth is underutilized until a big number of sources become active, as demonstrated from an average queue length very close to zero. Instead, when the Rate Controller is implemented, feedback signals are sent to the sources in order to keep the queue length close to the middle of the buffer. In fact, as shown in Fig. 5, where the average source bit rate and the relative MOS are reported, as the number of active sources increases, the average source bit

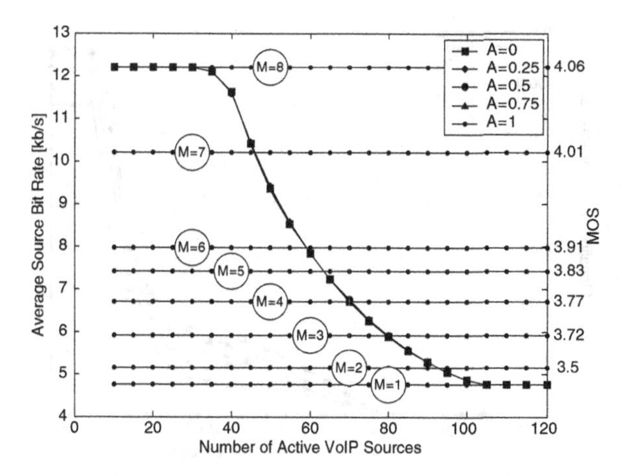


**Fig. 5.** Average source bit rate and MOS

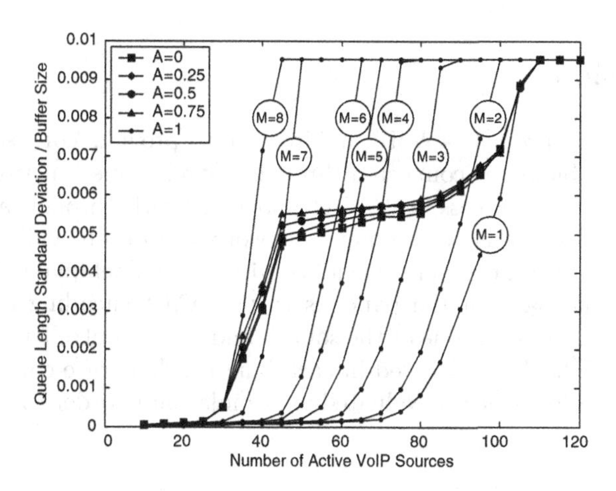


**Fig. 6.** Queue length standard deviation

rate decreases. In such a way, a better utilization of the bandwidth is done while the congestion is controlled.

In Fig. 6 we present the queue length standard deviation versus the number of active sources. Simulations results show that oscillation around the average queue length are very small when the Rate Controller is used. This means that the delay jitter, which is an important performance parameter for VoIP systems, is also small.

The counterpart of this mechanism is the overhead due to the feedback messages. Figure 7 shows that the signaling overhead increases with the number of active sources. Let us point out that Fig. 7, as well as Fig. 5, can be used to know the maximum number of calls the system should accept in order to guarantee a given maximum signaling overhead or, alternatively, a given quality in terms of MOS.

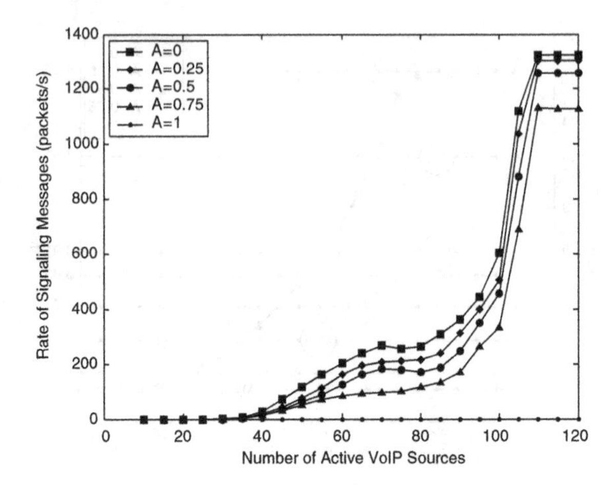


**Fig. 7.** Rate of signaling messages

## 4 Conclusions

In this paper we propose a VoIP architecture to provide VoIP service to an agglomerate of buildings connected through a wireless mesh network (WMN) to the Internet and accessing the telephone network through a VoIP Call Manager. The target is to maximize network utilization while minimizing congestion and losses, even in presence of wireless channel quality fluctuation. The idea at the basis of the paper is to use AMR encoding or any other multirate encoding technique at the source, and a rate controller implemented at the VoIP Multiplexer, located in each building. The rate control rules are defined, and performance is evaluated via simulation in order to achieve some hint to system design.

## Acknowledgment

This work has been partially supported by the Italian PRIN projects BORA-BORA and PROFILES.

## References

1. I. F. Akyildiz, X. Wang, and W. Wang, "Wireless mesh networks: a survey", *Computer Networks*, Volume 47, Issue 4, March 2005, pp 445–487.
2. G. Eneroth, G. Fodor, G. Leijonhufvud, A. Ràcz, and I. Szabò, "Applying ATM/AAL2 as a switching technology in third-generation mobile access networks", *IEEE Communications Magazine*, Volume 37, Issue 6, June 1999, pp 112–122.

3. J. M. Wozencraft and M. Horstein, "*Coding for two-way channels*", Technical Report 383, Research Laboratory of Electronics, MIT, Cambridge, MA, January 1961.
4. 3GPP TS 26.171: "AMR wideband Speech Codec; General description".
5. ITU-T Recommendation P.862, "*Perceptual evaluation of speech quality (PESQ), an objective method for end-to-end speech quality assessment of narrowband telephone networks and speech codecs,*" International Telecommunication Union, Geneva, Switzerland, February 2001.

# 20

# Toward the QoS Support in 4G Wireless Systems

A.L. Ruscelli[1] and G. Cecchetti[2]

[1] Scuola Superiore S. Anna, Pisa, Italy
a.ruscellii@sssup.it
[2] CNIT – Scuola Superiore S. Anna, Pisa, Italy
g.cecchetti@sssup.it

**Summary.** This paper presents a novel approach to support *Quality of Service* for wireless multimedia applications in the context of 4G wireless systems. Adopting a *Service Oriented Architecture*, it is inspired to *Open Wireless Architectures* (OWA), building a suitable framework over the top of the heterogeneous wireless MACs. It lets to enhance the existing QoS support provided by standard MAC protocols and it uses the *contract model* to guarantee QoS, taking into account the applications requests. It negotiates dynamically *Application Level Contracts* which will be translated seamlessly in *Resource Level Contracts* for the underlying network services. It receives the feedback by underlying network services to adjust the scheduling algorithms and policies to provide soft guarantees. The framework comprises *QoS Manager*, *Admission Control*, *Enhanced Scheduler*, *Predictor* and *Feedback System*. In particular, the QoS manager component is a middleware between applications and lower network layers and it is able to dynamically manage available resources under different load conditions in a transparent manner to application level.

## 1 Introduction

The increasing diffusion of multimedia communications involves the need to manage advanced and wideband multimedia services for which 4G systems provide a mix of concepts and technologies. Some of that are *evolutionary* because they are derived from 3G, while other are *revolutionary* because they are typical of this novel solution. In particular 4G networks improve the 3G networks approach suggesting a new wireless architecture suitable to the future wireless service provisioning. The latter will be characterized by global mobile access that implies the convergence of the wireless mobile and wireless access in an open, common, flexible and expandable platform. In this context 4G mobile technologies [1] find in *Open Wireless Architecture* (OWA) [2] or *Converged Broadband Wireless Platform*, the suitable model for realizing global mobile access, high quality of service, simple, seamless, automatic access to media services for voice, data, message, video, world-wide web, etc, utilizing an horizontal communication model. This architecture includes base-band signal processing, RF, Networking, OS and application parts so that the same end equipment can flexibly work in the wireless access domain as well as in the mobile cellular networks, with optimal spectrum efficiency and resource management. Moreover, the converged broadband wireless system necessarily has to take into account Quality of Service support. It is essential for several multimedia applications like VoIP, video conference call, audio and video streaming, contents distribution, Internet services and real time control services. The QoS support for OWA needs to be functional for integrated wired and wireless access modes using a common methodology and offering a differentiated service according to strict latency/throughput applications requirements, while the used medium offers time and space varying communication conditions. It turns out that the variability of available radio resources does not allow the network to provide hard QoS guarantees. Instead, the network must provide soft QoS guarantees constrained by a minimum channel quality. Some of these guarantees regards: *delay*, *delay jitter*, *packet loss ratio*, *throughput*, *bandwidth*. In particular the QoS provision must take into account the support done by each single access mode, however leaving space to build blocks for a full Quality architecture. In this context one trend is to use an adaptive QoS system with a relative QoS differentiation [3] based on different priority classes of Differentiated Service architecture to deliver multimedia data [4]. Another remarkable point of view is to introduce a cross-layer design with adaptive QoS assurance for multimedia transmissions [5,6].

In this article we present a novel framework to provide a comprehensive Soft QoS support for multimedia traffic streams, inspired both to cross-layer architecture idea and QoS differentiation and it allows to interface seamlessly multimedia applications with lower layers of a wireless network. The wireless application framework is composed by QoS Manager and Scheduling Subsystem (Admission Controller, Scheduler, Predictor and Feedback mechanism). We specifically focus on QoS Manager which is a middleware between applications and lower network layers and it is able to dynamically manage available resources under different load conditions in a transparent manner to application level. It accepts the different QoS requests from the various multimedia applications and it translates these in the specifics of each involved medium access standard, handling time-varying network conditions, heterogeneous traffic streams and link layer resources.

## 2 The Cross Layer Framework

The framework has a cross-layer architecture (Fig. 1) composed by a middleware for QoS, the *QoS Manager*, and the scheduling subsystem. The former transparently manages the communication levels for applications while the

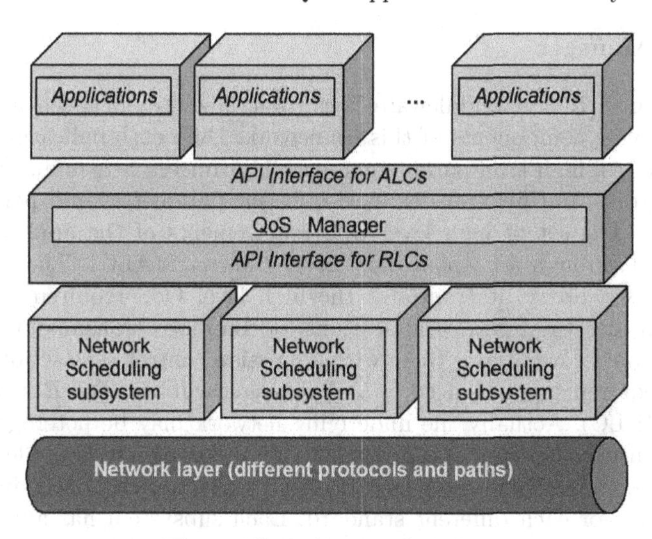


**Fig. 1.** Cross-layer architecture

latter consists of some building blocks to regulate the heterogeneous networks MAC layer. This subsystem may vary depending on the particular MAC(s) used (e.g. IEEE 802.11, IEEE802.16, MIMO, mobile public networks, wired networks, etc.). The QoS manager is independent both to application and MAC. Using this approach it lets one possible solution to convergence question, providing an open communication gateway architecture, which represents high level abstraction that lets practitioners to concentrate on the specification of the application requirements. Furthermore, it adopts a Service-Oriented Architecture which allows it relies mostly on the services provided and on the required applications. From the user perspective the framework allows an easy and simple access to multimedia services, hiding the complexity of the lower MAC levels of the different networks. The contract based scheduling implemented lets to the applications to dynamically specify its own set of complex and flexible execution requirements, written as a set of a service contracts for different resources, which are negotiated with the underlying implementation. To accept a set of contracts the QoS Manager has to check, as part of the negotiation, if it has enough resources to guarantee all the specified minimum requirements while keeping guarantees on all the previously accepted contracts negotiated by other application components. Eventually it adapts the requirements to available resources. If a result of this negotiation is accepted, the scheduling subsystem reserves enough capacity to guarantee the minimum requested resources and it reclaims any spare capacity available to share it among the different contracts that have specified their desire or ability for using additional capacity. The contract also contains Quality of Service tuning parameters that may be used by QoS manager.

## 2.1 QoS Manager

QoS manager [7] is a middleware layer that mediates between application and underlying components of this framework. Different applications specify different sets of high level parameters (e.g., Multimedia Streaming, VoIP, signaling protocol and file transfer have different parameters and performance indicators). The set of high level QoS requirements of the application will be specified through an *Application Level Contract* (ALC). The QoS manager acts as a proxy: it translates the high level QoS requirements of the application into the resource allocations, it computes transmission parameters values and it negotiates them with admission control. The set of low level resource requirements produced by QoS manager will be called *Resource Level Contract* (RLC). Actually the underlying network may be heterogeneous, it may vary in topology and standards, offering a completely variable scenario. For this reason, QoS manager has to interact with different scheduling subsystems, one for each different standard. Each subsystem has an admission control, a scheduler, a predictor and a feedback control. We can assume without lack of generality that each protocol does not interfere with other ones. When QoS manager interacts with a subsystem it provides the appropriate parameters and it takes into account the specific protocol used. We have implemented the QoS manager as a *"two-side" Application Program Inter-face* (API). The upper side interfaces applications while the bottom side interacts with scheduling subsystem. The applications can call the following functions of QoS manager:

```
int request_ALC (struct *ALCspec);
int modify_ALC (struct *ALCspec);
int cancel_ALC (struct *ALCspec);
```

An application uses `request_ALC` to negotiate an ALC with QoS manager. The latter checks the requirements specified with `struct *ALCspec through` *Admission Control subsystem* and returns the result 0 if the ALC is accepted, 1 if the ALC is modified or 2 if the ALC is rejected. If the ALC is modified, the application can cancel the current contract issuing `cancel_ALC`. At any time an application can request to modify the ALC by `modify_ALC` . This function behaves like `request_ALC`. To abort a contract an application use `cancel_ALC`, which should return 0 unless an error is occurred. The API bottom side is used by QoS manager to interact with the underlying levels. It consists of three functions for every protocol managed by QoS manager:

```
int request_RLC_proto (struct *RLCspec);
int cancel_RLC_proto (struct *RLCspec);
int get_RU_proto (struct *RU_proto);
```

where proto may be any supported network protocol (e.g. for 802.11e the function names are: `request_RLC_80211e` and `get_RU_80211e`). The first function is called by QoS manager to negotiate a RLC with the scheduling subsystem and it returns 0 if the RLC is accepted, 1 if the RLC is modified or 2 if the RLC is rejected. If the RLC is modified and ALC can still be satisfied, the QoS manager adjusts ALCspec and return it to the application requesting the corresponding contract, while if the RLC is modified but the ALC cannot be satisfied the QoS manager cancels the RLC through a `cancel_RLC_proto` and then notifies to the application that it cannot accept the ALC. RLCs are also canceled when ALC expires. The function `get_RU_proto` is used by QoS manager to query the scheduling subsystem about protocol resource utilization. The returning `RU_proto` lets the QoS manager to enhance its negotiation capability. Moreover QoS manager performs other actions useful to optimize QoS support:

- It adapts automatically the resource allocation to dynamic changes in the requirements of the application, tuning service parameters: when an application wants to change the contract profile, the QoS manager contacts again the corresponding admission control service and negotiates a new RLC. It perform a so-called adaptive resource allocation.
- It adapts dynamically the resource allocation in order to optimize the resource utilization without sacrificing on QoS requirements.
- It maintains as much as possible the resource allocation for each application as close the minimum that is needed to fulfill the ALC.
- If an overload occurs (e.g. due to varying network conditions or if a more important QoS request is received), it can decide to change one or more ALCs to degrade the QoS level of one or more applications by a call-back notification so that the application itself can adapt its QoS requirements.

## 2.2 The Scheduling Subsystem

The scheduling subsystem is composed by the admission control, the scheduler and the feedback mechanism.

*Admission control* verifies if there are sufficient resources for medium access to satisfy QoS manager requests. It computes the theoretical new bandwidth utilization and it checks if it is admissible without degradation of preexistent transmissions. The response is sent back to the QoS manager. If the instance request is successful a RLC is established and the QoS manager can communicate transmission parameters to the corresponding scheduler. The general admission test used is:

$$\sum_{i=0}^{N} \frac{Q_i}{P_i} \leq 1, \tag{1}$$

where $Q_i \equiv C_i/r_i$ is the average time budget of the medium which is reserved to the $i^{th}$ network station transmitting within each period $P_i$; $r_i$ is the physical bit rate assumed for admission control computations of the $i^{th}$ traffic stream, $C_i$ are the bytes transmitted during the $P_i$ and *Ulub* is least upper bound utilization factor computed for the worst-case available bandwidth. If there is not enough bandwidth to serve the new request, three different admission control policies exist which act as follows:

- *Saturation policy*: The highest possible budget is assigned to the task so that the total resource utilization does not exceed *Ulub*
- *Compression policy*: In respect of the established ALCs, all the RLCs are recomputed ("compressed") so that we can make new space for the new request
- *Reject policy*: The transmission is rejected

The *scheduler* manages each transmission for each admitted flow and it assigns dynamically both the period $P_i$ and transmission duration to follow the channel variability and streams characteristics. We propose a scheduler which can handle traffic stream (TS) with Soft Real Time guarantees [8] with special regard to VBR flows. VBR flows are supported by assigning transmission duration in agreement to the effective temporal demands. The assignment of $P_i$ is dynamic, so it lets to increase the transmission frequency of the applications having in queue traffic with tightening requirements of QoS. The scheduler is also able to reclaim the unused time of nodes which have exhausted their transmission before the end of their transmission duration and then it assigns that time to the nodes which have still useful data to transmit.

Delay or advance of the transmission with respect to the pre-agreed rate (in terms of bytes which have been anticipatively used or have not been transmitted by node) are formalized as the scheduling error $\varepsilon_i^{(k)}$, defined, at the $k$th time instant, as the difference between the cumulated bytes to transmit $z_i^{(k)} \equiv kC_i^{(k)}$ and the bytes actually transmitted $z_i^{(k)}$:

$$\varepsilon_i^{(k)} \equiv z_i^{(k)} - \overline{z}_i^{(k)} \tag{2}$$

The dynamic equation for the evolution of the scheduling error for the $i$th real-time data flow is:

$$\varepsilon_i^{(k+1)} \equiv \varepsilon_i^{(k)} + C_i^{(k)} - \gamma_i^{(k)} Q_i^{(k)} \tag{3}$$

where $\gamma_i^{(k)}$ is the actual channel speed.

The *feedback* mechanism senses the effective information acknowledged by nodes. It eventually uses the information provided by a predictor to vary transmission parameters of the scheduler in order to respect hard and soft deadlines. It is responsible to minimize the scheduling error.

The rapidity of this action can be improved turning on special weights $w_i$ for each traffic stream ($\mathrm{TS}_i$). The feedback system can compensate little variations of network conditions without the intervention of admission control to establish new RLCs. During *normal condition*, if

$$\sum_{i=0}^{N} \frac{Q_i}{P_i} \leq 1 \tag{4}$$

is kept, feedback system controls the scheduling error assigning:

$$\forall i,\ Q_i^{(k)} \triangleq \widetilde{Q}_i^{(k)} = \frac{C_i^{(k)} + \alpha_i \varepsilon_i^{(k)}}{\rho_i^{(k)}},$$

where $\widetilde{Q}_i^{(k)}$ is the required assigned budget to compensate the scheduling error, $\alpha_i \in ]0,1]$ and $\alpha_i \varepsilon_i^{(k)}$ is a fraction of the current scheduling error for each $TS_i$ and $\rho_i^{(k)}$ is the predicted channel speed at the physical layer.

During *overload condition* the allocated budget to each $TS_i$ is decreased. For example, if the feedback scheme uses a weighted distribution, for each $TS_i$, $\widetilde{Q}_i^{(k)}$ is decreased of an amount proportional to the weight $w_i$ assigning:

$$\forall i,\ Q_i^{(k)} \triangleq \widetilde{Q}_i^{(k)} - \frac{w_i \widetilde{Q}_i^{(k)}}{\sum_{j=1}^{N} w_j \widetilde{Q}_j^{(k)}} \left( \sum_{j=1}^{N} \widetilde{Q}_j^{(k)} - U_{lub} T_i \right),$$

where

$$\frac{w_i \widetilde{Q}_i^{(k)}}{\sum_{j=1}^{N} w_j \widetilde{Q}_j^{(k)}}$$

is the percentage of decreasing.

# 3 Conclusions

In this paper we have presented a description of a cross layer framework to integrate QoS support for 4G systems under time varying network conditions and different traffic specifications. It provides an interface to QoS support mechanisms for any applications with tightening guarantees and temporal boundaries. This framework lets applications establish contracts with QoS manager that administers the available resources from under-lying subsystems. The resulting QoS service is an improvement for applications running over wireless networks. The QoS manager acts as a proxy towards different network sub-systems which manage different wireless network protocols. Using this approach it lets one possible solution to convergence question and it takes a response to simultaneous need, in 4G systems, of ubiquity and diversity that imply flexibility and individuality, providing an open communication gateway architecture, which represents high level abstraction that lets practitioners to concentrate on the specification of the application requirements.


## Acknowledgments

This work has been carried out within the framework of the VICOM project founded by the Italian Ministry of University and Research (MiUR) and in part within FRESCOR EU project (Contract no. 034026).

## References

1. J. Shelper, "1g, 2g, 3g, 4g," *TechColumn*, April 2005
2. W. Lu and J. Hu, "Open wireless architecture – the core to 4g mobile communications," *CIC J. China Commun.*, April 2006
3. C. Dovrolis, D. Stiliadis, and P. Ramanathan, "Proportianal differentiated services: Delay differentiation and packet scheduling," *IEEE/CM Trans. Networking*, vol. 10, pp. 12–26, February 2002
4. J. Shin, "Quality-of-service mapping mechanism for packet video in diffentiated services network," *IEEE Trans. Multimedia*, vol. 3, pp. 219–231, 2001
5. A. Alwin, "Adaptive mobile multimedia networks," *IEEE Pers. Commun.*, vol. 3, no. 2, pp. 34–51, 1996
6. M. Mirhakkak, "Dynamic bandwidth management and adaptive application for a variable bandwidth wireless environment," *IEEE J. Select. Areas Commun.*, vol. 19, pp. 1984–1997, 2001
7. T. Cucinotta, L. Abeni, G. Lipari, L. Marzario, and L. Palopoli, "QoS management through adaptive reservations," *Real-Time Syst. J.*, vol. 29, no. 2-3, March 2005
8. J. Stankovic, K. Ramamritham, M. Spuri, and G. Buttazzo, "Deadline scheduling for real-time systems," K. A. Publishers, Ed., Boston, 1998

# 21

# A Scheduling Algorithm for Providing QoS Guarantees in 802.11e WLANs

G. Cecchetti and A.L. Ruscelli

CNIT – Scuola Superiore S. Anna, Pisa, Italy
g.cecchetti@sssup.it, a.ruscelli@sssup.it

**Summary.** In this paper we propose a scheduling algorithm for supporting Quality of Service (QoS) in an IEEE 802.11e network using the HCF Controlled Channel Access (HCCA) function. This is derived from Constant Bandwidth Server with Resource Constraints and adapted to wireless medium. It consists of a procedure to actually schedule transmission opportunities to HCCA flows with Quality of Service guarantees, in particular in the case of multimedia applications which present variable bit rate traffic.

## 1 Introduction

In recent years Wireless Local Area Networks (WLANs) are became very popular and the IEEE802.11 [1] has established as the world wide standard. At the same time, the continuous growth in the use of mobile devices that support multimedia applications and real-time services with strict latency/throughput requirements, such as multimedia video, VoIP (Voice Over IP), videoconference over a wireless channel, involves a great interest in the study of appropriate mechanisms to manage the wireless medium in order to achieve the expected *Quality of Service* (QoS). Recently approved IEEE 802.11e standard [2] for WLANs offers a complete set of primitives to provide delay guarantees while the previous IEEE 802.11b [3] was designated only for best effort services. However also IEEE 802.11e does not provide scheduling algorithms for packet transmission nor policies scheme for access control to the medium, leaving space to build blocks for a full Quality architecture.

Many research studies have evaluated the new standard employing analytical techniques [4] and simulations [5, 6], and they have demonstrated the usefulness of the proposed mechanisms of 802.11e. Subsequent works have proposed several scheduling algorithms to improve the QoS provisioning [7–10]. This improvement is necessary in particular in the case of VBR traffic for which the *reference* scheduler shows its limit. In fact it is particularly tailored for constant bit rate traffic.

In this paper we propose a new scheduling algorithm for the HCF Controlled Channel Access (HCCA) function of the IEEE 802.11e, namely the Wireless Constant Bandwidth Server with Resource constraints(WCBS-R). It does provide those flows that have been admitted to use the HCCA function with rate base guarantees. This algorithm is derived from real-time systems, and it is actually a modified version of Costant Bandwidth Server with Resource Constraints [11, 12]. The scheduling methodology adopted reserves a fraction of network bandwidth to each flow, assigning a suitable deadline to the server flow whenever the reserved time is consumed. Differently from the *reference* scheduler, WCBS-R is not based on periodic scheduling of fixed allocations but it manages dynamically the allocated capacity. Moreover the latter is made available for contention based access when it is not used by the HCCA flows. Through preliminary simulation results, we show that WCBS-R performs better than the reference scheduler, in terms of the capacity available for legacy DCF access, which is based on contention.

The rest of the paper is organized as follows. Section 2 introduces the IEEE 802.11e HCCA. In Sect. 4 we describe WCBS-R scheduling algorithm. Preliminary simulation results are discussed in Sect. 5 and conclusions are drawn in Sect. 6.

## 2 IEEE 802.11e Protocol Description

The new standard IEEE 802.11e introduces a new coordination function called the *Hybrid Coordination Function* (HCF) which multiplexes between two medium access modes: a distributed scheme called *Enhanced Distributed Channel Access* (EDCA) and a centralized scheme called *HCF Controlled Channel Access* (HCCA). To ensure compatibility whit legacy devices, the standard allows the coexistence of DCF and PCF with EDCA and HCCA.

### 2.1 Enhanced Distributed Channel Access

EDCA is a channel access mode which provides prioritized QoS and it enhances the original DCF by classifying traffic through the introduction of *Access Categories* (ACs), corresponding to different level of traffic priority. Each AC has its own transmission queue and its own set of channel access parameters. The most important ones are Contention Window (CWmin and CWmax), which sets backoff interval, and *Transmission Opportunity* (TXOP) limits which is the maximum duration for which a node can transmit after obtaining access to the channel. Using these parameters, when data arrives from higher layers it is classified and placed in the appropriate AC queue. Then an internal contention algorithm is used to calculate the total backoff time for each AC. The AC with the smallest backoff time wins the internal contention and uses this backoff value to contend externally for the wireless medium. Nodes with higher priority can access the channel earlier than other nodes and prioritized flows have the advantage of longer channel access with their TXOP.

## 2.2 HCF Controlled Channel Access

HCCA provides a centralized polling scheme to allocate guaranteed channel access to traffic flows based on their QoS requirements. It uses a QoS-aware *Hybrid Coordinator* (HC) which is usually located at the *QoS Access Point* (QAP) in infrastructured WLANs and it provides polled access to the wireless medium. In order to be included in the polling list of the HC, a *QoS Station* (QSTA) must send a QoS reservation request to the QAP, using the special QoS management frame, *Add traffic Stream* (ADDTS), which contains the Traffic Specification (TSPEC), which includes the following parameters: as *mean data rate* ($R_i$), *nominal Service Data Unit (SDU) size* ($N_i$), *minimum PHY rate* ($\Gamma_i$), delay bound ($D_i$) and *maximum service interval* ($MSI$).

HC aggregates every TSPEC of QSTA TSs and determines the values of parameters needed by the transmission itself: *Service Interval* (SI) and TXOP. SI is the time duration between successive polls for the node and it is a submultiple of the 802.11e beacon interval duration. TXOP is the transmission duration of each node based on the mean application data rates of its requested flows. Before the calculation of the latter parameters, QAP has to verify if the admission of each TS does not compromise the service guarantees of the already admitted TSs and, if the specified TS is accepted, QAP sends a positive acknowledgement which contains also the service start time that indicates the time from when the QSTA is allowed to transmit frames relative to considered TS.

When there are admitted QSTAs which desire to access the medium, the QAP listen to the medium itself and, if it is idle for a PIFS, HC gains control of the channel and, within the *Controlled Access Phase*(CAP), it polls a single QSTA at turn, according to its polling list, generated by a scheduler.

# 3 Related Work

Scheduling algorithms addressed to wireless networks have to take into account some limits due to wireless environment [13]. In particular, the wireless medium itself presents space and time varying characteristics, unlike what happens in wired networks, so wireless networks are subject to fast changes in *Signal to Interference plus Noise Ratio* (SINR) due to phenomena like *path loss, shadowing, multipath fading, signal attenuation* and *interference* [14].

This implies that the concept of fairness is difficult to apply. Furthermore, wireless resources (e.g. bandwidth and energy) are limited and that, jointly to need of lower computation complexity due to use of low-performance hardware, adds other constraints in the choice of algorithms.

Several studies has been done to verify performances of reference scheduler [5,6,15]. According to them, for every QSTA, fixed SI and TXOP based on mean values of the transmission parameters are useful for Constant Bit Rate TS, while they do not reflect the fluctuation of Variable Bit Rate TS. Particularly reference scheduler performances are evaluated using heterogeneous traffic stream like VoIP (G.711 codec), video stream (MPEG4 codec) and burst *best effort* data stream. Some alternative algorithms introduce the following features (a) variable SI and/or TXOP, (b) feedback based mechanism, (c) queue length model.

### 3.1 Deadline-Based Algorithms

The results obtained in the field of scheduling real-time tasks in a multiprogrammed environment has been adapted to the context of HCCA scheduling.

In [16] the authors propose the SETT-EDD scheduling algorithm which limits the amount of time during which the stations control the wireless medium, it improves the performance of the scheduler and it enhances its flexibility. It uses the mean TXOP as a guideline for allocating time and uses a token bucket scheme of time units or TXOP timer to allow nodes to vary their TXOP over time according to their needs. The TXOP timer of station $j$ increases at a constant rate equal to $TD_j/mSI_j$ (where $mSI_j$ is minimum SI of $j$th QSTA), which corresponds to the total fraction of time the station can spend in polled TXOPs. The TXOP timer has a maximum value equal to $MTD_j$ (where $MTD_j$ is the Maximum Time Duration of $j$th QSTA). The time spent by a station in a polled TXOP is deducted from the TXOP timer at the end of the TXOP. The station can be polled only when the value of the TXOP timer is greater than or equal to $mTD_j$, which ensures the transmission of at least one packet at the minimum PHY rate.

The authors also propose to change the service interval for each node based on the traffic profile and use Earliest Deadline First (EDF) to determine the polling order. If the due time to poll a station is $t$, the next poll shall be issued on a time $t'$ that satisfies the relation: $t + mSI < t' < t + MSI$. Time instant $t + mSI$ is the instant after which the next poll can be done, equivalent to the release time in the real-time scheduling theory. Time instant $t + MSI$ is the maximum time by which the next poll has to be done, or deadline time.

It has been shown that the proposed flexibility in the scheduler for voice and video traffic leads to significant reduction in average transmission delay (up to 33%) and packet loss ratio (up to 50%).

In [17] the authors combine the EDF algorithm with the Stack Resource Policy (SRP), which are efficient policies for scheduling real-time periodic tasks in a multiprogrammed environment. When a TS requests admission to the QAP, RTH computes a periodic timetable in which TSs are granted a fixed capacity. This allows TSs with different requirements to be scheduled efficiently, and therefore RTH admits more TSs than the sample scheduler. However, RTH cannot deal with VBR traffic.

### 3.2 Algorithms Based on Queue Length Estimation

Another approach to dealing with uplink VBR traffic is to estimate the length of uplink TS queues, and tune the length of TXOPs granted to them.

FHCF [18] tries to improve the fairness both of CBR and VBR flows by assigning variable TXOPs. These are computed using queue length. Actually FHCF is composed of two schedulers: the QAP scheduler estimates the varying queue length for each QSTA before the next SI and compares this value with the ideal queue length. The QAP scheduler uses a window of previous estimation errors for each TS in each QSTA to adapt the computation of the TXOP allocated to that QSTA. Because sending rate and packet size can change, this estimation can not be accurate. After this comparison QAP computes the additional requested time (positive or negative) for each TS of each QSTA and reallocates the corresponding TXOP duration. Then, the node scheduler located in each QSTA can redistribute the unused time among its different TSs since the TXOP is always allocated to a whole QSTA. It computes the number of packets to transmit in the TS and time required to transmit a packet according to its QoS requirements. Later, according to its allocated TXOP, it evaluates the remaining time that can be re-allocated. This is possible since each QSTA knows its TS queue size at the beginning of polling phase and it is able to estimate its queue length at the end of TXOP and the requested additional time for TS.

Performance study indicates that FHCF provides good fairness while supporting bandwidth and delay requirements for a large range of network loads and, because it uses to allocate TXOP the mean sending rate of VBR applications instead of the maximum sending rate usable for the standard HCF scheme, it may recover much time and more flows can accepted in HCCA. Furthermore, it is more efficient than the reference scheduler, admitting an higher number of traffic streams.

FBDS [19] assigns dynamically the TXOP according to queue length estimation while SI remains fixed. All the QSTAs which compose the communication system and its transmission queues are regarded as a system whose balance is perturbed by new incoming flows. The FBDS periodic scheduler, which uses HCF, behaves as a closed loop controller which restores this balance by bandwidth recovering. This is possible due queue length information sent by each QSTA through a 8-bit subfield of QoS Control Field. Moreover the closed loop system uses a discrete time model which permits to estimate queue length at beginning of new CAP phase and so it acts as compensation system against errors produced by channel perturbations not previewed by the scheduling algorithm.

This algorithm guarantees the delay bounds required by audio/video applications in presence of very broad set of traffic conditions and networks loads by using a control system action which ensures a maximum delay for queuing new frames.

# 4 WCBS-R Scheduling Protocol

The QAP schedules traffic streams using an algorithm derived from the soft real-time scheduling literature, with regard to the Constant Bandwidth Server with Resource Constraints (CBS-R) scheduling algorithm [12]. The CBS algorithm was modified to suit the needs of wireless traffic and named WCBS-R. It can be defined as follows:

1. A WCBS-R is characterized by a capacity $c_i$ and by an ordered pair $(Q_i,T_i)$, where $Q_i$ is the maximum capacity and $T_i$ is the service interval of the $TS_i$. The ratio $U_i = Q_i/T_i$ is denoted as the $TS_i$ bandwidth. At each instant, a fixed deadline $d_i$ is associated with the $TS_i$. At the beginning $d_i = 0$.
2. Each served transmission chunk $H_{j,k}$ is assigned a dynamic deadline $d_{j,k}$ equal to the current $TS_i$ deadline $d_i$.
3. Whenever a $TS_i$ is served for transmission, the capacity $c_i$ is decreased by the same amount.
4. When $c_i = 0$, the $TS_i$ capacity is recharged at the maximum value $Q_i$ and a new $TS_i$ deadline is generated as $d_i = d_i + T_i$. Notice that there are no finite intervals of time in which the capacity is equal to zero.
5. A WCBS-R is said to be *active* at time $t$ if there are pending transmissions (remember the budget $c_i$ is always greater than 0); A WCBS-R is said to be *idle* at time $t$ if it is not *active.*
6. When a transmission of $TS_i$ arrives and the WCBS-R is *active* the request is enqueued in a queue of pending transmissions according to a given (arbitrary) discipline (e.g., FIFO).
7. When a flow $TS_i$ arrives and the WCBS-R is *idle*, if $c_i \geq (d_i - r_i)U_i$ the scheduler generates a new deadline $d_i = r_i + T_i$ (where $r_i$ is the arrival time of $TS_i$) and $c_i$ is recharged at the maximum value $Q_i$, otherwise the scheduler generates a new deadline $d_i = \max(r_i + T_i, d_i)$ and the capacity becomes $c_i = c_i + (d_i^{new} - d_i^{old})U_i$.
8. When a $TS$ finishes, the next pending transmission, if any, is served using the current capacity and deadline. If there are no pending transmission, the scheduler becomes *idle*.
9. At any instant, a $TS$ is assigned the last deadline generated by the scheduler.
10. Whenever a served flow $TS_i$ tries to access a critical section, if $c_i < \xi_i$ (where $\xi_i$ is the duration of the *longest critical section* of $TS_i$ such that $\xi_i < Q_i$ ), a capacity replenishment occurs, that is $c_i = c_i + Q_i$ and a new scheduler deadline is generated as $d_i = d_i + T_i$.

## 4.1 Admission Control

Let $B_i$ denote the maximum duration of the transmission time of $TSs$ with period longer than $TS_i$. A sufficient condition for a set of $n$ TS each one characterized by a period $T_i$ and capacity $C_i$, to be schedulable is as follows:

$$\frac{B_i}{T_i} + \sum_{j \leq i} \frac{Q_i}{T_i} \leq 1, \quad \forall i : 1 \leq i \leq n. \tag{1}$$

So, when admitting a stream $i$, the QAP has to calculate its $Q_i$ and $T_i$, and to check if (1) holds. Given the TSPEC for $i$ we define:

$$Q_{min} := \lceil \frac{R_i \cdot T_i}{N_i} \rceil, \quad Q_{max} := \lceil \frac{\Pi_i \cdot T_i}{M_i} \rceil$$

where $\Pi_i$ is the peak data rate, $M_i$ is the maximum SDU size for the $i$th TSPEC. For WCBS-R, we use: $Q_i = Q_{min} + CWF(Q_{max} - Q_{min})$. For $T_i$ we use the MSI. The QAP keeps track of the allocated capacity, and when doing admission control it checks if a new stream would require more capacity than the system can provide.

If it can be admitted and it is a downlink stream no more actions than updating the currently used capacity have to be performed, otherwise, if it is an uplink one, the stream is added to the polling list, with a poll time $p_i$ equal to the current time, so that it will be polled as soon as possible, on the next call to the scheduler.

### 4.2 Enqueueing a Packet

When a packet arrives, the QAP has to check if its associated stream i was already active. If it was not, it has to check if the remaining $c_i$ can be given to the stream without exceeding the $Q_i/P_i$ utilization of the medium, otherwise it has to postpone the deadline of the stream, replenishing its capacity.

### 4.3 Dequeueing a Packet

When in a CAP the QAP has to chose the next packet to send, if first updates the status of the stream being served, changing its capacity as needed, and updating its deadline if necessary. Then it checks if there are polling streams that can be added to the active list (i.e., their $p_i$ is passed,) changing their state and requeueing them if necessary.

It then requeues the active stream if it has switched to a polling or idle state or if it is no more the one with the earliest deadline, selecting the next task in EDF order.

If there are no active streams a CP is started. If the selected stream i is an uplink one the corresponding station is given a TXOP of $c_i$, otherwise the packet to be sent is extracted from the QAP queues.

## 5 Experimental Results

In this section we analyze WCBS-R through simulation.

**Table 1.** MAC/PHY simulation parameters

| Parameter | Value |
|---|---|
| SIFS (μs) | 10 |
| PIFS (μs) | 30 |
| DIFS (μs) | 50 |
| SlotTime (μs) | 20 |
| PHY header(μs) | 192 |
| Data rate (Mb s$^{-1}$) | 11 |
| Basic rate (Mb s$^{-1}$) | 1 |
| Bit error rate (b s$^{-1}$) | 0 |

### 5.1 Simulation Settings

The physical layer parameters are those specified by the High Rate Direct Sequence Spread Spectrum (HR-DSSS) [3], also known as 802.11b, and are reported in Table 1.

We focus on the system performance in ideal conditions so we assume that the channel is error-free, while MAC level fragmentation and multirate support are disabled. Furthermore we assume that all nodes can directly communicate with each other. Therefore, the hidden node problem and the packet capture are not taken into consideration and the RTS/CTS protection mechanism is disabled.

We have implemented the proposed W-CBS in the ns-2 network simulator [20], using the HCCA implementation framework described in [21]. Then we compared the results with respect of reference IEEE 802.11e standard scheduler. The analysis has been carried out using the method of independent replications. Specifically we ran independent replications of 600 s each with 100 s warm-up period until the 95% confidence interval is reached for each performance measure. Confidence intervals are not drawn whenever negligible. Then we compared the results with respect of reference IEEE 802.11e standard scheduler.

### 5.2 Admission Control Analysis

We first evaluated the performance of W-CBS in terms of the admission control limit.

Figure 1 shows the number of admitted videoconference TSs, as a function of the number of admitted VoIP G.711 TSs. In both cases, the sample scheduler curve lies significantly below the WCBS-R curve. This behavior confirms that the sample scheduler cannot efficiently accommodate TSs with different TSPECs. In fact, firstly, it polls TSs with $\Delta i >$ SI more often than needed, by setting the scheduling duration to the smallest TS period. Secondly, it overestimates the capacity needed by TSs.

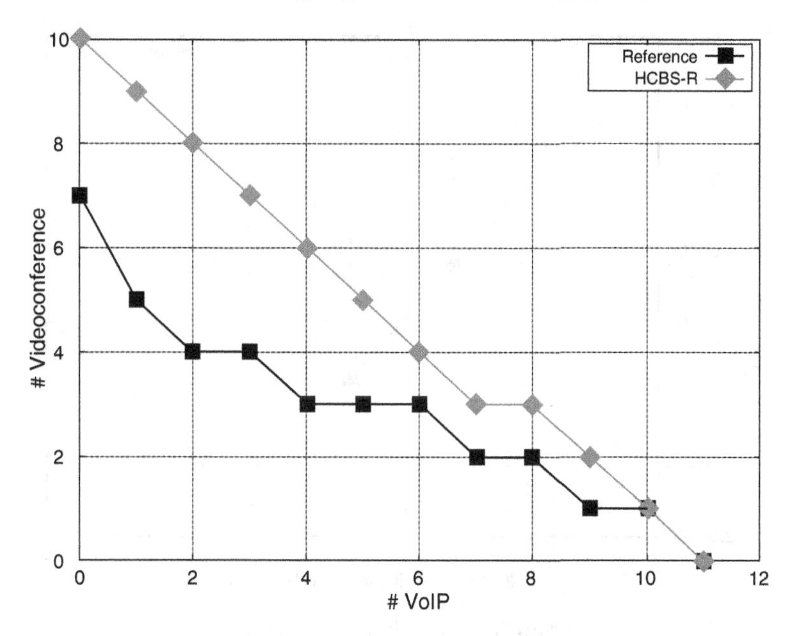


**Fig. 1.** Admission control. Number of admitted videoconference TSs against the number of admitted VoIP G.711 TSs

### 5.3 Data Throughput

We evaluate a scenario with four stations with mixed CBR and VBR traffic. To do so, we set up an increasing number of QSTAs, from 0 to 4, each having a bi-directional VoIP TS and bi-directional Video Conference TS. The delay bound of VoIP is set to 20 ms and that of VC TSs to 33 ms.

The Fig. 2 shows the throughput achieved by stations with data traffic against the number of stations with bi-directional VoIP and VC sessions. Stations with data traffic operate in asymptotic conditions, i.e. they always have a frame to transmit. The packet length of data traffic is constant and equal to 1,500 bytes. If there are not any stations with CBR and VBR TSs, the data throughput is maximum and WCBS-R behaves in a very similar way to the reference scheduler. Otherwise, if there are TSs with significantly different delay bound requirements, such as the VoIP and VC TSs, the MAC overhead of the reference scheduler is higher than that with framework scheduler and, therefore, the throughput achievable by data traffic is much lower.

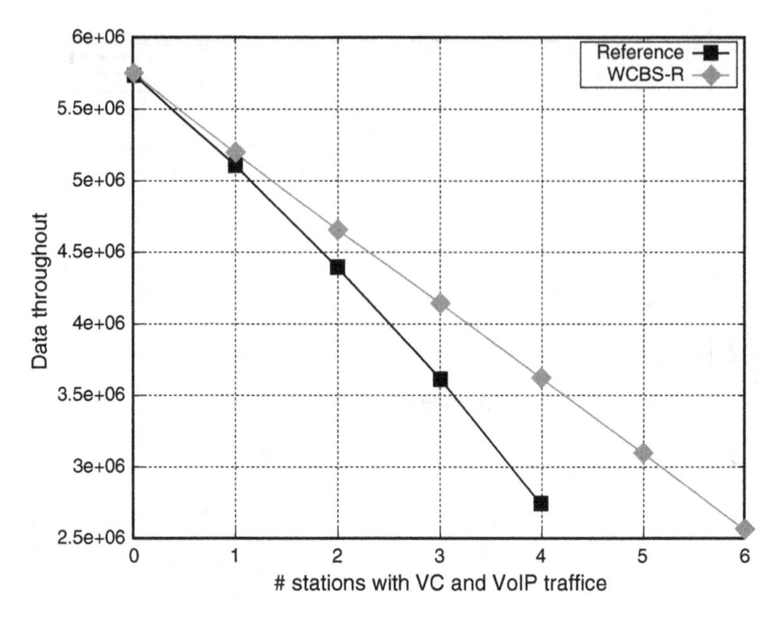


**Fig. 2.** Throughput of stations

## 6 Conclusions

In this paper we have defined a new scheduling algorithm alternative to the reference scheduler to integrate a QoS support in IEEE 802.11e wireless networks with soft real-time guarantees. The scheduling algorithm, named WCBS-R, is based on CBS-R Real-Time algorithm which permits to dynamically manages the medium resources and it supports variable packet size and variable bit rate traffic streams. The simulation analysis shows that WCBS-R efficiently services QoS traffic with different delay bounds, thus saving capacity which is made available to contention-based data traffic and outperforming the reference scheduler.

## References

1. *Wireless LAN Medium Access Control(MAC) and Physical Layer (PHY) Specification*, IEEE Std. 802.11, 1997
2. *802.11E-2005 IEEE Standard for Information technology Telecommunications and information exchange between systems Local and metropolitan area networks Specific requirements Part 11: Wireless LAN Medium Access Control (MAC) and Physical Layer (PHY) specifications: Amendment 8: Medium Access Control (MAC) Quality of Service Enhancements*, IEEE Std. 802.11e, 2005
3. *Wireless LAN Medium Access Control (MAC) and Physical Layer (PHY) Specification: Higher-Speed Physical Layer Extension in the 2.4 GHz Band*, IEEE Std. 802.11b, 1999

4. H. Zhu and I. Chlamtac, "An analytical model for IEEE 802.11e EDCF differential services," in *Proc. 12th International Conference Computer Communication and Networks*, Dallas, TX, October 2003
5. A. Grilo and M. Nunes, "Performance evaluation of IEEE 802.11e," in *Proc. PIMRC 2002*, vol. 1, Lisboa, Portugal, September 2002, p. 511–517
6. J. Cowling and S. Selvakennedy, "A detailed investigation of the IEEE 802.11e HCF reference scheduler for VBR traffic," in *13th International Conference On Computer Communications And Networks (ICCCN 2004)*, Chicago, US, October 2004
7. H. Fattah and C. Leung, "An overview of scheduling algorithms in wireless multimedia networks," *IEEE Wireless*, vol. 9, no. 5, pp. 76–83, October 2002
8. S. Lu, V. Bharghavan, and R. Srikant, "Fair scheduling in wireless packet networks," *IEEE/ACM Trans. Net.*, vol. 7, no. 4, pp. 473–489, August 1999
9. S. Tsao, "Extending earliest due-date scheduling algorithms for wireless networks with location dependent errors," in *Proc. IEEE VTC. 2000*, Boston, MA, September 2000
10. A. Grilo, M. Macedo, and M. Nunes, "A service discipline for support of IP QoS in IEEE 802.11 networks," in *Proc. PWC 2001*, Laapenranta, Finland, August 2001
11. L. Abeni and G. Buttazzo, "Integrating multimedia applications in hard real-time sistems," in *IEEE Real Time-Systems Symposium*, December 1998, pp. 4–13
12. M. Caccamo and L. Sha, "Aperiodic servers with resource constraints," in *IEEE RealTime Systems Symposium*, December 2001
13. Y. Cao and O. L. Li, "Scheduling algorithms on broad-band wireless networks," *Proceeding of the IEEE*, vol. 89, pp. 76–87, January 2001
14. Y. L. Boudec and T. Thiran, "A short tutorial on network calculus i: Fundamental bounds in communication networks," in *IEEE ISCAS*, Geneva, Switzerland, May 2000, pp. IV–93–IV–96
15. S. Mangold, S. Choi, P. May, O. Klein, G. Hiertz, and L. Stibor, "IEEE 802.11e wireless LAN for quality of service," in *Proc. European Wireless (EW2002)*, 2002
16. A. Grilo, M. Macedo, and M. Nunes, "A scheduling algorithm for QoS support in IEEE 802.11e networks," *IEEE Wireless Communications*, pp. 36–43, june 2003
17. C. Cicconetti, L. Lenzini, E. Mingozzi, and G. Stea, "Efficient provisioning of real-time qos guarantees in ieee 802.11e wlans," in *European Wireless 2006*, Athens, April 2006
18. P. Ansel, Q. Ni, and T. Turletti, "An efficient scheduling scheme for IEEE 802.11e," in *Proc. Modeling and Optimization in Mobile, Ad Hoc and Wireless Networks*, 2004
19. A. Annese, G. Boggia, P. Camarda, L. A. Grieco, and S. Mascolo, "Providing delay guarantees in IEEE 802.11e networks," in *59th IEEE Semiannual Vehicular Technology Conference, VTC Spring*, 2004
20. *Network Simulator 2*, http://www.isi.edu/nsnam/ns/
21. C. Cicconetti, L. Lenzini, E. Mingozzi, and G. Stea, "A software architecture for simulating ieee 802.11e hcca," in *3rd IPS MoMe*, Warsaw (Poland), March 2005

22

# Mobility Management QoS Measures Evaluation for Next Generation Mobile Data Networks

Kumudu S. Munasinghe and Abbas Jamalipour

School of Electrical and Information Engineering University of Sydney, Australia
kumudu@ee.usyd.edu.au, a.jamalipour@ieee.org

**Summary.** An evaluation of Quality-of-Service (QoS) measures in mobility management for interworked Next Generation Mobile Data Networks (NGMNs) is presented in this paper. The interworking framework used for the QoS analysis is a novel approach for NGMNs. This enables multiple cellular networking technologies such as the Universal Mobile Telecommunications System (UMTS) and CDMA2000 technology to interwork with Wireless Local Area Networks (WLANs) under a common platform. The proposed framework exploits the IP Multimedia Subsystem (IMS) as a universal coupling mediator for real-time session negotiation and management and Mobile IP (MIP) for IP mobility management. The latter part of the paper uses an analytical model for analyzing the QoS measures related to mobility management and uses an OPNET based model for verification.

## 1 Introduction

Ubiquitous data services and relatively high data rates across heterogeneous networks could be achieved by interworking 3G cellular networks with WLANs. Thus WLANs can be considered as a complementary technology for 3G cellular data networks as well as a compulsory element of the future NGMN [1]. Having identified the importance of such an interworking mechanism, we have currently proposed a solution for achieving session continuation during a vertical handoff between WLAN and UMTS networks [2]. However, the proposed interworking mechanism does not fully achieve the motive of a NGMN since global roaming and interoperability beyond one cellular system to another is not supported. Therefore, the first half of this paper outlines our contribution towards designing a common platform for coupling CDMA2000, UMTS, and WLAN networks as an extension to our earlier works.

The novelty of our previous architecture for WLAN-UMTS interworking is that it uses the 3GPP's IMS as an arbitrator for real-time session negotiation and management [3]. Since 3GPP2 also adopts a similar concept as 3GPP by introducing an IMS in the CDMA2000 Core Network (CN), the IMS can be deemed as a top candidate as a universal coupling mediator [4]. However, since there are substantial differences between 3GPP and 2GPP2 CNs in regards to the level of contribution of the IMS in relation to mobility and session management, a major overhaul in the existing architecture is required. The second half of this paper focuses on analytically modeling the mobility management QoS aspect within the packet switching service domain for the extended framework. QoS measures such as vertical handoff delay, transient packet loss, and signaling cost are modeled and compared against our previous works. Furthermore, an OPNET based simulation platform is constructed for verification of handoff delay, packet loss, and jitter.

The reminder of this paper is organized as follows. The next section presents an overview on IMS based mobility and session management. Followed by this is the proposed architecture, which describes how the IMS has been used as a universal coupling mediator for interworking. The next sections are on analytically modeling, simulation based verification, and finally the concluding remarks.

## 2 IMS-Based Mobility and Session Management

The UMTS Release 5 within its CN, introduced the IP Multimedia Subsystem [3]. As illustrated in Fig. 1, it consists of the essential requirements for controlling of multimedia sessions and provisioning of IP multimedia services for a UMTS network [3].

The key elements of the IMS framework are the Call State Control Functions (CSCFs) and the Home Subscriber Server (HSS). A CSCF can be loosely defined as a Session Initiation Protocol (SIP) proxy server and the HSS as a master database for user profiles. The CSCFs can be identified as follows; Proxy-CSCF, Interrogating-CSCF, and Serving-CSCF. The P-CSCF is the first SIP proxy receiving a SIP request. The P-CSCF forwards session requests to the S-CSCF via the I-CSCF. The task of the I-CSCF is to select

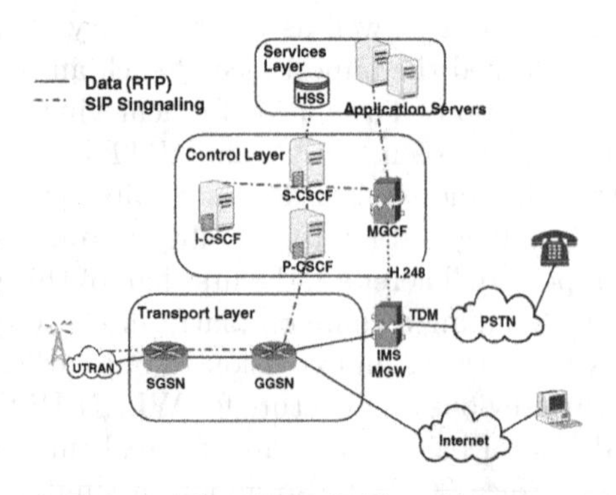


**Fig. 1.** The 3GPP-IMS architecture

the appropriate S-CSCF by checking with the HSS. The S-CSCF is the actual SIP server that eventually performs the user registration and handles session control for the IMS network. The Media Gateway Control Function (MGCF) interconnects with circuit switched networks via the corresponding IMS Media Gateway (IMS-MGW).

SIP is the core protocol chosen by the 3GPP for signaling and session management within the IMS [5]. The extensible nature of SIP has also been utilized by the 3GPP for incorporating additional features to suit the IMS [6]. Since 3GPP uses an IMS (or Application Layer) based mobility and session management approach, SIP is used to implement all four mobility aspects as defined by IMT-2000; namely, personal, terminal, session, and service mobility [7].

The general packet data support portion of the 3GPP2's Multi Media Domain (MMD) is known as the Packet Data Subsystem (PDS) and the entity that comprise the multimedia session capability is collectively known as the IMS [4]. As illustrated in Fig. 2, the initial release of the 3GPP2-IMS is based on 3GPP's IMS Release 5 specifications. However, there exist considerable differences within the 3GPP2-IMS specifications [8]. Firstly, 3GPP2 uses MIP for IP mobility (i.e., terminal mobility) management and SIP for session mobility management where as 3GPP exclusively uses SIP for all types of mobility management.

Secondly, although 3GPP mandates IMS over IPv6, 3GPP2 does not make such a specification. Therefore, either MIPv4 [9] over IPv4 or MIPv6 [10] over IPv6 can be used. If MIPv4 is deployed, the Packet Data Switching Node (PDSN) of the home network may act as the MIP Home Agent (HA) or MIP Foreign Agent (FA) as the Mobile Node (MN) moves form one PDSN to another. Since 3GPP2's standard for CDMA2000 does not fully support inter-PDSN mobility for IPv6, the proposed design will primarily be based on MIPv4 [11]. However, transition of this model to MIPv6 can be easily done by addressing user authentication, address allocation and enabling the MN to perform the MIPv6 update procedures [12].

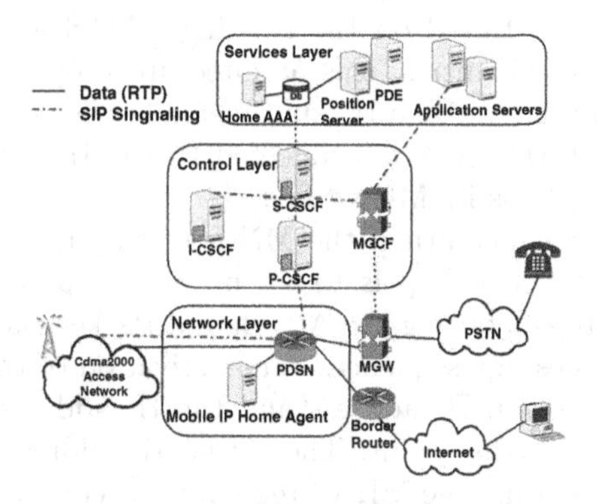


**Fig. 2.** The 3GPP2-IMS architecture

Although it is of lesser importance, other differences between 3GPP and 3GPP2 IMS can be briefly stated as follows; 3GPP2's Home AAA and its database being equivalent to 3GPP's HSS and the P-CSCF and PDSN do not need to reside in the same network. The other differences are codecs, QoS procedures, S-CSCF/P-CSCF/HSS interfaces, authentication procedures to name a few.

# 3 Interworking Architecture

The proposed architecture can be seen as a universally applicable, joint MIP-SIP/IMS approach for mobility and session management in NGMNs. Nevertheless, the aim of this approach is not to develop a hybrid version of the IMS by integrating MIP features (i.e., a S-CSCF with an integrated MIP-HA) as suggested in [13]. Depending on the version of MIP used, few variations for this signaling framework may exist. The approach used in our design works on a pure IPv4 platform. The main reasons for using MIPv4 is that it eliminates the complexity of IP addresses managing and enables IP mobility management (by transporting single/static IP address) in such a way that node mobility is transparent to the layers above. As a result, the IMS is able to provide session mobility for a roaming MN without any disruption of IP connectivity thus, enabling seamless service continuity.

According to the 3GPP2's IMS standard IMS signaling can be implemented over IPv4. Nevertheless, since the IMS is primarily a framework for SIP based session control, even though the original specification mentions otherwise, 3GPP's IMS can also be successfully implemented over IPv4. However, it is worth noting that route optimization is used for overcoming triangular routing. Another important motive for using an all IPv4 based platform is the fact that network operators are coming to a sad realization that despite of the initial hype, the migration towards IPv6 is slowly becoming more and more practically unrealistic.

Our proposed internetworking architecture, with signaling and data routes, is illustrated in Fig. 3. Based on the existing UMTS and CDMA2000 mobility management mechanisms, this architecture proposes a framework for NGMN's for providing real-time IP multimedia services. As per the illustration on Fig. 3, the UMTS CN is connected to the IP network through the GGSN, which also acts as its MIPv4 FA.

Prior to IP address acquiring, the MN goes through the system acquisition procedures. The next step is to set up a data pipeline. This must be completed in two-steps by using the Attach and Packet Data Protocol (PDP) context activation message sequences. The activation of the MN's PDP context does not allocate an IP address and the IP address field of the PDP context is not filled at this point. The actual IP address allocation for the MN is initiated by sending the MIPv4 registration requests to it's HA via the

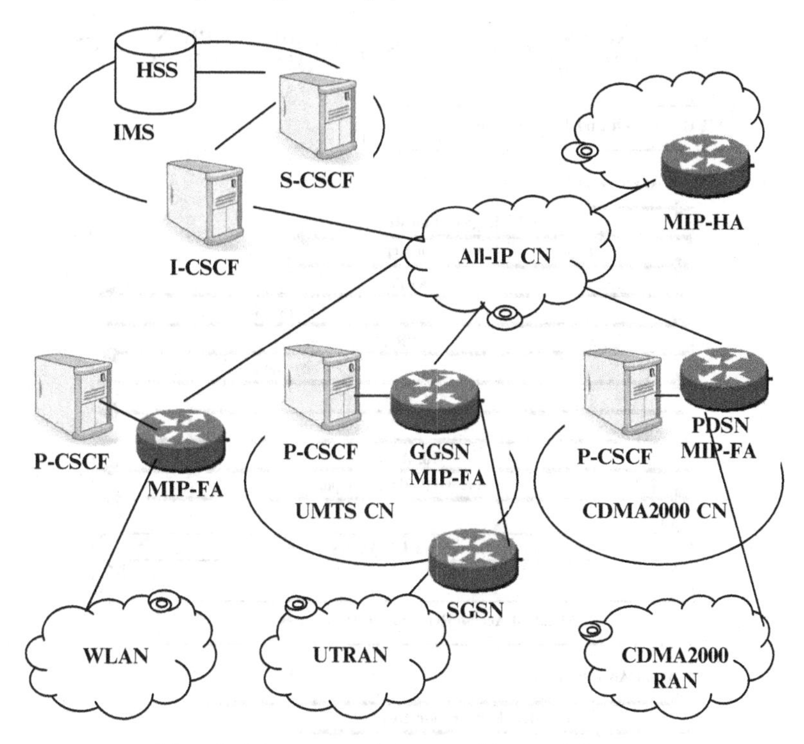


**Fig. 3.** The proposed interworking architecture

GGSN (MIP-FA) as specified under 3GPP's [14]. When the GGSN (MIP-FA) receives this information the MN gets its home IP address assigned.

Now the MN is able to identify the P-CSCF for the registration with the IMS. Prior to establishing a SIP session, the MN requires performing a service registration function to let the IMS know its location. The MN acts as a SIP client and sends a SIP registration message to its home system through the P-CSCF. Once authorized, a suitable S-CSCF for the MN is assigned and its subscriber profile is sent to the designated S-CSCF.

After the activation of the PDP context and the service registration, the MN is ready to establish a media/data/call session. As illustrated in Fig. 4, the sequence of the SIP session origination procedure can be described as follows. The mobile origination procedure is initiated by a SIP INVITE message sent form the UMTS interface of the source MN. Within the Session Description Protocol (SDP) body of this SIP INVITE contains a request to follow the precondition call flow model. This is important because some clients require certain preconditions (that is, QoS levels). Next, the destination responds with a 183 Session Progress containing the required information of media streams and codecs applicable for this session. This is acknowledged by a PRACK request which is reciprocated by a 200 OK response generated by the destination. Next an UPDATE request is sent by the source containing

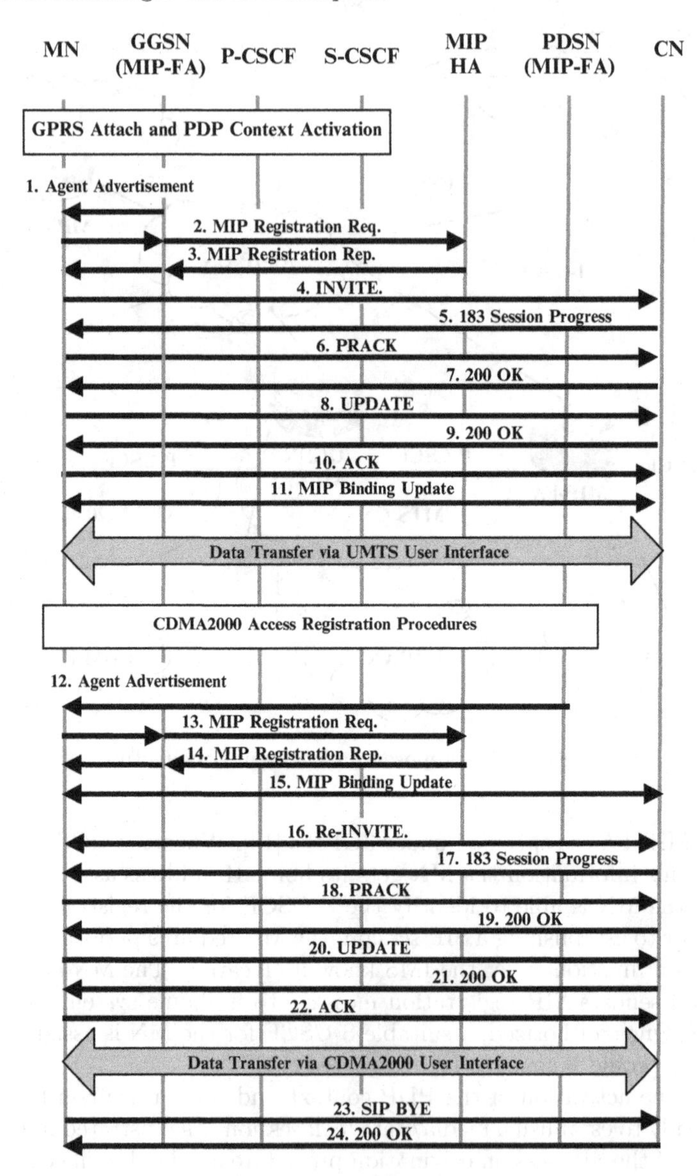


**Fig. 4.** MIP-SIP signaling flow

another SDP offer confirming that resources are reserved. Once the destination receives the UPDATE request, it generates a 200 OK response. Once this is done, the MN can start the media/data flow and the session will be in progress (via the UMTS interface).

When this MN roams between CDMA2000 and UMTS systems intersystem roaming takes place. The message flow for an intersystem roaming from UMTS to CDMA2000 can be described as follows. Firstly the standard CDMA2000 link layer access registration procedures are performed. Next the CDMA2000 interface performs the MIP registration procedures with the PDSN (MIP-FA) as explained previously. This is when the PDSN (MIP-FA) forwards this request to the MIP-HA and the HA assigns the home IP address to the new CDMA2000 interface. Lastly the exchanging of a MIP Binding Update message between the MN and the CN for avoiding triangular routing [9].

The next stage is the taking place of the IMS-SIP session handoff procedures. This requires sending a SIP Re-INVITE (with same Call-ID and other identifiers corresponding to the ongoing session) to the destination SIP UAC. Followed by this is a resource/preconditions reservation for the CDMA2000 interface. Once this is successfully done the new session flow can be initiated. It is important to note that until such time that the new data flow is initiated via the CDMA2000 interface, the data flow via the UMTS interface remains active. Thus the model follows the make-before-break handoff mechanism as proposed in our previous works [2]. Finally, the UMTS interface performs the implicit detach procedures with the SGSN and the PDP context is deactivated. Intersystem roaming form CDMA2000 to UMTS can also take place in a similar manner. Furthermore, since this design is an extension to our UMTS-WLAN interworking platform, WLAN-CDMA2000 roaming can also be accommodated within this architecture in a similar manner.

## 4 Analytical Modeling

An analytical model is devised for evaluating the proposed MIP-SIP based scheme against our previous works (i.e., the pure SIP scheme) to analyze the QoS metrics and measures involved in mobility management. More precisely, QoS metrics such as handoff delay, total packet loss, and Signaling overhead are analyzed. The derivations of the numerical results have been omitted in the following sections for brevity.

### 4.1 Handoff Delay

A standard vertical handoff delay during mid-session mobility consists of the following subprocedures (or delays); $D_1$ = Link layer HO delay, $D_2$ = movement detection delay, $D_3$ = address allocation delay, $D_4$ = session reconfiguration delay, and $D_5$ = packet re-transmission delay.

The vertical handoff delay/s at the network layer (and above) are calculated independent of the link layer delay $D_1$ and mainly consist of $D_3$ and $D_4$. According to our proposed architecture of MIP-IMS based vertical handoff, there is no DHCP related address allocation; hence it can be argued that the main contributor for network layer based vertical handoff delay is $D_4$. The session re-configuration delay ($D_4$) mainly consists of MIP handoff delay ($D_{MIP}$) and IMS based session handoff delay ($D_{IMS}$)

$$D_4 = D_{MIP} + D_{IMS} \,. \tag{1}$$

Let $D(S, H_{a-b})$ denote the end-to-end transmission delay of a message of size $S$ sent from $a$ (an MN away) to $b$ via wireless and wired links. Thus, $D(S, H_{a-b})$ can be expressed as follows [15]:

$$D\left(S, H_{a-b}\right) = \left[\frac{S}{B_{wl}} + L_{wl}\right] + \left[\frac{S}{B_w} + L_w\right] \times H_{a-b} + \left[H_{a-b} + 1\right] \times L_{proc} \,, \tag{2}$$

where, $S$ is the average size of a signalling message, $H_{a-b}$ is the average number of hops between $a$ and $b$, $B_{wl}$ is the bandwidth of the wireless link, $B_w$ is the bandwidth of the wired link, $L_{wl}$ is the latency of the wireless link, $L_w$ is the latency of the wired link, and $L_{proc}$ is the processing delay at node. Therefore, by using (2), $D_{MIP}$ and $D_{IMS}$ for the architecture given by Figs. 3 and 4 can be expressed as in (3) and (4), respectively.

$$\begin{aligned} D_{MIP} = {} & 2[D(S_{MIPReg}, H_{CDMA-FA}) + D(S_{MIPReg}, H_{FA-HA}) \\ & + D(S_{MIPBU}, H_{CDMA-CN})] \end{aligned} \tag{3}$$

$$\begin{aligned} D_{IMS} = {} & D(S_{Re-INVITE}, H_{CDMA-CN}) + D(S_{183-SP}, H_{CDMA-CN}) \\ & + D(S_{PRACK}, H_{CDMA-CN}) + D(S_{OK}, H_{CDMA-CN}) \\ & + D(S_{UPDATE}, H_{CDMA-CN}) + D(S_{OK}, H_{CDMA-CN}) \\ & + D(S_{ACK}, H_{CDMA-CN}) + \Delta \,, \end{aligned} \tag{4}$$

where, $\Delta$ is additional IMS (application layer) latency due to HSS lookup process. Now, by using (2), for comparison purposes $\mathrm{D}_{IMS}$ for the pure SIP approach architecture given by [2] can be expressed as in (5)

$$\begin{aligned} D_{IMS} = {} & D(S_{Refer}, H_{UMTS-CDMA}) + D(S_{202Accept}, H_{UMTS-CDMA}) \\ & + D(S_{INVITE}, H_{CDMA-CN}) + D(S_{183-SP}, H_{CDMA-CN}) \\ & + D(S_{Notify}, H_{UMTS-CDMA}) + D(S_{PRACK}, H_{CDMA-CN}) \\ & + D(S_{OK}, H_{CDMA-CN}) + D(S_{UPDATE}, H_{CDMA-CN}) \\ & + D(S_{OK}, H_{CDMA-CN}) + D(S_{ACK}, H_{CDMA-CN}) \\ & + D(S_{OKNotify}, H_{UMTS-CDMA}) + \Delta \,. \end{aligned} \tag{5}$$

It is worth reminding that make-before-break handoff is applied in the proposed handoff scenarios, which helps compensate for large handoff delays. However, for purpose of a complete analysis of the vertical handoff delay, the standard straight forward case of break-before-make handoff scenario is used.

### 4.2 Packet Loss

The total packet loss (*Pkt_loss*) during a session can be defined as the sum of all lost packets during the vertical handoff while the MN is receiving the downlink data packets. It is assumed that the packet loss begins when the L2 handoff is detected and all in-flight packets are lost during the vertical handoff time. Thus, it can be expressed as follows [15]:

$$Pkt\_loss = \left[\frac{1}{2T_{ad}} + D_4\right] \times \lambda_d \times N_m \ , \qquad (6)$$

where, $T_{ad}$ is the time interval between MIP agent advertisements, $\lambda_d$ is the downlink packet transmission rate, and $N_m$ is the av. number of handoffs during a session. $N_m$ is known as $t_s/t_r$, where $t_r$ is av. network resident time and $t_s$ is av. call (session) connection time.

### 4.3 Signaling Overhead

The resultant signaling overhead of mobility management during vertical handoff can be analyzed as follows. The signaling overhead is the accumulative traffic load on exchanging signaling messages during the MN's communication session. The overhead incurred by a message can be defined as [16]

$$O_{message} = S_{message} \times H_{a-b} \ . \qquad (7)$$

Thus the total signaling overhead incurred by vertical handoffs during a given data session for the MIP+SIP scenario can be computed as follows:

$$O_{MIP+IMS} = \lambda_m \left[\sum_{i=1}^{n_1} \left(S_{MIP-i} \times H_{(a-b)-i}\right) + \sum_{i=1}^{n_2} \left(S_{IMS-Invite-i} \times H_{(a-b)-i}\right)\right] ,$$

$$O_{MIP+IMS} = \left[\lambda_m \sum_{i=1}^{n_3} \left(S_{IMS-ReInvite-i} \times H_{(a-b)-i}\right)\right] \frac{\lambda_s}{\lambda_m} + \lambda_m \left[\sum_{i=1}^{n_1} \left(S_{MIP-i} \times H_{(a-b)-i}\right) + \sum_{i=1}^{n_2} \left(S_{IMS-Invite-i} \times H_{(a-b)-i}\right)\right] , \qquad (9)$$

where $n_1$, $n_2$, and $n_3$ represent the number of messages involved in each handoff/message sequence. Furthermore, $\lambda_m$ is the average network mobility rate of a MH and $\lambda_s$ is the av. call (session) arrival rate. $\lambda_s/\lambda_m$ is known as call-to-mobility rate (CMR). Now, the total signaling overhead incurred by vertical handoffs during a given data session for the Pure SIP scenario can be computed as follows:

$$O_{IMS} = \lambda_m \sum_{i=1}^{n_1} \left(S_{IMS-Invite-i} \times H_{(a-b)-i}\right) + \lambda_s \sum_{i-1}^{n_2} \left(S_{IMS-ReInvite-i} \times H_{(a-b)-i}\right), \tag{10}$$

$$O_{IMS} = \left[\lambda_m \sum_{i=1}^{n_2} \left(S_{IMS-ReInvite-i} \times H_{(a-b)-i}\right)\right] \frac{\lambda_s}{\lambda_m} + \lambda_m \sum_{i=1}^{n_1} \left(S_{IMS-Invite-i} \times H_{(a-b)-i}\right). \tag{11}$$

## 5 Numerical Results and Analysis

The following numerical results are generated using MIPv4 and 3GPP-SIP messages. Table 1 shows the typical MIPv4 and SIP message sizes and other related parameters. MIPv4 values are based on [17] and IMS-SIP values are based on [18]. The relative distances in hops are illustrated in Fig. 5. Other related parameters have been partly obtained form [15, 16] to maintain consistency.

Based on these assumptions, the following analytical results are derived for (1), (6), and (9) for the scenarios of a MIP-SIP based vertical handoff and a pure SIP based vertical handoff. Figure 6 illustrates the behavior of the vertical handoff delay against the wireless link bandwidth. The behavioral trends of the graphs indicate that the vertical handoff delay is largely reduced in the MIP-SIP scheme in contrast with a pure IMS-SIP based scheme. These reductions result from the use of rather short MIP messages for handling IP mobility instead of the IMS related SIP Refer method.

**Table 1.** MIPv4, IMS-SIP message sizes and values of other parameters

| MIPv4 message | Size (Bytes) | IMS-SIP | Size (Bytes) | Parameter | Value |
|---|---|---|---|---|---|
| Agent solicit. | 67 | INVITE | 736 | $B_{wl}$ | 16 K–54 Mbps |
| Agent advert | 28 | Re-INVITE | 731 | $B_w$ | 100 Mbps |
| Reg. request | 60 | 183 Sess. Prog | 847 | $L_{wl}$ | 2 ms |
| Reg. reply | 56 | PRACK | 571 | $L_w$ | 0.5 ms |
| Binding update | 66 | 200 OK | 558 | $L_{proc}$ | 0.001 s |
| Binding ack. | 66 | UPDATE | 546 | $\lambda_m$ | 10 h$^{-1}$ |
| | | ACK | 314 | CMR | 0.1–10 |
| | | REFER | 750 | $\Delta$ | 100 ms |
| | | 200Accepted | 550 | $T_{ad}$ | 1 s |
| | | NOTIFY | 550 | $\lambda_d$ | 64 Kbps |
| | | OK NOTIFY | 550 | | |

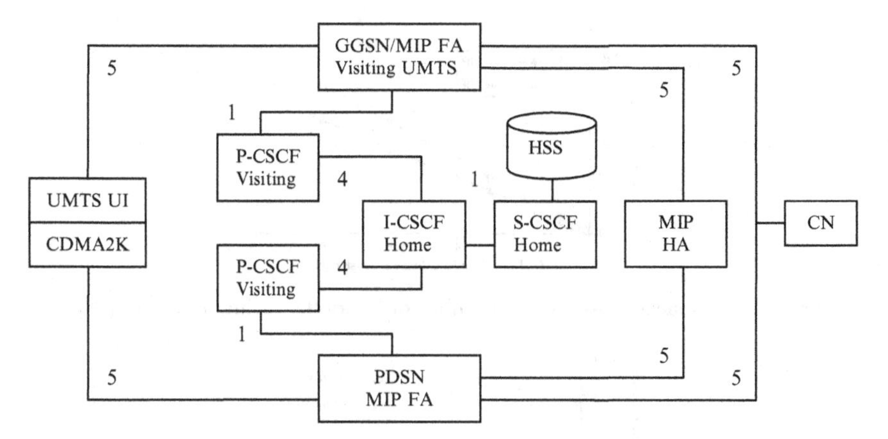


**Fig. 5.** Relative distances in hops

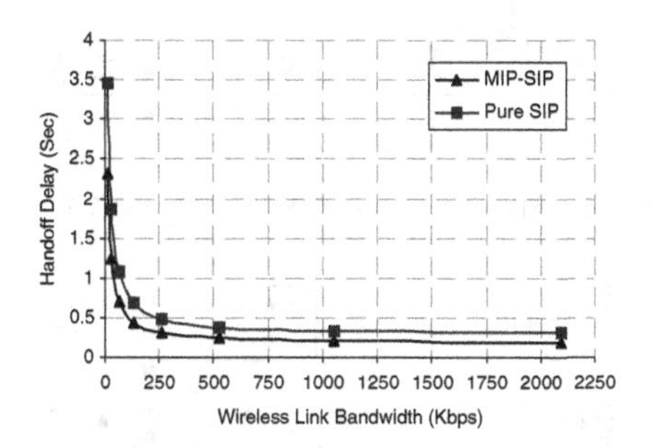


**Fig. 6.** Vertical handoff delay vs. wireless link bandwidth

Furthermore, the MIP-SIP method has a relatively less number of application layer/IMS based processing latencies, which also contributes to an overall low vertical handoff delay. Another interesting observation is that when the bandwidth of the wireless link increases form 0 to 2 Mbps the handoff delay decreases exponentially. It also indicates that for relatively wider bandwidths (say, beyond 512 Mbps onwards) the handoff delay decrease becomes relatively slower. Therefore, this indicates that vertical handoff delays cannot be simply reduced by purely increasing the wireless link bandwidth.

Figure 7 demonstrates the total packet loss during the vertical handoff time as the number of handoffs per session is increased. According to (6), the packet loss during a vertical handoff is relatively proportional to the vertical handoff delay. Therefore, since the pure SIP approach has a relatively higher handoff delay, the resultant packet loss is always greater than in the case of the MIP-SIP method. Another interesting observation of the plots in Fig. 7 is that the two graphs diverge form one another as the number of handoffs per session increases. The reason for the pure SIP graph to diverge from the MIP-SIP graph are the application layer based additional IMS related latencies.

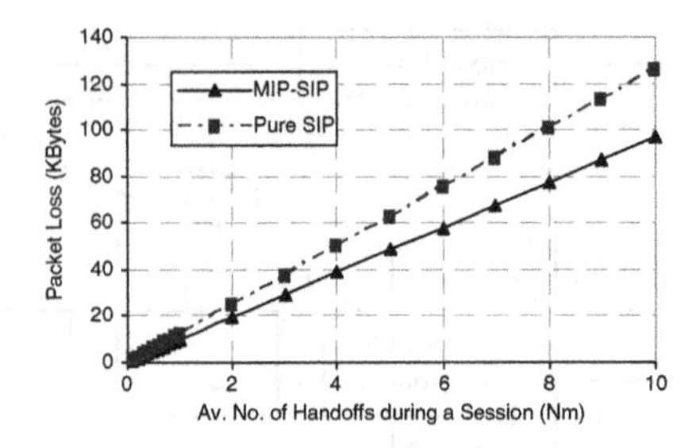


**Fig. 7.** Total packet loss vs. av. number of handoffs during a session ($N_m$)

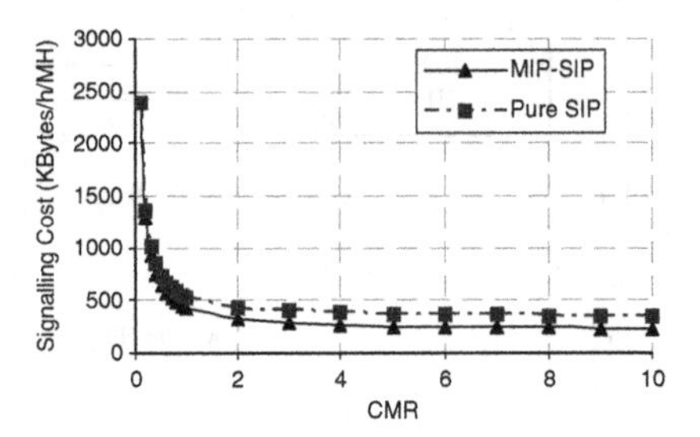


**Fig. 8.** Signalling cost vs. CMR at $\lambda_s = 5\,\text{h}^{-1}\,\text{MH}^{-1}$

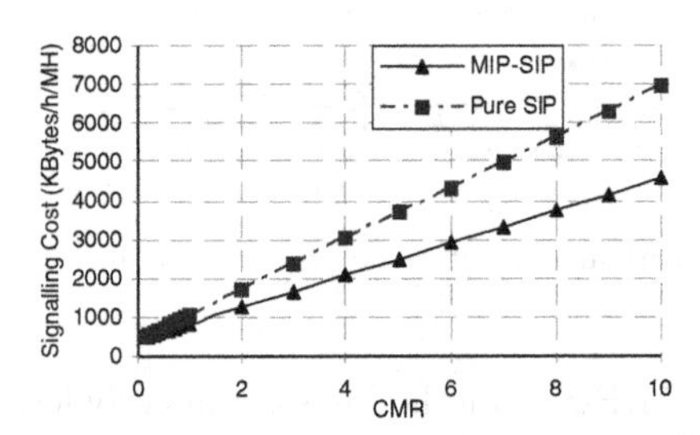


**Fig. 9.** Signalling cost vs. CMR at $\lambda_m = 10\,\text{h}^{-1}\,\text{MH}^{-1}$

These additional latencies substantially contribute to increase handoff delays as the number of handoffs per session increases and eventually the packet loss.

Figures 8 and 9 illustrate the behavior of the total mobility management related signaling overhead/cost against CMR at $\lambda_s = 5\,\text{h}^{-1}\text{MH}^{-1}$ and $\lambda_m = 10\,\text{h}^{-1}\text{MH}^{-1}$ respectively. That is, in Fig. 8, the call (session) arrival rate ($\lambda_s$) is fixed and in Fig. 9, the network mobility rate ($\lambda_m$) is fixed. In both cases, MIP-SIP approach shows a clear-cut reduction in the overall signaling cost in contrast to the pure SIP approach. Figure 8 also indicates that the signaling cost of the pure SIP method is approximately over 30% than the MIP-SIP approach for relatively high CMR values.

According to the graph in Fig. 8, the signaling cost reduces exponentially with the increase of CMR since the mobility rate ($\lambda_m$) declines when the session arrival rate ($\lambda_s$) is constant. Conversely, the graphs in Fig. 9 indicate that the signaling cost increases linearly as the CMR increases since the session arrival rate ($\lambda_s$) increases when the mobility rate ($\lambda_m$) is constant. This also leads to a very interesting conclusion. That is, more signaling overhead is incurred during the vertical handoff than at the time of the initial session setup.

## 6 Simulation Results and Validation

In order to validate the above numerical results and analysis, a network simulation scenario is modeled using OPNET Modeler 11.5. Since OPNET's standard SIP components do not address the specifications of the IMS, substantial modifications are required for the development of a fully functional SIP-IMS model. Our newly developed SIP-IMS model is an enhanced version of the basic IMS-SIP signaling model, which is currently available under the contributed models library of the OPNET University Program [19]. Followed by this, below the IMS architecture a MIP v4 framework is constructed for handling IP mobility at the network layer (and limiting the IMS exclusively for session mobility management).

The next task is creating an All-IP based heterogeneous network with MIP and SIP signaling as illustrated in Fig. 3 with the relative distance hops of Fig 5. Once such a network is created, simulations can be made for observing vertical handoffs for the proposed MIP-SIP mobility management framework over a heterogeneous network. According to Fig. 3, since IMS and MIP protocols are implemented in the CN, their behavior can be considered to be independent to the underlying networks (i.e., to WLAN, UMTS, or CDMA2000). Taking the above facts and limitations of OPNET 11.5 into consideration, a simple All-IP heterogeneous test bed is created by interworking a UMTS network with a WLAN. Using the above described platform, measurements are collected for investigating QoS measures for a joint MIP-SIP vertical hand-off (say, from UMTS to WLAN in this case).

The average vertical handoff delay obtained by the OPNET model for a single VoIP session over a 2 Mbps wireless link bandwidth for the MIP-SIP model is 190 ms, and the same for a pure SIP based mechanism is 302 ms. Thus the handoff delay for the pure SIP approach is greater by 37% in comparison to the MIP-SIP mechanism. A similar trend can also be observed from the curves illustrated by Fig. 6. Similarly the packet loss metric obtained from the OPNET simulation platform behaves similar to the numerical results of packet loss. Thus the average packet loss for a single handoff during a single session (i.e., when $N_m = 1$), for a 64 Kbps link for MIP-SIP mechanism is 10 KBytes and pure SIP mechanism is 12.5 KBytes. This is also closely inline with the results of Fig. 7. Another interesting point about our simulation results is that, to a certain extent, they are inline with results published for the case of horizontal handoff delays [16, 20].

Additionally to the QoS metrics investigated through analytical modeling, the OPNET model investigated the jitter metric (i.e., the variation in the interarrival delay). As the simulation results indicate, the jitter is 28 ms and 47 ms for joint MIP-SIP and pure SIP based approaches for the case of a single VoIP session. This is also a particularly encouraging observation. Since a jitter rate below 50 ms is expected to provide acceptable voice quality in real-time VoIP applications, the above readings indicate the likelihood of maintaining satisfactory (if not seamless) levels of voice communications during the vertical handoff process [21].

## 7 Conclusions

This paper presented a mobility management QoS evaluation for a unified mobility and session management framework for NGMNs. This model enabled internetworking between UMTS, CDMA2000, and WLAN networks where the IMS acted as a unified session controller at the application layer and MIP handled IP mobility management at the network layer. The QoS evaluation was based on an analytical modeling and an OPNET based network simulation model was used for verification of these results. The results indicated that despite the additional overheads imposed by SIP signaling, this new approach could be successfully used for interworking in NGMNs. The QoS investigation further indicated that the proposed mobility management mechanism has the potential of transferring a session with acceptable levels of service continuation.

## Acknowledgments

This work is supported by an ARC Linkage Project in collaboration with SingTel Optus.

## References

1. M. Buddhikot, et al., "Integration of 802.11 and third-generation wireless data networks," in *Proceedings of the IEEE INFOCOM*, San Francisco, California, Apr. 2003.
2. K. Munasinghe and A. Jamalipour, "A 3GPP-IMS based Approach for Converging Next Generation Mobile Data Networks," in *Proceedings of the IEEE ICC*, Glasgow, UK, Jun. 2007.
3. 3GPP, "IP Multimedia Subsystem (IMS)," 3GPP TS 23.228 Version 6.10.0 Release 6, 2005.
4. 3GPP2, "All-IP Core Network Multimedia Domain," 3GPP2 X.S0013-002-A v1.0, Nov. 2005.
5. J. Rosenberg, et al., "SIP: Session Initiation Protocol," RFC 3261, 2002.

6. F. G. Marquez, M. G. Rodriguez, T. R. Valladares, T. de Miguel, and L. A. Galindo, "Interworking of IP multimedia core networks between 3GPP and WLAN," *IEEE Wireless Communications*, vol. 12, pp. 58–65, 2005.
7. N. Banerjee, K. Basu, and S. K. Das, "Hand-off delay analysis in SIP-based mobility management in wireless networks," in *Proceedings of the International Parallel and Distributed Processing Symposium*, Nice, France, Apr. 2003.
8. A. Mahendran and J. Nasielski, "3GPP/3GPP2 IMS Differences," 3GPP2, S00-20020513-030, 2002.
9. C. Perkins, "IP Mobility Support for IPv4," IETF RFC 3344, 2002.
10. D. Johnson, C. Perkins, and J. Arkko, "Mobility Support in IPv6," IETF RFC 3775, 2004.
11. 3GPP2, "Wireless IP Network Standard," 3GPP2 1 PP2 P.S0001-B, 2002.
12. D. Choi, et al., "Transition to IPv6 and support for IPv4/IPv6 interoperability in IMS," *Bell Labs Technical Journal*, vol. 10, pp. 261–270, 2006.
13. J.-W. Jung, R. Mudumbai, D. Montgomery, and H.-K. Kahng, "Performance evaluation of two layered mobility management using mobile IP and session initiation protocol," in *Proceedings of the IEEE GLOBECOM*, San Francisco, California, Dec. 2003.
14. 3GPP, "Combined GSM and Mobile IP Mobility Handling in UMTS IP CN," 3GPP TR 23.923 v 3.0.0, 2000.
15. S. C. Lo, G. Lee, W. T. Chen, and J. C. Liu, "Architecture for mobility and QoS support in all-IP wireless networks," *IEEE JSAC*, vol. 22, pp. 691–705, 2004.
16. Q. Wang and M. A. Abu-Rgheff, "Interacting mobile IP and SIP for efficient mobility support in all IP wireless networks," in *Proceedings of the IEE International Conference on 3G Mobile Communication Technologies*, London, UK, Oct. 2004.
17. H. C. Fathi, S.; Prasad, R., "Mobility management for VoIP: evaluation of mobile IP-based protocols," in *Proceedings of the IEEE ICC*, Seoul, Korea, May 2005.
18. D. Pesch, M. I. Pous, and G. Foster, "Performance evaluation of SIP-based multimedia services in UMTS," *Computer Networks*, vol. 49, pp. 385–403, 2005.
19. A. H. Enrique Vazquez and Jose Ignacio Fernandez, "SIP-IMS Model for OPNET Modeller," OPNET University Program Contributed Models, 2005.
20. S. Zeadally, F. Siddiqui, N. DeepakMavatoor, and P. Randhavva, "SIP and mobile IP integration to support seamless mobility," in *Proceedings of the IEEE PIMRC*, Barcelona, Spain, Sept. 2004.
21. M. J. Karam and F. A. Tobagi, "Analysis of the delay and jitter of voice traffic over the Internet," in *Proceedings of the IEEE INFOCOM*, Anchorage, Alaska, Apr. 2001.

# 23

# Wireless Resource Allocation Considering Value of Frequency for Multi-Band Mobile Communication Systems

Hidenori Takanashi[1], Rihito Saito[1], Dorsaf Azzabi[1], Yoshikuni Onozato[1], and Yoshitaka Hara[2]

[1] Graduate School of Engineering, Gunma University 1-5-1 Tenjincho, Kiryu, Gunma, 376-8515, Japan
takanasi@nztl.cs.gunma-u.ac.jp, saito@nztl.cs.gunma-u.ac.jp, azzabi@nztl.cs.gunma-u.ac.jp, onozato@nztl.cs.gunma-u.ac.jp

[2] Mitsubishi Electric Corporation, Information Technology R&D Center 5-1-1 Ofuna Kamakura Kanagawa, 247-8501, Japan
Hara.Yoshitaka@cs.MitsubishiElectric.co.jp

**Summary.** Now we face the problem of a scarce bandwidth resource since we have nearly exhausted useful low frequency resources because of explosive growth of cellular phones. In order to provide wide coverage and connectivity among people, we introduce the "Value of frequency" and propose wireless resource allocation scheme based on the value of frequency for the multi-band mobile communication system, which can support both high data rate services and wide service coverage using frequency resources. The simulation of the multi-band mobile communication system demonstrates effectiveness of our wireless allocation scheme. It is shown that both the degree of dissatisfaction and the blocking probability decrease for some system parameters while assessing the degree of dissatisfaction.

## 1 Introduction

The spectrum requirements for communication purposes will increase by as much as 200–300% up to 2010 [1]. The resource of frequency spectrum is limited. Therefore in order to cope with the radical improvement of the frequency usage, technologies such as multi-radio access, broadband access techniques, software defined radio, smart antennas, multi-hop schemes have been heavily investigated. On the other hand, economic and regulatory consequences of dynamic spectrum access technologies and management have been studied [1]. However and whatever technologies are employed, wireless resources are limited. Then users have to share the precious resources.

Multi-band mobile communication systems [3] can provide wide service coverage area for much larger number of users than conventional one-band system while utilizing different propagation characteristics of the different frequency bandwidths. Frequency bandwidths are managed by a base station. The highest frequency band is first checked and if it is not available, the lower frequency band is checked successively until an available frequency band is found. Otherwise, the request is blocked.

Every frequency has different properties. Lower frequency is easy to handle but cannot provide so much higher data rate and consequently, high capacity. Higher frequency may carry higher data rate but smaller coverage compared to lower frequency. Since the wireless resource is limited and is also a common property, we would like to share it among many users to pursue a fair usage. We consider the lower frequency bandwidth has the higher value in use. The higher the frequency is, the more the bandwidth is utilized, and the easier to be allocated. Therefore we set the value of frequency on the basis of the lowest frequency bandwidth. Thanks to the development of wireless technologies, it is possible to use frequency bandwidth with variable coding rates and variable data rates. We propose a wireless resource allocation scheme considering the value of frequency so that the same value of frequency is shared among users in the multi-band mobile communication systems. In the proposed scheme, the amount of channel capacity is assigned according to the value of frequency. Then the allocated bandwidth of a frequency band is proportional to the value of the frequency.

Human–wireless network interaction is done by using wireless resources. Bandwidth is allocated for users to transmit and receive information. Since frequency is a scarce resource, it is not true that we can use bandwidth as much as we want. Usability [2] can be measured by the perceived efficiency or elegance when users use wireless resources. When the number of users increases, the amount of bandwidth allocated for a user decreases. From the view point of communication services users' perceived usability may decrease. In this paper we would like to increase the degree of sharing while satisfying the usability at some allowable level. Usability can be defined in several ways. It is difficult to define it since human beings feel different ways depending on the things they are doing.

In this paper we take a multi-band mobile communication system as an example. Users will perceive the services through wireless resource of frequencies. Because of limited resources of frequencies, users may encounter blocking of usage. Increasing the number of users may cause deterioration of the communication services for which users may feel dissatisfaction. Compared to the conventional method, we assess the degree of dissatisfaction using simplified model. It is shown that both the degree of dissatisfaction and the blocking probability decrease for some system parameters. Effectiveness of our wireless allocation method is demonstrated through the simulation of the multi-band mobile communication system.

## 2 Bandwidth Allocation Based on the Value of Frequency

The strategy to apportion small coverage bands prior to wide coverage bands is effective in the multi-band mobile communication systems [2]. The conventional method allocates the bandwidth to the user on equal-sized channels. Therefore, even if the number of channels in the high frequency belt increases, the users in heavily crowded spectrum condition can not benefit from it because they cannot connect to this high frequency belt. Thus, the increase in high frequency belt isn't a fundamental solution to the channel blocking. To increase the number of channels with a low frequency belt becomes a key for reducing the blocking rate. We introduce the concept of "value of frequency" in order to adjust the amount of the allocations of the frequency band. The frequency allocation method is, then, examined by value of frequency in the multi-band mobile communication method.

The value of frequency of the band with the highest value of frequency is assumed to be $VF_1$. The value of $f_1$ is provided like $VF_1$ considering the characteristics of frequency $f_1$. Values of $f_2, \ldots, f_i$ are provided like $VF_2, \ldots, VF_i$ similarly. When the value of each frequency is examined, it is used for scaling the unit of bandwidth allocation to be adaptable in wireless resource allocation.

In the multi-band mobile communication method, the number of accommodations is expanded by considering the value of frequency in allocating this frequency. The low frequency number belt has twice the value of the high frequency belt if $VF_2 : VF_1 = 1 : 2$. In the bandwidth allocated in each terminal, the low frequency number belt is equal to half the number of the high frequency belt.

Figure 1 depicts the conventional method of channel allocation for each frequency band $f_1, f_2$, and $f_3$. The capacity of each frequency is defined by $C_1, C_2$, and $C_3$, where $C_1 < C_2 < C_3$. Each channel is allocated with equal size of frequency bandwidth.

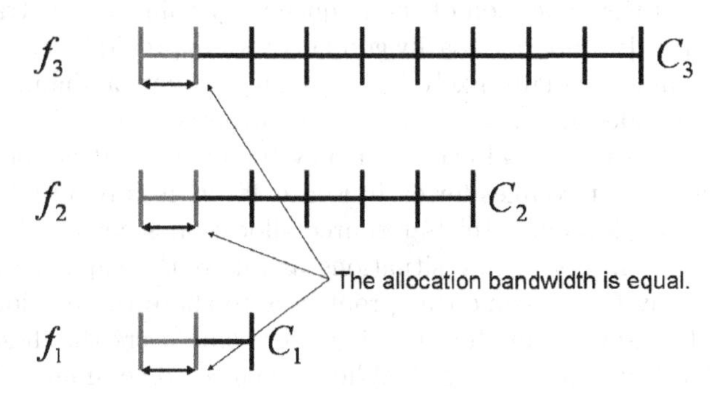


**Fig. 1.** Conventional method of bandwidth allocation

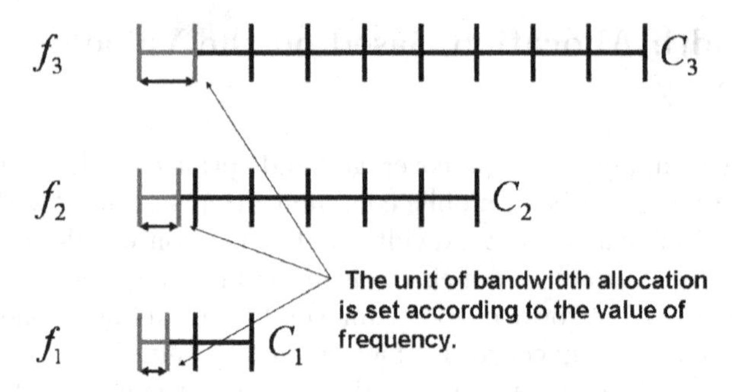


**Fig. 2.** Bandwidth allocation depending on the value of frequency

Figure 2 represents the proposed method. The value of frequency is assumed to be $VF_3 : VF_2 : VF_1 = 1 : 1.5 : 2$, and the ratio of allocated bandwidths is given by $f_3 : f_2 : f_1 = 1 : 0.75 : 0.5$. The channels have equal bandwidths in the same frequency belt. The unit of bandwidth allocation decreases as the frequency decreases because the value is considered in the allocation. The number of channels of low frequency number belts increases. Assigned bandwidth for each user in the different frequency belt has the bandwidth inverse proportion to its value of frequency. This is an important point of the proposed method.

In the conventional method, all frequency belts have the same units of allocated bandwidth. In our approach, every frequency belt has its own allocated bandwidth that is proportional to its capacity. Frequency belts have, thus, been accommodated to treat more users by increasing the number of their channels according to their value of frequency.

## 3 Human–Wireless Resource Interactions for Soft Adaptability

The influence of the allocation of the frequency by value on the transmission rate is very large. It is not necessarily good in order to provide services for users to allocate channels equally likely by the same amount of channels. Various services are provided in wider service areas so that the required bandwidths are varying. The situation of each user may be different. It is not affordable to preserve wireless resources for each user as far as it is required. Therefore it is necessary to develop wireless resource allocation schemes to assign precious wireless resources in any situations as far as the capacity is allowed. One possible way to overcome this problem is to share the precious wireless resources in fine grained modularity. Then for some users the dissatisfaction with this allocation may be caused. When a connected communication is set, the dissatisfaction is generated in the following two ways:

1. Dissatisfaction with blocking(DB): a user may feel when a new incoming connection is blocked
2. Dissatisfaction with connection(DC): a user may feel when it is connected with the low frequency number belt with high value

The dissatisfaction that the user may feel varies depending on each user. Then, to express these variations, we consider the value of dissatisfaction as random numbers from some predefined range.

### 3.1 DB: Dissatisfaction when Blocking

The minimum value of the dissatisfaction that the user may feel when a blocking occurs is defined by $DB_{\min}$. The maximum value is defined by $DB_{\max}$. Both $DB_{\min}$ and $DB_{\max}$ are real numbers. The dissatisfaction when the blocking occurs takes the random number from the range of $[DB_{\min}, DB_{\max}]$.

### 3.2 DC: Dissatisfaction when Connecting with Low Frequency Number Belt

The dissatisfaction when connecting with low frequency number belt takes the random number from the range of $[DC_{\min}, DC_{\max}]$. The minimum value of the dissatisfaction that the user may feel when connecting with low frequency number belt is assumed to be $DC_{\min}$. The largest value is assumed to be $DC_{\max}$. Both $DC_{\min}$ and $DC_{\max}$ are real numbers. This is because it is expected to influence the data rate as the allocation of the bandwidth decreases as the number of frequency decreases.

Let us take the case of two frequency bands. In Fig. 3, we depict the change of dissatisfaction for $VF_1/VF_2$. Thereafter, $VF_1/VF_2$ is denoted as $x$. The user doesn't express dissatisfaction if $x = 1$. In this case, the minimum and the maximum values of dissatisfactions are 0. If $VF_2$ is much smaller than $VF_1$, the allocated bandwidth for the low frequency belt may decrease too much. Then the user may feel almost the same dissatisfaction as blocking because the provided service is not enough. Therefore we assume, $DC_{\min} = DB_{\min} DC_{\max} = DB_{\max}$. We define $\alpha$ as a very large value of $x$. $DC_{\min}$ and $DC_{\max}$ of the dissatisfaction for a given $x$ are obtained from the Fig. 3

$$DC_{\min} = (DB_{\min} - 1)(x - 1) \quad \text{for} \quad 1 < x < \alpha, \tag{1}$$

$$DC_{\max} = (DB_{\max} - 1)(x - 1) \quad \text{for} \quad 1 < x < \alpha\,. \tag{2}$$

In order to grasp what kind of dissatisfactions users may feel, we calculate the dissatisfaction with blocking (DB) when a new incoming connection is blocked and the dissatisfaction with connection (DC) raised by (1) and (2).

A user may feel dissatisfied when the new communication demand is blocked. This is expressed by $UD_b$. The dissatisfaction, when connecting with

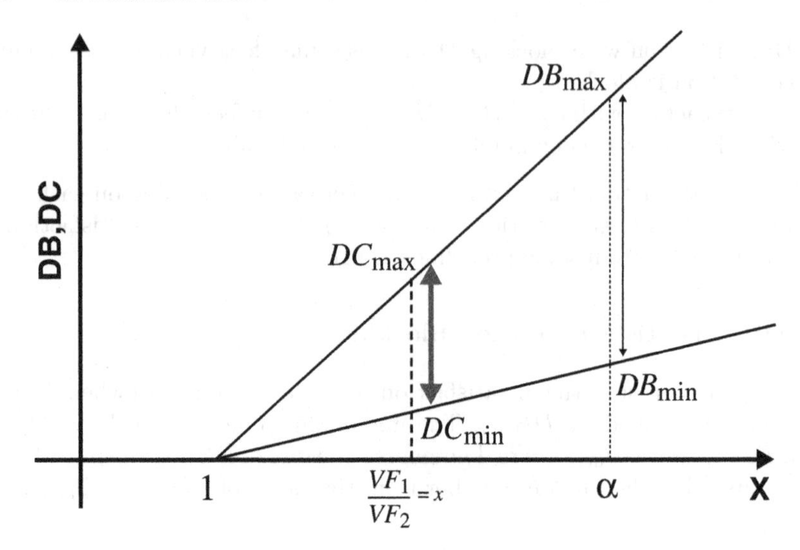


**Fig. 3.** Range of the change of dissatisfaction by the change in value

the low frequency number belt, is expressed by $UD_c$. The sum of all dissatisfactions is denoted as D

$$\mathrm{D} = (\text{Probability of blocking})UD_b + (\text{Probability of connected})UD_c\ .$$

## 4 Our Proposed Scheme for the Two-Band System

We consider the two-band mobile communication system [3] where $f_2$ is high frequency bandwidth and $f_1$ is low frequency bandwidth. The two-band mobile communication can provide wide service coverage area for much larger number of users than conventional one-band system while utilizing different propagation characteristics of the two frequency bandwidths. In Fig. 4, coverage rates of the two-band mobile system are shown.

Coverage rate of $f_1$ is defined by $R_1$, where

$$R_1 = \{(\text{ShadedArea}) + (\text{WhiteArea})\}/(\text{TotalArea}).$$

Coverage rate of $f_2$ is defined by $R_2$, where

$$R_2 = (\text{ShadedArea})/(\text{TotalArea}).$$

Two frequency bandwidths are managed by a base station. The higher frequency band is first checked and if it is not available, the lower frequency band is checked. If an available frequency band is not found, the request is blocked.

Human–wireless network interaction in the two-band mobile communication system is shown in Fig. 5. In Fig. 5, the system side depicts how the new request is

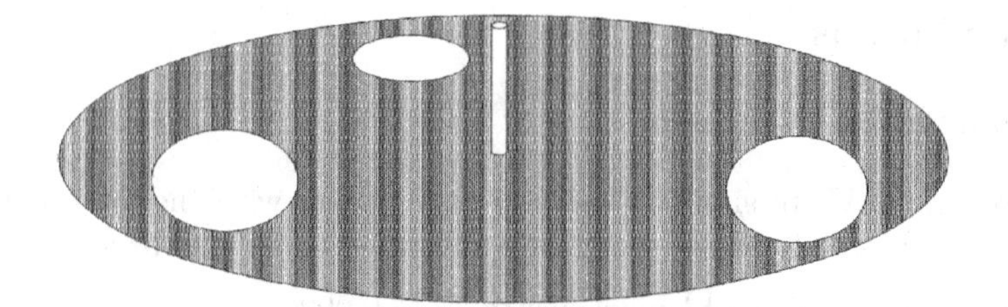

**Fig. 4.** Coverage rates $R_1$ and $R_2$

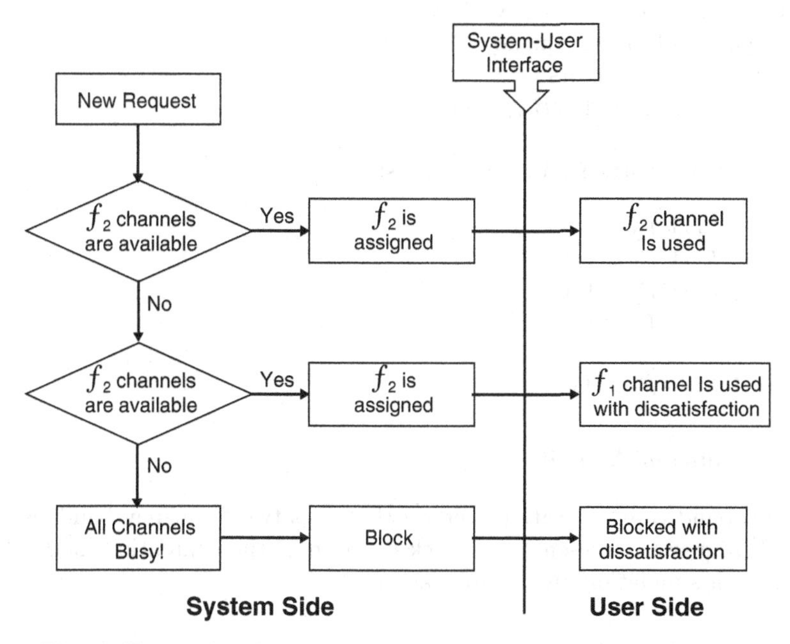


**Fig. 5.** Human–wireless network interaction in the two-band system

handled focusing on channel allocation in the system. The user side of Fig. 5 shows how users behave against the result of the channel allocation. Dissatisfaction occurs for the case when $f_2$ channel has not been available and $f_1$ channel is allocated or the request is blocked. Users would like to use $f_2$ channel since it has wide coverage and it is easy to handle. The base station may assign it as far as it is available. If it is occupied, the base station has to assign $f_2$ channel. Users may feel dissatisfaction for this assignment. The user dissatisfactions are so variable that we assume dissatisfaction as random variables.

In the conventional allocation scheme [3], Hertz is the basic unit of allocation. However we notice that frequency has different value. In the proposed scheme, we reduce Hertz to the same denomination for each frequency bandwidth, and we use this denominated frequency as the basic unit of allocation. For example, the value of frequency $f_2$ is 1.4 means that the ratio of allocated bandwidths, $f_2 : f_1$, is 1.4:1.

# 5 Simulations

## 5.1 Simulation Parameters

We use MATLAB to simulate the proposed system with the following parameters. We calculate the blocking rate and the dissatisfactions as follows:

- Coverage area rate of low frequency $R_1 = 0.9999$
  Coverage rate of high frequency $R_2 = 0.97$
- Capacity of low frequency $C_1 = 5$
  Capacity of high frequency $C_2 = 300$
- 300,000 new requests
- $DB_{\min} = 0.6$, and $DB_{\max} = 1.2$
- $\alpha = 4$
- Trial value patterns for $x$ as follows:

  $VF_1/VF_2 = 1$
  $VF_1/VF_2 = 1.2$
  $VF_1/VF_2 = 1.4$
  $VF_1/VF_2 = 1.6$
  ...
  $VF_1/VF_2 = 4.0$

## 5.2 Simulation Algorithm

A multi-band communication method that uses two frequency bands is simulated. Figure 6 represents the block diagram of the simulation model. The abstract of simulation algorithm is as follows:

STEP1 Set value $x$, and format the channel according to $x$.
STEP2 Dissatisfaction in the low frequency number belt is set.
STEP3 $f_2$ channel is allocated if it is possible to accommodate. (GO TO STEP7)
STEP4 It is not possible to accommodate in $f_2$, and possible to accommodate in $f_1$. (GO TO STEP6)
STEP5 The block is generated if it is not possible to accommodate in $f_1$. (GO TO STEP6)
STEP6 Total sum of the dissatisfaction is obtained. (GO TO STEP7)
STEP7 End

## 5.3 Simulation Results

Simulations for various values of frequencies are done. When the values of frequency $f_2$ are 1.0, 1.4, and 1.8, the blocking probability and the degree of dissatisfaction is depicted for various input traffic in Figs. 7, 8, and 9, respectively.

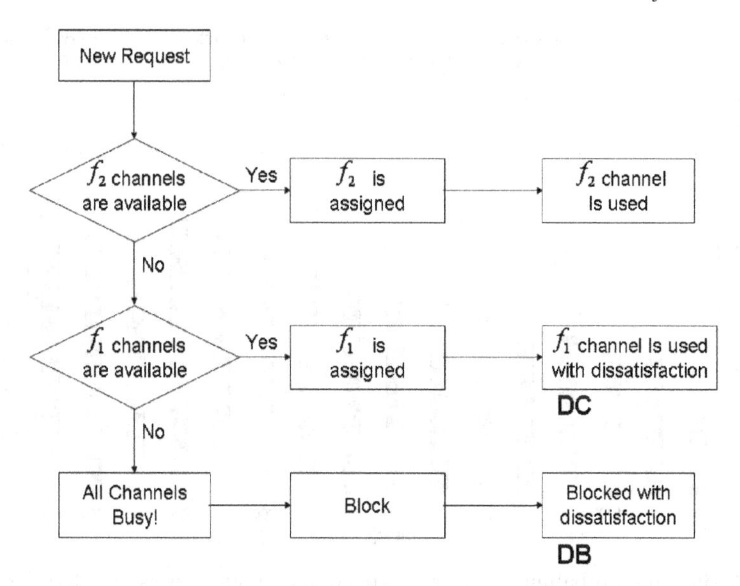


**Fig. 6.** Block diagram of the simulation model

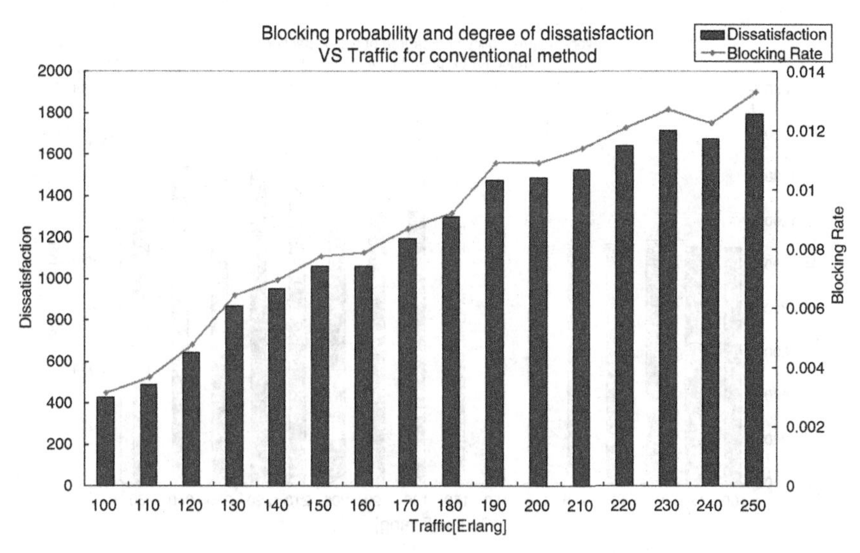


**Fig. 7.** Blocking probability and the degree of dissatisfaction vs. input traffic when the value of frequency is 1.0 (conventional)

The conventional method is shown in Fig. 7. In the conventional method, the denomination of frequency allocation unit has not been executed. So the dissatisfaction accompanied with connection does not occur. The dissatisfaction is only caused by blocking. Figure 7 shows that the dissatisfaction increases as the blocking probability increases. In Fig. 8, as traffic increases, the blocking

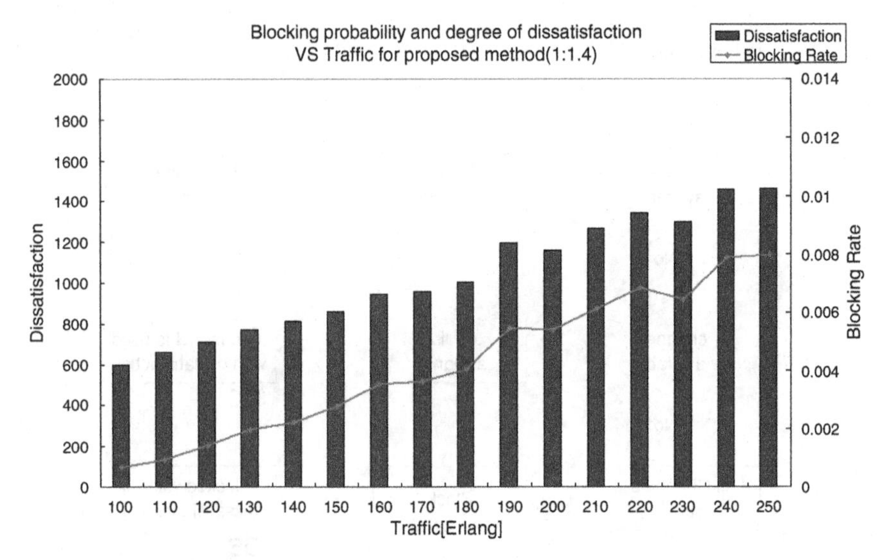


**Fig. 8.** Blocking probability and the degree of dissatisfaction vs. input traffic when the value of frequency is 1.4

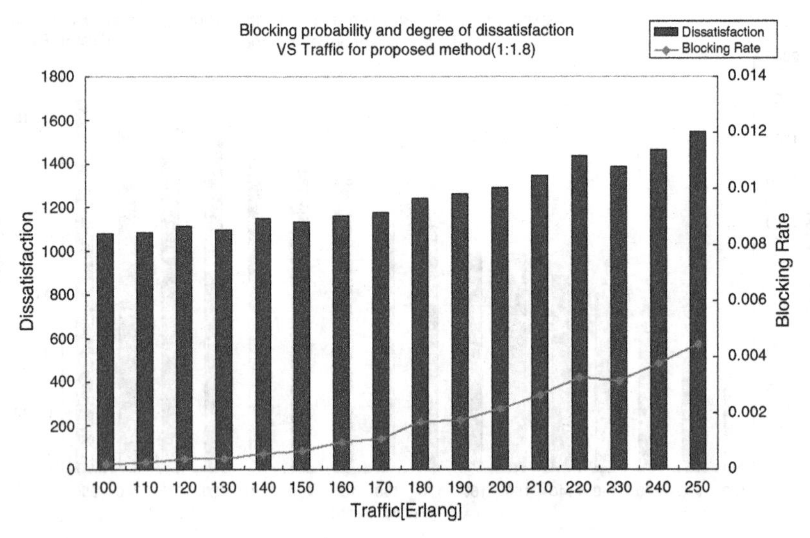


**Fig. 9.** Blocking probability and the degree of dissatisfaction vs. input traffic when the value of frequency is 1.8

probability and the degree of dissatisfaction increase while the degree of dissatisfaction is kept less than 1,300. Our proposed method may decrease the blocking probability thanks to the frequency denomination. Therefore we can get lesser degree of dissatisfaction. However when the value of frequency $f_2$ is 1.8, too much fragmentation caused by the higher value of frequency can

**Table 1.** Channel capacity for various values of frequency satisfying the degree of dissatisfaction is less than 1,300

| Value | 1 | 1.2 | 1.4 | 1.6 | 1.8 | 2.0 |
|---|---|---|---|---|---|---|
| Channel capacity | 180 | 190 | 230 | 220 | 200 | NA |

increase the degree of dissatisfaction with connection even though the blocking probability becomes much smaller.

Table 1 gives the channel capacity satisfying the degree of dissatisfaction is less than 1,300. Our proposed allocation scheme gives more channel capacity than the conventional method. In our model of multi-band mobile communication when we set the value of frequency to 1.4, 127% of channel capacity can be obtained for the same level of dissatisfaction.

## 6 Conclusion

We introduced the value of frequency to proportionate the allocated bandwidth units to each frequency belt. Because of the denominated frequency, the proposed wireless resource allocation scheme may adjust to the frequency coverage area and capacity, in various situations.

From our simulation, an appropriate balance of the two frequency bandwidths is achieved to support both high data rate services and wide coverage within the allowable level of dissatisfaction for the two-band mobile system employing the proposed scheme.

## References

1. Fredrik Berggren, Olav Queseth, Zander, Borje Asp, Christian Jonsson, Peter Stenumgaard, Niklas Z. Kviselius, Bertil Thorngren, Urban Landmark, and Jonas Wessel, "DynamicSpectrum Access Phase 1:Scenarios and research challenges," 23 September 2001, TRITA-S3-RST-0407
2. Jakob Nielsen and Hoa Loranger, Prioritizing Web Usability, New Riders Publishing, 2006
3. Yoshitaka Hara and Kazuyoshi Oshima, "Multiband mobile communication system for wide coverage and high data rate," IEICE Trans. Commun. vol. E89-B, no. 9, pp.2357–2457, September 2006

# Session 5

# Wireless Sensor Networks

24

# Wireless Sensor Networks and SNMP: Data Publication Over an IP Network

L. Berruti, L. Denegri, and S. Zappatore

DIST, University of Genoa, Via Opera Pia 13, I-16145 Genova, Italy
luca.berruti@cnit.it, denegri@dist.unige.it, sandro.zappatore@unige.it

**Summary.** The availability of a wide range of cheap single chip computers, and of integrated radio-transceivers operating in the ISM band, has significantly encouraged the design and deployment of efficient wireless sensor networks (WSNs). This paper presents the original design and implementation of a multi-hop WSN, based on a Microchip MCU and a Cypress RF device, able to publish the data gathered from the field through the facilities offered by SNMP agents. The algorithm adopted to manage the network is discussed, and a possible application scenario is illustrated. A working prototype based on the hardware platform, software and protocol described in this paper has been deployed and tested, and some of the obtained results are shown.

## 1 Introduction

Wireless Sensor Networks (WSNs) [1, 2] are clearly the fastest and cheapest way to collect data from a wide area, but they must be subjected to some constraints:

- Nodes of the WSN must be cheap enough, because their number can be very high to cover very wide outdoor areas or very complex indoor environments
- A protocol must be used that allows the WSN to self configure to avoid an expensive and time consuming manual setup; the protocol has to configure the WSN at start-up, to manage periodical wanted or unwanted network changes and to route data from source nodes to one or several sinks
- To cover wide areas, the WSN must implement a multi-hop routing protocol, using suitable structures, such as mesh, clusters, trees or others, to organise the nodes

Furthermore, data collected using the WSN must be provided to final users who not always can be near the WSN sinks, but who can easily find a PC connected to the Internet nowadays. This paper describes how to design and develop a working protocol and the algorithm that fulfils the above constraints and presents a framework that allows the collection of data and their delivery via the Simple Network Management Protocol (SNMP) [3, 4] over an IP network such as the Internet. However, it should be noted that the novelty of the paper is not in the algorithm itself (which is based on the approaches presented in [5, 6]), but rather in highlighting and solving the architectural problems involved in deploying the sensor network, and in publishing data gathered by sensors through the facilities offered by SNMP agents. The paper is organized as follows. Section 2 describes the elements of our WSN. The algorithm managing the network, as well as the packet structures, are discussed in Sect. 3. Section 4 presents the original approach used to publish collected data over the Internet, while Sects. 5 and 6 illustrates the preliminary results and a possible application scenario, respectively. Finally, in the last section, the conclusions are drawn.

## 2 WSN Elements

Figure 1 sketches our sensor network and the related "gateway" to an IP infrastructure. The core elements of the network are the nodes: in our implementation, the hardware platform has been designed with off-the-shelf elements,

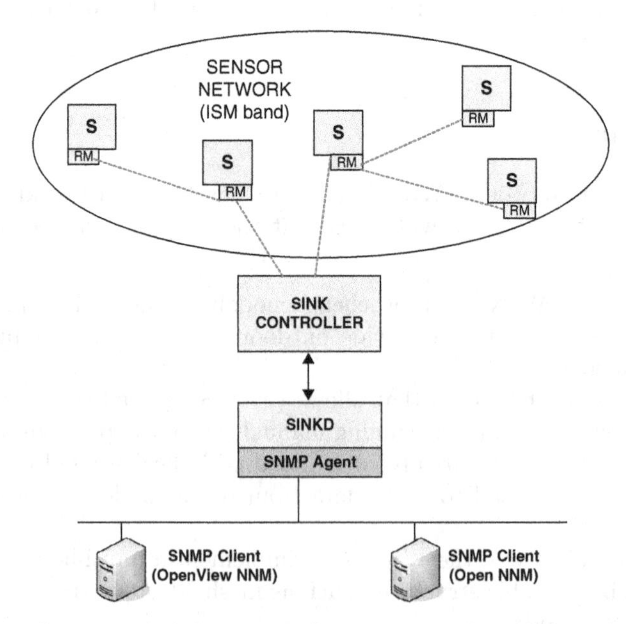


**Fig. 1.** Overall structure of the proposed sensor network publishing data by means of a SNMP agent

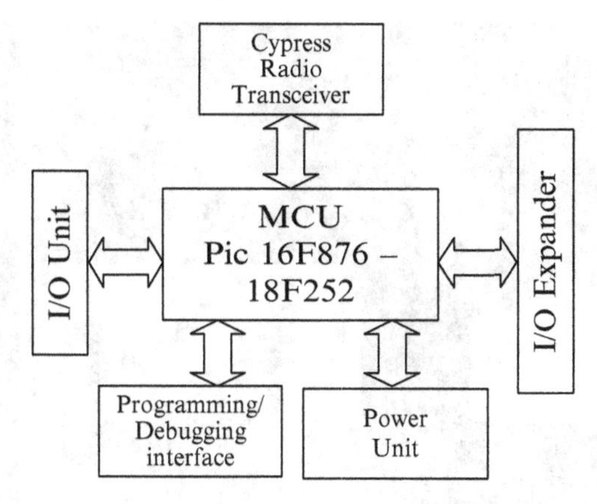


**Fig. 2.** Functional blocks of a WSN node

and this allows to build nodes that are cheap, but have good computational and communication capabilities.

Thereby, during the design phase, much attention was paid in order (1) to adopt commercial components, (2) to choose integrated circuits providing a PDIP/SOIC package (in this manner any possible part replacement is quite simple), (3) to exploit a RF transceiver operating in the free ISM band (2.4 GHz), able to internally spread and serialize/de-serialize data to be transmitted, and (4) to keep the size of the node board as small as possible, thus permitting an easy deployment of the WSN.

The functional blocks of a node are depicted in Fig. 2.

The core element is represented by the MCU, a PIC 16f876 [7] (or PIC 18f252, that is pin-to-pin compatible with the previous one): this latter communicates, via a SPI, with a radio transceiver, based on the Cypress CY-WUSB6935 chip [8]. In our implementation, we adopted the Cypress based "High speed multi-channel transceiver," produced by Aurel S.p.A., an Italian medium enterprise, specialized in RF modules design [9].

Moreover, by means of the SPI, it is possible to connect the MCU with an (optional) I/O expander chip, thus increasing the overall I/O capabilities of the WSN node. The Programming/Debugging interface provides an ICD tap, which allows the user to upload/debug the firmware directly on the MCU. The Power Unit (PU) monitors the battery status and, if an external power supply is available (for instance, a solar panel), the PU controls and regulates the battery charge. The I/O unit consists of a very simple interface, which permits the MCU to acquire signals from the field. Although the MCU can handle eight digital (viz. on–off) inputs, eight digital outputs and four analog input channels, the onboard circuitry permits to manage only one digital input, one digital output, and one analog input channel: to achieve more I/O capacity an I/O expander board must be plugged in the sensor main board.

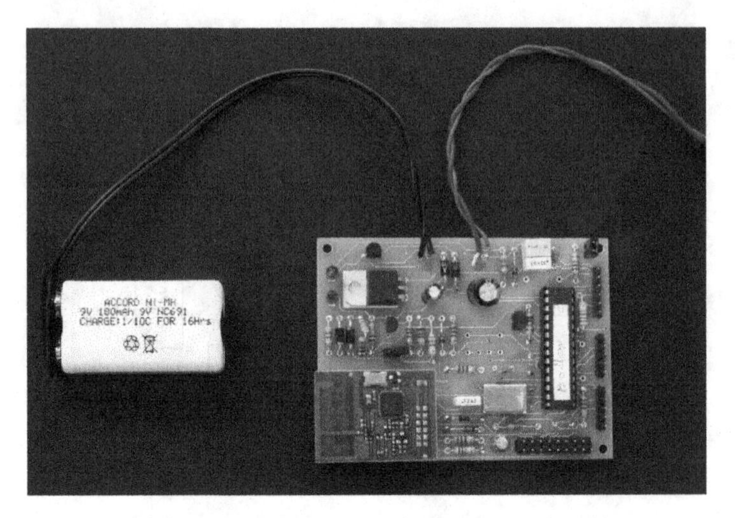


**Fig. 3.** Main board of the implemented wireless sensor

This design choice is motivated by the fact that, in general, a node has to monitor/control only one signal. Furthermore, the design of an expander board can be carried out independently of the development of the sensor main board, since the expander board and its field interface can be tuned to the specific electrical characteristics of the signals to be monitored.

Figure 3 presents the board of a sensor developed within our research activity.

The various protocol layers of the WSN, as well as the driver handling the RF transceiver, the software portions managing the power unit and the signal acquisition and signal conditioning are implemented on the MCU.

At the "centre" of the network there is the sink, which includes two main elements: a network sink controller (NSC), and a sink manager (SM).

The hardware and firmware of the former are similar to those of a common sensor node, while the latter consists of a single board ARM computer (booting Linux), produced by Technologic Systems [10].

The NSC actually manages the sensor network, while the SM publishes data gathered from the sensors by means of an SNMP agent. To this aim, the SM continuously communicates with the NSC in order to get any information acquired from the field, and to send data toward the sensors to properly set their digital outputs. The SM stores information to/from the WSN in an internal real-time database, which is also accessed by an SNMP agent that, in turn, exposes data according to a MIB defined structure. The functional blocks of the sink are depicted in Fig. 4.

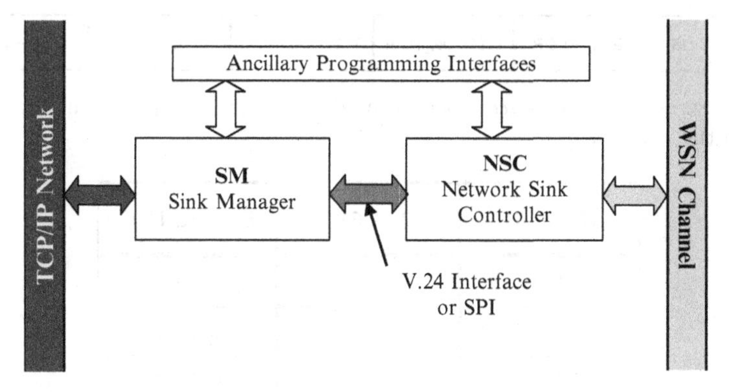


**Fig. 4.** Functional blocks of the sink

## 3 Protocol and WSN Operation

The protocol implemented can be considered an Energy Efficient Route protocol and, specifically, a Minimum Hop Route protocol [5]: this solution can be considered a good trade-off between power saving and hardware/firmware requirements. It can also be included in flooding or gossiping techniques, though some of the problems solved by these approaches have been faced using different solutions.

The basic idea of the protocol is to divide the operation of the network in two phases: network management mode and normal network operation mode, which consists of actual data exchange.

### 3.1 Network Management

The first phase allows the automatic configuration of the network after its deployment. The configuration is periodically updated in order to take into account any possible topology change. The protocol machine controlling each node starts in a listening state and it continuously monitors the channel activity on a hard coded frequency. The sole active node in the network is the sink, which advertises data regarding the status of the network by broadcasting *beacon* packets, whose structure is shown in Fig. 5.

The field SEND_ADDR consists of a unique identifier (ID) of the sender node (each node has a unique ID derived from the manufacturer identifier, a number laser-fused in the RF transceiver). Since every *beacon* is always originated by the sink, the first *beacon* packet contains as SEND_ADDR, the ID of the sink. The DEST_ADDR is a broadcast address while the BEACON_HDR field consists of a byte that specifies the packet type. The other relevant fields of this packet are: CHAN, HOP, TTL and FLAG. CHAN indicates the number of the channel that will be used for the normal network operations (it is chosen by the sink among the most quiet channels). The channel used for

| SEND_ADDR | DEST_ADDR | BEACON_HDR | TTL(7) FLAG(1) | HOP | CHAN |
|---|---|---|---|---|---|

**Fig. 5.** Structure of a *beacon* packet used to advertise topological information through the network

| NODE ID | HOP COUNT | EXPIRE TIME |
|---|---|---|
| ID 1 | HOP 1 | EX_TIME 1 |
| ID 2 | HOP 2 | EX_TIME 2 |
| ID 3 | HOP 3 | EX_TIME 3 |

**Fig. 6.** Routing table of a node: each entry consists of the node ID, the number of hops needed to reach the sink through the node ID, and the expire time

the normal operation mode always differs from the one used during the start up phase. HOP is initialized with zero by the sink and is used to measure, in hop, the distance of a node from the sink. TTL is a seven bit time-to-live counter, employed to limit the number of hops a packet can be transmitted through, and finally, FLAG is a switch bit: it is set during the first *beacon* transmission, it is reset during the second transmission, and so on. In other words, even *beacons* are marked with "0" and odd ones with "1."

Upon receiving a *beacon*, the sensors add a new entry in their routing tables (see Fig. 6), and record the sender ID and the number of hops needed to reach the sink. The entries of the routing table are soft state, so they also initialize the expire time at the maximum value.

*Beacons* can reach nodes from different paths and, consequently, with different delays: therefore, upon receiving the first *beacon*, each node continues to listen to the channel for a certain amount of time (*listening window*) before choosing the best route toward the sink among those in its routing table. Moreover, after a randomly generated time delay, each node generates a new *beacon* packet, where the SEND_ADDR is replaced with the ID of the node and the HOP field with the minimum number of hops (incremented by 1) needed to reach the sink, on the basis of the Hop Count field contained in the routing table of the node. The random delay aims at desynchronizing the transmission of packets that otherwise could collide. This delay plus the *listening window* time is called *beacon window*.

Once all the tasks of the network management phase are completed, the nodes switch to the channel devoted to the normal network operation mode. In this mode, the sink periodically broadcasts the *beacon* packet on the normal operation channel: the *beacon* is flooded through all the nodes exploiting the same procedure used during the network management phase, the routing tables are refreshed and their entries accordingly updated or deleted if the

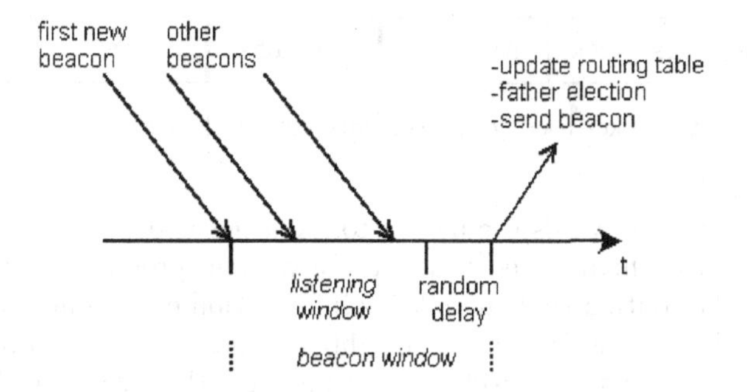


**Fig. 7.** Time line of a *beacon window*

expire-time counter reaches zero. Changes in network topology are managed as follows: if an (old) path does not exist anymore, the related entry is deleted after a certain amount of time, because it is no longer refreshed by new *beacons*. Vice versa, if a new and better path is found, for example because a (new) node has entered the network, the new available path is notified, by means of *beacons*, to all the nodes belonging to the new path. Obviously, the routing tables of these nodes are congruently updated, in order to reflect the new topology. The FLAG bit, set or reset by the sink after sending a *beacon* packet, is used by the nodes to distinguish a new *beacon* from an old one, where "old" refers to a *beacon* sent during an old (previous) transmission. This allows to avoid loop creation: within a *beacon window*, only *beacons* with the same flags are accepted, while *beacons* with flags equal to the one of the previous *beacon window*, but outside the *beacon window*, are rejected. In fact, these *beacon* packets are probably clones of an old *beacon* packet. If some *beacons* cause a loop, owing to a pathologic situation, the TTL control mechanism prevents their neverending transmission, thus avoiding energy wastes and possible network congestions.

At the end of the network management phase, only local information on the network topology is available at the sensors: each node knows only its neighbours and, among these ones, it chooses the node nearest to the sink.

Figure 7 shows events during the *beacon window*: *beacons* arrivals and actions performed by the protocol algorithm.

Finally, if a new node wants to enter the network in the normal operation mode, it periodically scans all the channels until it receives a data packet or a *beacon* packet. Then, upon receiving the next *beacon*, the new node initializes its routing table and begins to operate.

### 3.2 Data Forwarding

Whenever a node has to send a *data packet* to the sink, it addresses the packet to its father node, that is the node with the minimum hop distance in the routing table. In turn, upon receiving the data packet, the addressed

| SEND_ADDR | DEST_ADDR | DATA_HDR | SRC_ADDR | DATA | DATA |
|---|---|---|---|---|---|

**Fig. 8.** Data packet used to deliver data collected by a sensor to the sink

node immediately forwards the packet to its father node, and so on up to the sink. It should be highlighted that the forwarding process could be further improved if the routing table maintains information concerning, not only the neighborhood (viz. the IDs of the neighbor nodes), but also other kinds of data, such as the battery level: in this manner, the selection of the route toward the sink could be made by weighting different metrics. The structure of a *data packet* is shown in Fig. 8.

The SEND_ADDR specifies the ID of the node which the packet is forwarded to, the DEST_ADDR is the ID of the node that is currently handling the data packet, while SRC_ADDR specifies the ID of the node which originated the packet. Finally, the field DATA contains the actual payload.

## 4 Delivering Data on an IP Network

Data received from the sink node have to be gathered and delivered: this task is done by using the SNMP [11]. This choice is motivated by the fact that many SNMP applications were developed in order to monitor and supervise electronic devices and systems. Generally, these applications provide effective tools to simply manage and archive incoming data. Furthermore, these tools are able to notify possible alarm conditions to human personnel or to a support system, which, in turn, can attempt to automatically issue appropriate commands.

As previously mentioned, in our sensor network implementation, the sink includes two major components: the SNC and the SM. Specifically, the latter consists of an embedded system based on an ARM9 architecture, linked to the NSC via a SPI or RS232 interface. The SM launches a sink daemon (viz the *sinkd*) and a SNMP agent [12] on top of a GNU/Linux operating system.

The *sinkd* collects data acquired by the peripheral nodes and archives the related information in a real-time data base. Once a SNMP request (a so called SNMP GET) is received by the SNMP agent, the latter browses the real time data base in order to properly reply to the request. Vice versa, whenever a SNMP client has to send a command to a specific node (for instance, to switch on a digital output), it issues a SNMP SET command that will be handled by the SNMP agent. In turn, the agent will forward (after proper coding and packet assembling) the request to the addressed node by exploiting the *sinkd* APIs.

In this manner, the entire WSN is "published" according to the SNMP standard, where all the data are organized in a Management Information Base (MIB) [13]. The design of the latter is based on the IF-MIB [14], a MIB commonly used to describe the network interfaces of a generic system: in our case, each sensor node plays the role of a sort of interface attached to the sink.

The information related to each sensor is grouped inside a table called *snTable*. A specific sensor is univocally identified within the system by means of a table index (*snIndex*).

In the following the definition of the *snIndex* and the *snTable* according to ASN.1 notation are shown:

```
– SensorIndex contains the semantics of snIndex and
– should be used for any objects defined in other
– MIB modules that need these semantics.
SensorIndex ::= TEXTUAL-CONVENTION
    DISPLAY-HINT "d"
    STATUS      current
    DESCRIPTION
            "A unique value, greater than zero, for
            each sensor in the managed system. It is
            recommended that values are assigned
            contiguously starting from 1."
    SYNTAX      Integer32 (1..2147483647)
– the Sensors table
– The Sensors table contains information on the
– entity's sensors.
snTable OBJECT-TYPE
    SYNTAX      SEQUENCE OF snEntry
    MAX-ACCESS not-accessible
    STATUS      current
    DESCRIPTION
            "A list of sensor entries. The number of
            entries is given by the value of
            snNumber."
    ::= { sensors 2 }
```

Each table entry (*snEntry*) contains properties related to a general purpose sensor.

```
snEntry OBJECT-TYPE
    SYNTAX              snEntry
    MAX-ACCESS not-accessible
    STATUS              current
    DESCRIPTION
                "An entry containing management
                information applicable to a
                particular sensor."
```

```
    INDEX        { snIndex }
    ::= { snTable 1 }
snEntry ::=
                "The value of the first digital input
                read from the sensor.
                A value of on (1) means high logical
                level.
                A value of off (2) means low logical
                level."
    ::= { snEntry 4 }
snOutDigital OBJECT-TYPE
    SYNTAX INTEGER { on (1), off (2) }
    MAX-ACCESS read-write
    STATUS           current
    DESCRIPTION
                "This object is used to change the status
                of the first digital output of the
                sensor.
                A value of on (1) pulls up the digital
                output to high logical level.
                A value of off (2) pushes down the digital
                output to low logical
                level."
    ::= { snEntry 5 }
    SEQUENCE {
        snIndex                      SensorIndex,
        snDescr                      DisplayString,
        snInAnalog                   Integer32,
    snInDigital                  INTEGER,
        snOutDigital                 INTEGER
    }
snIndex OBJECT-TYPE
    SYNTAX           SensorIndex
    MAX-ACCESS read-only
    STATUS           current
    DESCRIPTION
                "A unique value, greater than zero, for
                each sensor. It is recommended that
                values are assigned contiguously
                starting from 1."
    ::= { snEntry 1 }
snDescr OBJECT-TYPE
    SYNTAX           DisplayString (SIZE (0..255))
    MAX-ACCESS read-only
    STATUS           current
```

```
    DESCRIPTION
                "A textual string containing information
                about the sensor."
    ::= { snEntry 2 }
snInAnalog OBJECT-TYPE
    SYNTAX          Integer32 (0..65535)
    MAX-ACCESS read-only
    STATUS          current
    DESCRIPTION
                "The value of the first analog input read
                from the sensor."
    ::= { snEntry 3 }
snInDigital OBJECT-TYPE
    SYNTAX INTEGER { on (1), off (2) }
    MAX-ACCESS read-only
    STATUS          current
    DESCRIPTION
```

Since a network can include sensors of different types at the same time, the MIB previously described has been extended by adopting a tree representation, where each branch specifies aspects unique to the sensor considered. A similar approach permits to focus the attention on the specific hardware features of each sensor (i.e. the type of input signals the sensor can acquire, as well as the physical quantities which must be recorded). Just to clarify the concept, Fig. 9 shows how the description of a generic sensor (*snTable*) can be extended in order to better describe a sensor (*termalTable*) devoted to collect data regarding the temperature of a certain system. Each entry of the *termalTable* contains two indexes, called *termalIndex* and *termalSnIndex*. The former is used to identify the sensor inside the *termalTable* (it plays the same role of *snIndex* for the *snTable*). The latter is a reference, which permits to extract the information contained in the *snTable*.

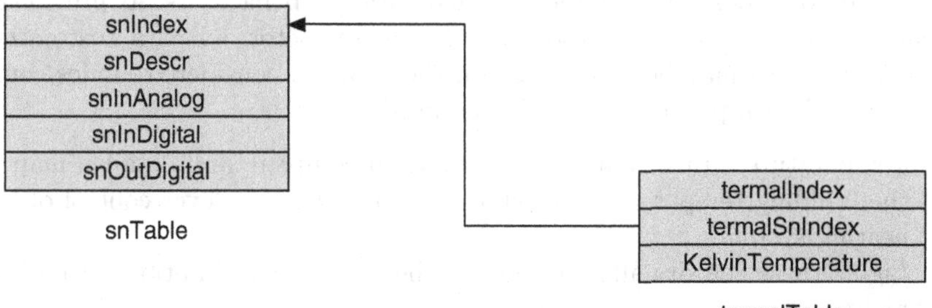


**Fig. 9.** General and specific representation of a sensor derived from an IF-MIB

## 5 Preliminary Results

During node development, some preliminary performance tests of the radio-transceivers has been carried out. In particular, coverage area and interference immunity have been considered. The test bed was a floor of a building, with a number of contiguous offices and laboratories.

The goal was to estimate the average coverage area and the capability of rejecting cochannel and adjacent-channels interferences.

A first set of tests proved that using the maximum transmission power (15 dBm) the coverage area extends to about 30 m in every direction from the node; decreasing the transmission power to −14 dBm reduces the coverage area to about 7–8 m. Long distance measurements, performed under a quasi free space condition, highlighted a maximum transmission range of about 500 m.

A good interference immunity has been proved by generating, with another radio-transceiver, signals at the same frequency or on an adjacent channel. Cochannel tests were performed using different pseudo-codes, while adjacent-channel tests were carried out using the same pseudo-code for the main and the interfering signal.

## 6 Application Scenario

The system described in this paper can represent a very innovative tool to remotely control buildings or structures where complex devices are placed (i.e. a server farm). The nodes of the WSN should be divided in two sets:

- Sensor nodes: they measure analog or digital signals coming from the field or from specific devices
- Actuator nodes: they interact with the field by changing the state of some control signals

The flexibility of the developed hardware platform allows an easy customization of nodes with different characteristics or behaviours. As previously mentioned in Sect. 2, each sensor node has a connector, which a customized expansion board may be plugged to. In the scenario considered, nodes could be equipped with probes in order to monitor:

- Temperature: in this manner, users have a complete and detailed map of the building temperature, which can be useful for a clever control of the heating system
- Light: these sensors allow to reduce the energy consumption during the night
- Smoke/gas escapes: the WSN represents a valid tool to detect starting blazes and immediately adopt specific countermeasures

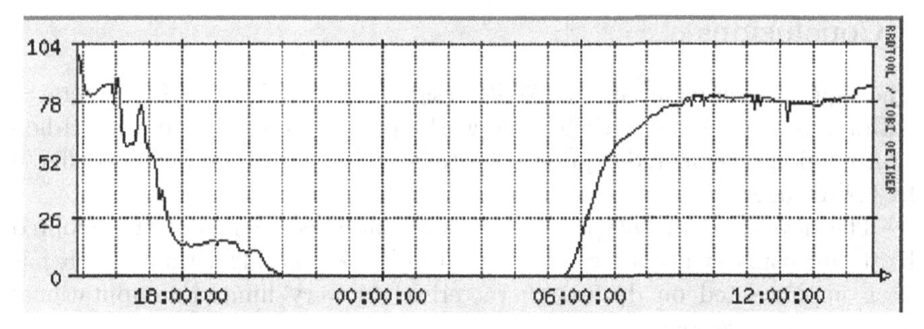


**Fig. 10.** Behavior of daily measured light intensity

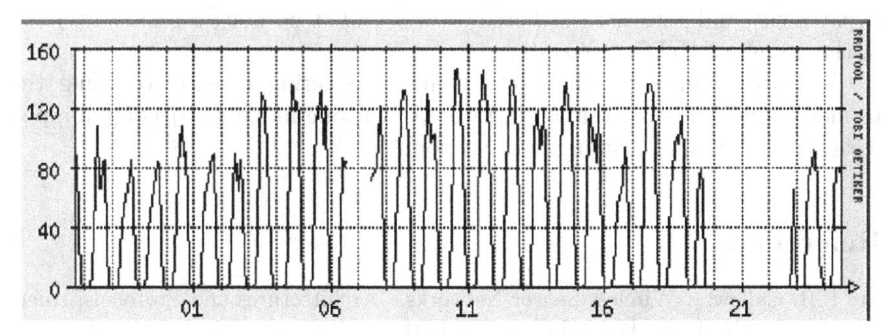


**Fig. 11.** Behavior of monthly measured light intensity

Obviously, the different sensor types do not need to really be different nodes of the WSN, since a specific characterization can be obtained by an appropriate expansion board.

On the other hand, actuator nodes have to interact with devices, by switching them on or off, or regulating some of their parameters.

All these nodes can be added to the WSN independently and at different times, because the network is able to reconfigure itself whenever new nodes enter the network. This is a great advantage compared to wired systems, especially if a remote control system is deployed after the building construction.

Data gathered by sensors are routed to the sink and then published via a SNMP agent. In this manner, a common Network Node Manager (NNM), for instance the commercial NNM belonging to HP-Openview suite [15], can be fruitfully exploited in order to manage anomalous situations or alarms.

Alternatively, open-source SNMP tools can be employed to record data and to graphically represent them. Figures 10 and 11 report two graphics showing light intensity measured by a sensor daily and monthly, respectively. The plots are obtained by means of an ad-hoc customization of the well-known MRTG monitor tool [16]. The gaps in Fig. 11 correspond to maintenance periods, during which the network was down.

## 7 Conclusions

The paper presented an original WSN designed and implemented from scratch within our research activity. The network operates according to a multi-hop transmission scheme and exposes collected data by exploiting facilities offered by SNMP agent.

The developed prototype clearly shows the effectiveness of the adopted hardware components as well as the efficiency of the algorithm, which has been implemented on devices characterised by very limited computational and memory resources.

The network was experimentally tested within the building housing our department, and data were gathered by means of an ad-hoc customized version of the popular MRTG tool suite.

A possible future application we are considering consists of using the framework here described as a platform for tracking moving objects (equipped with sensors) within a structured environment.

## References

1. E.H. Callaway, Wireless Sensor Networks – Architectures and Protocols, Auerbach Publications, Boca Raton, FL, 2004.
2. I.F. Akyildiz, W. Su, Y. Sankarasubramaniam, E. Cayirci, A Survey on Sensor Networks, IEEE Communications Magazine, 40(8):102–114, 2002.
3. J. Saperia, SNMP at the Edge: Building Effective Service Management Systems, McGraw-Hill Professional, New York, 2002.
4. D.R. Mauro, K.J. Schmidt, Essential SNMP, O'Really, 2005.
5. W.R. Heinzelman, J. Kulik, H. Balakrishnan, Adaptive Protocols for Information Dissemination in Wireless Sensor Networks, Proceedings of the 5th Annual ACM/IEEE International Conference on Mobile Computing and Networking, Seattle, USA, 1999.
6. W.R. Heinzelman, A. Chandrakasan, H. Balakrishnan, Energy-Efficient Communication Protocol for Wireless Microsensor Networks, Proceedings of the 33rd Hawaii International Conference on System Sciences, vol 8, 2000.
7. See for further information http://www.microchip.com.
8. Cypress, "CYWUSB6934: WirelessUSB™ LS 2.4 GHz DSSS Radio SoC", Datasheet, available on line at http://www.cypress.com.
9. See for further information http://www.aurelwireless.com.
10. See for further information http://www.embeddedarm.com.
11. J.D. Case, M. Fedor, M.L. Schoffstall, J. Davin, Simple Network Management Protocol (SNMP), IETF, RFC 1157, 1990.
12. See for further information http://net-snmp.sourceforge.net.
13. K. Mc Cloghrie, M.T. Rose, Management Information Base for Network Management of TCP/IP-Based Internets, IETF, RFC 1156, 1990.
14. K. McCloghrie, F. Kastenholz, Evolution of the Interfaces Group of MIB-II, IETF, RFC 1573, 1994.
15. See for further information http://h20229.www2.hp.com.
16. See for further information http://oss.oetiker.ch/mrtg/.

# 25

# Self-Localization of Wireless Sensor Nodes By Means of Autonomous Mobile Robots

Andrea Zanella, Emanuele Menegatti, and Luca Lazzaretto

University of Padova, Department of Information Engineering (DEI)
Via Gradenigo 6/B, 35131 Padova, Italy
zanella@dei.unipd.it, emg@dei.unipd.it, luca.lazzaretto@ieee.org

**Summary.** In general, a wireless sensor network consists of a large number of low-cost, static nodes that organize themselves in order to deliver events notification to a sink node in a multi-hop fashion. Typically, nodes are battery driven and are limited in terms of processing, storing and communication capabilities. On the contrary, an autonomous mobile robot is an expensive object, equipped with advanced interfaces and capable of performing complex tasks. The complementary capabilities of these two technologies can be integrated in a synergetic manner not only to enhance the performance of each single system, but also to create novel applications and services.

In this paper, we will describe the RAMSES2 project, which aims at investigating the potential benefits resulting from the integration of WSNs and AMRs. As case study, we present and analyze the first experimental results concerning the self-localization problem, by which a wireless sensor node, placed in an unknown location, infers its own position by processing the information received by an AMR that moves in its proximity, thus acting as mobile beacon. The advantage of using mobile beacons for localization in WSN has been already reported in literature. However, most of the previous work refers to outdoor scenarios, while in this paper we report results obtained in a typical indoor environment. The localization problem in indoor environment represents a challenging benchmark to check the functionalities of the RAMSES2 hybrid platform and, despite the project is still in a very initial stage, the first results are promising and call for further investigation of this novel and interesting domain.

## 1 Introduction

Recently, the research on Wireless Sensors Network (WSN) has enlarged its scope by considering network components with some mobility capabilities. Such mobile agents can increase the system performance in different ways, for example by moving over the area covered by the static sensor nodes for collecting data from the peripheral sensors, replacing exhausted nodes, synchronizing the nodes, or, in general, increasing the energy efficiency and the network lifetime. Most of the mobile nodes used by the WSN community are low cost and small size robots, such as MICAbot, CostBots, Robomote, Millibots. These robots have very limited autonomous capabilities and, in general, they do not carry additional sensors or more powerful computational and communication units.

RAMSES2,[1] a research project recently funded by the University of Padua (Italy), gives a further boost to the paradigm of WSN with mobile nodes by proposing the integration of the WSN with sophisticated (and expensive) Autonomous Mobile Robots (AMR), having advanced mobility, processing and communication capabilities.

A possible application in this context consists in using the WSN to monitor some strategic areas, in order to prevent some natural or artificial disasters (e.g., earthquakes, fire, land/snow-slide, chemical or nuclear emergencies) by signaling alarming deviations from standard values of environmental parameters (e.g., temperature, humidity, pressure, etc.). When a dangerous event is detected by the WSN, it may activate a team of AMRs that react in the most suitable way. In case of fire detection, for example, the AMR squad could query the WSN to locate the zone containing the fire. AMRs equipped with fire extinguishers can coordinate their action in order to maximize their ability to extinguish the fire, while other AMRs can utilize ad hoc communication channels to deliver high-quality video images of the zone interested by the event to a control centre.

Leveraging on the complementarity of WSN and AMRs, RAMSES2 project aims at developing strategies to enhance the performance of both systems and enable novel functionalities, in particular in the context of surveillance and rescue. The integration of AMRs and WSNs is still an unexplored field that, on the one hand, offers very exciting development perspectives and, on the other hand, raises several scientific challenges. Among the plethora of open research problems in this area, RAMSES2 project will focus only on a limited number of topics that are of particular interest for their scientific and practical spin-off. More specifically, the project will address the problems of WSN-aided motion planning for AMRs, on the one hand, and on AMR-aided network maintenance for WSN, on the other hand.

## WSN-Aided Motion Planning

Motion planning involves determining the motions of mobile nodes so that they reach a goal state by optimizing a criterion, such as the minimization of the distance traveled without colliding into obstacles. This research area has been widely studied in robotics in the last decades and many successful algorithms and techniques have been developed and used in many applications. Furthermore, reactive techniques for autonomous navigation toward a goal avoiding obstacles have been also widely studied and applied in robotics. In

[1] RAMSES2: integration of Autonomous Mobile robots and wireless sensor networks for Surveillance and rescue

the framework of WSN with mobile nodes, motion planning is an important issue that has not been sufficiently addressed yet. Existing motion planning algorithms require an environment model, i.e., a map, and a kinematic/dynamic model of the mobile node. If the environment model is unknown or subject to unexpected variations, as in case of fire, it is necessary to apply map-building techniques which require suitable sensors. Therefore, cooperation between AMR and WSN can improve the localization/tracking procedures of mobile nodes. More specifically, the project will investigate the possibilities of using anchor nodes, i.e., static nodes that know their own spatial coordinates, to reduce the uncertainty in the position estimation of AMRs, thus improving the performance of the navigation algorithm.

Also, the project will investigate how to apply reactive techniques used in robotics to guide the mobile nodes toward a spot of interest (for instance, the WSN node location in which a fire has been detected or in which there is a hole in the connectivity). In this case, the mobile platform moves reacting to sensorial stimulus of the WSN. The problem here is to identify the most effective interaction pattern between the mobile robots and WSN, in order to reach the goal. For instance, a random walk algorithm using gradient descent might be applied to guide the mobile node to a focus of interest (e.g., the highest peak of temperature in the building) [1, 2], while diffusion-based path planning can be applied when it is known that the quantities of interest in the system are generated via a diffusion process [3]. This algorithm assumes that a network of mobile sensors can be commanded to collect samples of the distribution of interest. These samples are then used as constraints for a predictive model of the process. The predicted distribution from the model is then used to determine new sampling locations (for instance, to identify connectivity holes).

## AMR-Aided Network Maintenance for WSN

The use of mobile nodes (i.e., the AMRs) allows energy savings in the static nodes (e.g., by carrying some of the traffic burden). Mobile node can improve or repair the communications in the WSN. Additionally, the advanced features of the mobile robots can be exploited to provide basic services to the WSN devices, such as calibration, positioning and so on. In this paper, in particular, we present the first results obtained in the context of self-localization of indoor wireless sensor node by means of beacon messages broadcasted by an AMR. In other words, the mobile robot, which is capable of estimating with good accuracy its own position in a given area, is used as a *mobile beacon* to reduce the uncertainty on the position estimated by the wireless sensor nodes placed in unknown locations in the area.

In the recent period, several works have tackled the problem, proposing the use of one or more mobile nodes to improve the localization of static nodes of the network. Pathirana et al. estimated the position of the nodes of the network measuring the received signal strength (RSSI) of the radio messages received by a mobile node mounted on a Lego Mindstorm robot [4]. Here, a Robuts Extended Kalman Filter (REKF)-based state estimator solves the localization given the RSSI measurements. Shenoy and Tan present a system in which a mobile robot can be navigated to the event location by the un-localized sensor nodes, which in turn are finely localized by the mobile robot [5]. The localization is achieved by building bounding boxes representing the communication range of the mobile node. Accumulating this information, while the robot moves, shrinks the bounding box and then increase the accuracy of the localization of the nodes. Unfortunately, this system was implemented in simulation only.

Our approach is very similar to the one presented by Sichitiu et Ramadurai in [6]. They used PDAs equipped with IEEE 802.11 communication as static target nodes, while the mobile element was a remotely controlled truck carrying a PDAs and a GPS sensor. The position estimation is based on the different measurements of the RSSI received by the static nodes while the transmitting mobile node is moving. Each node applies Bayesian inference to refine its position estimation at each received message.

We realized a similar architecture, with the main difference that we used actual sensor devices as static unlocalized nodes and a self-made autonomous mobile robot as mobile beacon. Also, our experiments have been performed indoor, which is recognized as a much more hostile environment for self-localization schemes [6, 7].

The remaining of this paper is organized as follows. Section 2 briefly describes the set-up used to perform the experiments. In Sect. 3, the channel model is described and characterized on the basis of the collected measurements. The same measurements are, then, used to obtain the results presented in Sect. 4, which is aimed at evaluating the potential advantages of the hybrid system with respect to static WSN in solving the self-localization problem. Finally, Sect. 5 concludes the paper.

## 2 Experimental Set-Up

The RAMSES2 testbed merges the equipments and competences of SIGNET (Special Interest Group on Networking) and IAS-Lab (Intelligent and Autonomous Systems), two research labs of the Dep. of Information Engineering, at the University of Padova.

At this stage, the testbed consists of a self-made autonomous mobile robot, based on the Pioneer 2 ActivMedia platform, and a WSN composed of a dozen of EyesIFX sensor nodes, produced by Infineon Technologies.

The sensor nodes, intensively used by SIGNET to test protocols and algorithms for classical Wireless Sensor Networks [8, 9], can be programmed and powered via USB, thus permitting easy interconnection with other digital devices. Each board is equipped with a radio interface that provides 19.2 kbps transmission rate by using an FSK modulation in the 868.3 MHz band. The platform is fitted with light and temperature sensors. Furthermore, a *Received Signal Strength* (RSS) circuit can be used to have a measure (RSSI) proportional to the received signal strength.

The AMR provided by the IAS-Lab runs Linux OS with on top the middleware Miro [10]. Miro is a distributed object oriented framework for mobile robot control, based on CORBA (Common Object Request Broker Architecture). The robot computational power is provided by a standard ATX motherboard with an Intel 1.6 GHz Intel Pentium 4 with 256 MB RAM and a 160 GB hard disk. The native on-board sensors are an omnidirectional camera, composed of a standard CCD camera and a convex omnidirectional mirror, and the odometers connected to the two driven wheels. At the present stage, the omnidirectional camera is not used and the position of the robot is estimated by integrating over time the odometric information. The robot is also equipped with a PCMCIA IEEE 802.11g wireless Ethernet card to communicate with other robots and with external computers.

In order to enable the communication between the AMR and the WSN, an EyesIFX sensor node has been connected to the robot via the USB port on the ATX motherboard. Then, we extended the Miro class called Server, creating a new class called EyesService that manage the communication between the AMR and the on-board EyesIFX node.

The architecture has been tested by running a simple experiment in an empty corridor of 4.5 m × 10 m (ceiling was approx 4 m height). Figure 1a shows a snapshot of the location. The robot was programmed to move along three parallel straight lines in the corridor, as shown by the cross-marked line in Fig. 1b. The average travelling speed was approximately $0.24\,\mathrm{m\,s^{-1}}$. The robot coordinates were continuously updated by using the odometric information and passed every 50 ms to the on-board EyesIFX node which, in turn, propagated the message through the EyesIFX radio interface. Ten static sensor nodes, named from $S1$ to $S10$, were placed in the area to form an incomplete lattice, as shown by the red bullets in Fig. 1b. Nodes were

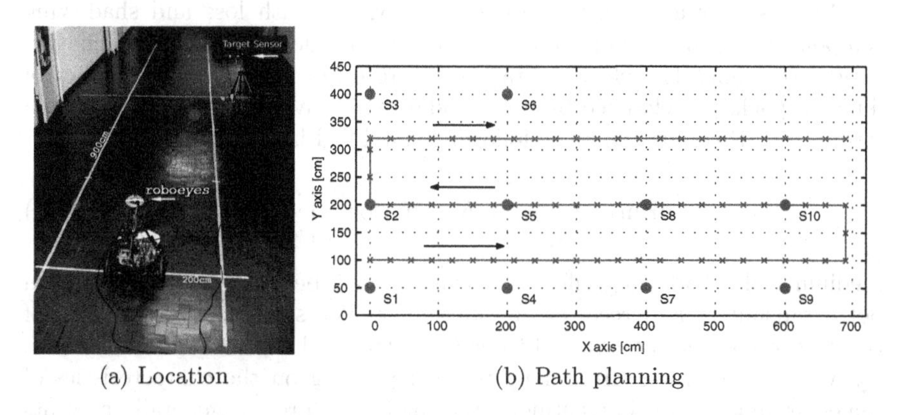


(a) Location (b) Path planning

**Fig. 1.** Experimental set-up

programmed to receive the robot messages, read the robot coordinates and store them in a table, together with the RSSI value measured during the packet reception.

The data collected during the experimental campaign have been first used to characterize the parameters of a simple radio channel model, as explained in Sect. 3. The model, then, has been used to derive a simple RSSI-based distance estimate that, in turn, was used in a multilateration algorithm to estimate the position of the static nodes, as it will be explained later on in this paper.

## 3 Channel Modelling and Characterization

Our first goal is to determine a suitable radio channel model for our environment. According to [11], an accurate modelling of the indoor channels is difficult to obtain. In any case, we aim at developing solutions that are as much as possible independent of the specific environment in which they are tested. Therefore, we consider a simple path loss channel model, in which the generic $i$th node, placed at distance $d_i$ from the transmitter, receives a signal with power $P_i$ (in dBm) given by:

$$P_i \text{ (dBm)} = P_{Tx} + K - 10\eta \log_{10}\left[\frac{d_i}{d_0}\right] + \Psi_i + \alpha_i(t) . \tag{1}$$

In (1), $P_{Tx}$ is the nominal transmission power (in dBm), $K$ is a unitless constant that depends on the environment, $d_0$ is a reference distance for the antenna far field, and $\eta$ is the path loss coefficient. The term $\Psi_i$ denotes the random attenuation due to shadowing, while $\alpha_i(t)$ counts for the fast fading effect. Typically, shadowing is almost constant over long time periods, while fast fading shows rapid fluctuations, so that packets received in different time epochs likely experience equal shadowing, but almost independent fading.

When sender and receiver are stationary, the path loss and shadowing components are practically time-invariant, while the fast fading attenuation varies over time. Therefore, averaging the received power over a number of different packets exchanged by static nodes we can average out the fast fading term in (1), thus obtaining the following simplified law:

$$P_i \text{ (dBm)} \simeq P_{Tx} + K - 10\eta \log_{10}\left[\frac{d_i}{d_0}\right] + \Psi_i . \tag{2}$$

Medium-scale shadowing effect, however, cannot be easily eliminated when both transmitter and receiver are stationary. The statistical distribution of this factor is generally assumed to be Gaussian, with zero mean and variance $\sigma^2_{\Psi_i}$ whose value range from 4 up to 12 depending on the characteristics of the environment [11]. Furthermore, the shadowing process generally presents spatial correlation, although in this study we will assume the shadowing terms to be independent and identically distributed. The effect of shadowing in RSSI-based localization schemes can be mitigated by increasing the number of locations from which beacon messages are sent. In this case, we would ideally obtain a pure path loss model:

$$P_i \text{ (dBm)} \simeq P_{Tx} + K - 10\eta \log_{10} \left[ \frac{d_i}{d_0} \right] . \tag{3}$$

In static WSNs, the number of beacons that can be deployed in a given area is limited to few units, due to cost and/or practical constraints. Therefore, the RSSI-based localization mechanisms in static WSNs are usually affected by large errors and provide very poor performance, in particular in indoor environments [9]. The use of an AMR alleviates the problem by dramatically increasing the number of beaconing locations, which is somehow equivalent to the deployment of a very large number of *virtual* static beacons.

As a matter of fact, our experimental setting allowed us to collect about 2000 RSSI measurements for each static sensor, from a large number of different beacon locations, with minimum effort. In order to determine the propagation model parameters from the set of collected data we first need to convert the integer RSSI measures into dBm values. According to the EyesIFX datasheet, the nominal relation between received signal power (in dBm) and RSSI is shown in Fig. 2 for the case with Low Noise Amplifier active (upper curve) and switched off (bottom curve). In our experiment, the transmission power of the mobile node, mounted on the AMR, was set to the nominal value of +5 dBm at the antenna connector. With this setting, we noticed that the coverage range was comparable with the size of our indoor test-bed

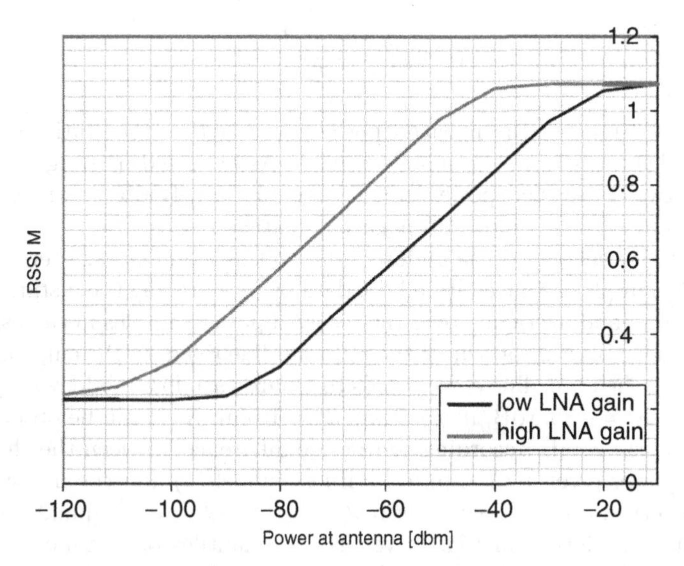


**Fig. 2.** RSSI (V) vs received power (dBm) with an without amplification gain

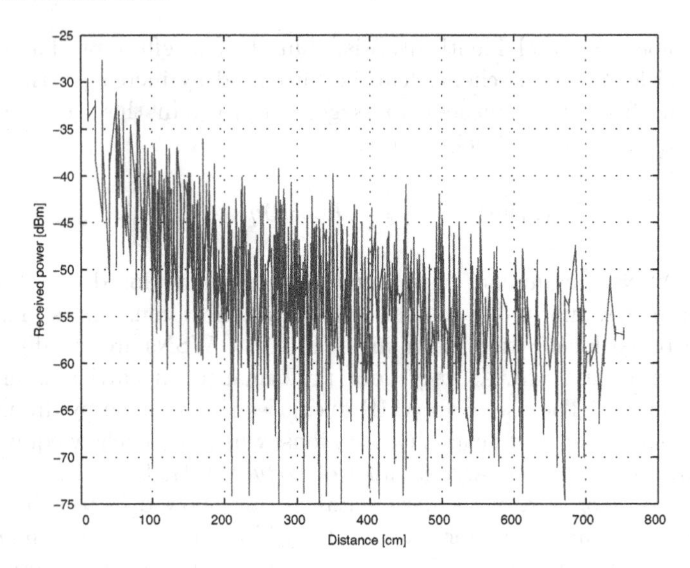


**Fig. 3.** Received power (dBm) vs distance (cm)

environment. Therefore, we switched off the LNA at the receivers, in order to have the RSS circuit likely working in the linear region. Before being accessible, the RSSI value is converted in digital form by a linear ADC with 16 bits, which maps the RSSI voltage into an integer in the interval $[0, 2^{16} - 1]$. Therefore, the relation between measured RSSI and received power level (in dBm) in the linear region is given by the following equation:

$$P_i\,(\mathrm{dBm}) = \frac{RSSI}{14} - 114\,. \tag{4}$$

Figure 3 reports the received power measured by the static sensors for different distances from the mobile beacon. As expected, the logarithmic relation between received power and distance is largely obscured by random variations due to the indoor propagation characteristics.

In order to determine the underlying path loss model parameters $K$, $\eta$ and $d_0$, as given in (3), we have filtered the data by partitioning the distance axis in slots of 10 cm and averaging over the received power samples in each slot. The result is shown in Fig. 4a, where the red solid line refers to the empirical data, while the dashed blue line corresponds to a pure path loss model with parameters $P_{Tx} + K = -30.5\,\mathrm{dB}$, $\eta = 1.5$ and $d_0 = 10\,\mathrm{cm}$. As it can be observed, the fitting is fairly good. According to (2), the difference between the theoretical received power given by the pure path loss model (3) and the measured values shall return the shadowing term $\Psi_i$. The QQ-plot of the quantiles of such a difference is plotted in Fig. 4b versus the quantiles of a standard Normal distribution. The quantile–quantile plot reveals that, as expected, the error

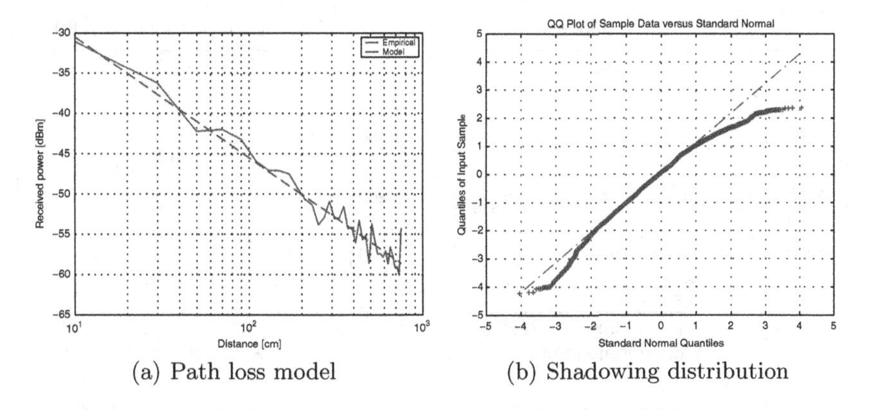


(a) Path loss model

(b) Shadowing distribution

**Fig. 4.** Propagation model matching

samples distribution is fairly close to a Normal distribution. Mean and standard deviation can be estimate from the empirical samples, resulting equal to $\mu_\Psi \simeq -0.0348 \pm 0.0860\,\text{dB}$ and $\sigma_\Psi \simeq 6.339 \pm 0.0614\,\text{dB}$, respectively, where the range corresponds to the 95% confidence interval.

# 4 Static Nodes Localization

We have applied the well-known multi-lateration algorithm [12] to estimate the position of the static sensors in the area. The algorithm has any node compute its own position by intersecting the circles centered on the positions occupied by the robot and radius equal to the estimated distance between the robot and the node itself. The distance has been obtained from the measured RSSI, by using the aforementioned path loss model. Ideally, the intersection should be a single point on a surface, but due to channel and environment impairments, this intersection as a matter of facts identifies an area where the node is likely to be found. In practice, the environment has been represented by an occupancy grid quantized in cells of 20 cm $\times$ 20 cm. Each cell is assigned a weight that is initially set to zero. Every time a static node receives a packet, it extracts the transmitter coordinates from the packet payload and reads the RSSI measured during the packet reception. The RSSI is used to get an estimate of the transmitter distance $d$, by reverting (3). Then, the node increments by one the weight of all the cells at distance $d$ from the transmitter. The cell that scores the maximum weight is elected as target node location.

Figure 5 shows an example of the positioning obtained by applying the multi-lateration algorithm to different subsets of the complete data set collected during the experiment, by each static node. More specifically, the plot in the upper-left corner reports the real positions of the static node. The other three plots, in clock-wise order, report the position estimates obtained by each sensor node by using only a subset of $N = 1,280$, $N = 160$ and $N = 40$

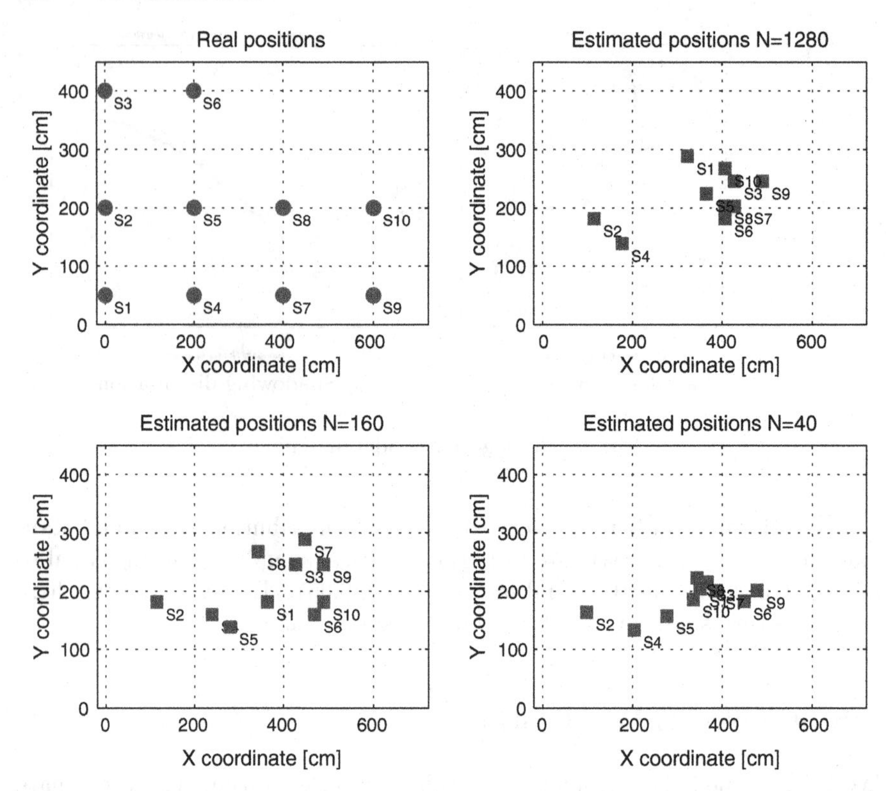


**Fig. 5.** Example of self-localization result with different number $N$ of (random) virtual beacons

RSSI samples, respectively, randomly picked from the entire data set. At a first glance, we can notice that the localization error, defined as the distance between the real and estimated node location, is rather large, in particular for the nodes in proximity of the area borders. Furthermore, increasing the number of RSSI samples apparently does not yield any significant benefit.

In Fig. 6, the results have been obtained by selecting the $N$ highest RSSI readings out of the complete data set collected by each sensor node. As it can be observed, in this second case the nodes positioning is much more precise than in the previous case. Also, we can notice that the localization improves as the number of samples reduces. This result reveals that the simple localization mechanism considered in this paper is sensitive to the ranging errors, which are more relevant when considering lower RSSI values due to the logarithmic nature of the path loss model.

These results confirm that localization in indoor environments is a challenging task, which requires more sophisticated solutions.

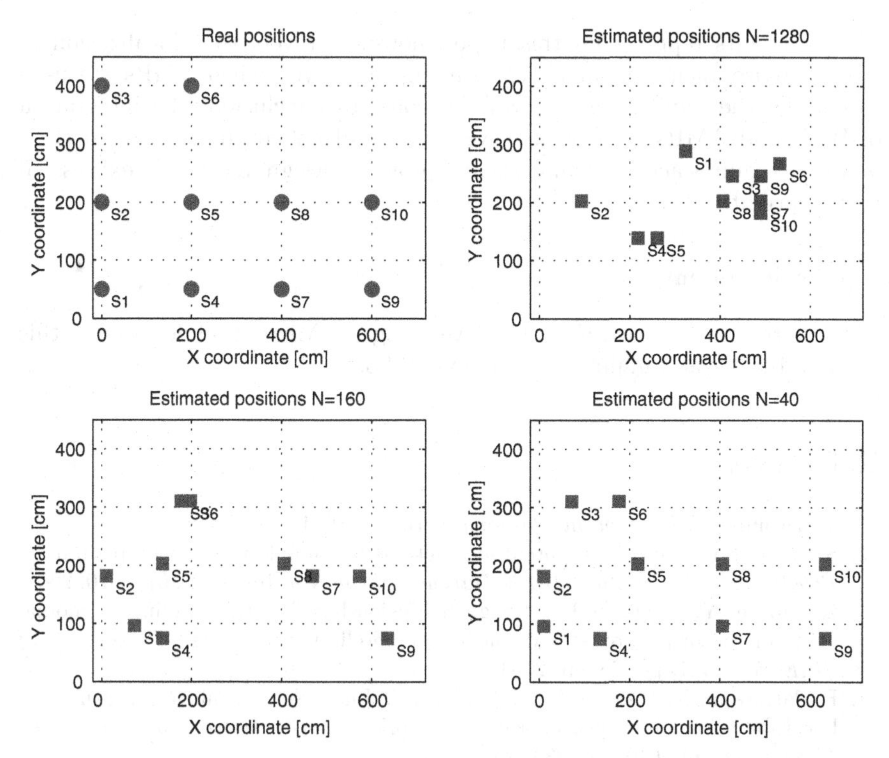


**Fig. 6.** Example of self-localization result with different number $N$ of (ordered) virtual beacons

# 5 Conclusions

In this paper, we have presented the RAMSES2 project, whose aim is to investigate the potentialities of the interaction between wireless sensor networks and autonomous mobile robots. As a first case study, we have considered the well-known self-localization problem for wireless sensor nodes.

Most of classical localization mechanisms based on RSSI measurements show very poor performance when deployed in static real-world network, for several reasons. A primary problem is represented by the shadowing, which introduces long-term random variations to the distance-vs-RSSI law dictated by the path-loss model. Another relevant problem in self-localization algorithms is the loose calibration of RSSI circuits and transmitter potentiometers of the sensor nodes. Both such problems can be alleviated when an AMR is used as reference node, acting as *mobile beacon*. In fact, any time the AMR transmits a beacon message from a different position it acts as a sort of virtual static beacon. The high number of virtual beacons permits to mitigate the effect of shadowing. Furthermore, since the beacon messages are all transmitted by the same node, calibration problems are avoided.

The results reported in this paper, however, reveal that localization in indoor environment remains a challenging task, even when AMRs are used to enlarge the number of (virtual) beacons. In conclusion, the integration of WSN and AMRs appears as a very powerful networking paradigm that, however, requires accurate analysis and protocol design in order to express all its potentialities.

**Acknowledgments**

We wish to thank Enrico Frigo for developing the Micro modules used in this work and to fix and maintain the mobile robot.

## References

1. J. Latombe, *Robot Motion Planning*, Springer, 1991.
2. S. Ge and Y. Cui, "New potential functions for mobile robot path planning," *Robotics and Automation, IEEE Transactions on*, vol. 16(5), pp. 615–620, 2000.
3. K. Moore, Y. Chen, and Z. Song, "Diffusion-based path planning in mobile actuator-sensor networks(MAS- net): some preliminary results," *Proceedings of SPIE*, vol. 5421, pp. 58–69, 2004.
4. P. Pathirana, N. Bulusu, A. Savkin, and S. Jha, "Node localization using mobile robots in delay-tolerant sensor networks," *IEEE Transactions on Mobile Computing*, vol. 4(3), pp. 285–296, 2005.
5. S. Shenoy and J. Tan, "Simultaneous localization and mobile robot navigation in a hybrid sensor network," *Intelligent Robots and Systems, 2005.(IROS 2005). Proceedings. 2005 IEEE/RSJ International Conference on*, vol. 1, 2005.
6. M. Sichitiu and V. Ramadurai, "Localization of wireless sensor networks with a mobile beacon," *Mobile Ad-hoc and Sensor Systems, 2004 IEEE International Conference on*, pp. 174–183, 2004.
7. N. B. Priyantha, H. Balakrishnan, E. Demaine, and S. Teller, "Mobile-Assisted Localization in Wireless Sensor Networks," in *IEEE INFOCOM*, Miami, FL, March 2005.
8. R. Crepaldi, A. Harris, A. Scarpa, A. Zanella, and M. Zorzi, "Signetlab: deployable sensor network testbed and management tool," in *SenSys '06: Proceedings of the 4th international conference on Embedded networked sensor systems.* New York, NY, USA: ACM, pp. 375–376, 2006.
9. R. Crepaldi, A. Harris, A. Scarpa, A. Zanella, and M. Zorzi, "Testbed implementation and refinement of a range-based localization algorithm for wireless sensor networks," in *3rd IEE Mobility Conference 2006*, Oct., pp. 25–27, 2006.
10. H. Utz, S. Sablatnog, S. Enderle, and G. Kraetzschmar, "Miro-middleware for mobile robot applications," *Robotics and Automation, IEEE Transactions on*, vol. 18(4), pp. 493–497, 2002.
11. A. Goldsmith, *Wireless Communications.* Cambridge University Press, 2005.
12. J. Hightower and G. Borriello, "A survey and taxonomy of location system for ubiquitous computing," University of Washington, CSE Dept., Seattle, WA 98195, *Tech. Rep.* UW-CSE 01-08-03, Aug. 2001.

# An Experimental Study of Aggregator Nodes Positioning in Wireless Sensor Networks

Laura Galluccio, Alessandro Leonardi, Giacomo Morabito, and Sergio Palazzo

Dipartimento di Ingegneria Informatica e delle Telecomunicazioni
University of Catania, Italy
laura.galluccio@diit.unict.it,alessandro.leonardi@diit.unict.it
giacomo.morabito@diit.unict.it,sergio.palazzo@diit.unict.it

**Summary.** Wireless sensor networks are composed of devices characterized by limited energy and processing capabilities. This aspect becomes crucial in case of applications where a lot of redundant data go to the sink wasting precious bandwidth and energy resources. In the recent past, some solutions aimed at supporting reduction in energy consumption at network nodes through the use of data aggregation have been proposed. In this paper we perform an experimental study of data aggregation for wireless sensor networks with a focus on aggregator node positioning. In particular, we evaluate a data aggregation techniques which gathers a variable number of packets at fixed time intervals, and then provide, through simulations, some interesting results on advantageous positioning of aggregator nodes.

## 1 Introduction

Wireless sensor networks are characterized by limited processing, storage and energy capabilities. This aspect becomes critical for network lifetime in case of propagation of numerous messages from the sensors up to the sink. As an example when sensors are disseminated in the interest area to monitor a given phenomenon, they either periodically or upon occurring an anomalous event, notify the sink with appropriate messages. Accordingly, network is overloaded by possible redundant packets alerting and/or notifying about the same event. In this case the main drawback which could be met is related to the higher energy consumption at those network nodes which, depending on their location, relay a multitude of packets to the sink. As a consequence of the intense network traffic, also congestion can be frequently met at these nodes. In the recent past, some solutions aimed at supporting reduction in energy consumption at network nodes through the use of data aggregation have been proposed [1, 2]. Data aggregation is a technique which allows to fuse together data coming from multiple sources and destined to the sink. This approach implies the need for appropriately positioning aggregator nodes in order to reduce energy consumption and congestion due to collisions met at certain bottleneck nodes. However, it is well known that the problem of optimal location of network aggregators is a NP hard problem [3] and, accordingly, only suboptimal solutions can be considered. In this paper we will perform a preliminary study of data aggregation for wireless sensor networks with a focus on aggregator nodes positioning. To this purpose, let us preliminarily observe that two different approaches for data aggregation can be distinguished. The first is based on letting aggregator nodes fuse a preestablished number of data packets, once collected at their buffer. However, this technique does not allow to control the delay in the aggregation process because the time when the aggregated packet will be sent depends on the time-varying statistics of the received traffic. Another approach is based on aggregating a variable number of data packets but at fixed time intervals. This second scheme is convenient under the perspective of controlling the delay, but turns out to be suboptimal in that it is not possible to always aggregate a controlled number of packets which could thus result in a waste of resources. In this paper we focus on the latter scenario where aggregation is performed at fixed time intervals.

The rest of this paper is organized as follows. In Sect. 2 we show in detail the aggregation policy used. Section 3 provides some results obtained through simulations so as to derive some insights into positioning of aggregator nodes throughout the network. Finally, in Sect. 4 some conclusions and considerations on future work are drawn.

## 2 Aggregation Policy

Aggregation is an operation performed inside the network in order to reduce the amount of data packets forwarded to the sink. Aggregation implies the need for some particular nodes, denoted as aggregators, in charge of performing the necessary processing to fuse together the information carried by many data packets so that only few aggregated packets are sent to the sink. The aggregated packet is a new packet which, inside its payload, contains the information carried by the packets collected.

To make aggregation more efficient, in case of event-based applications, it is possible to exploit the correlation among the data carried by different packets, thus reducing the size of the aggregated packet's payload.

Let us call $t_0$ the time instant when the aggregator node starts receiving data packets from the neighbor nodes. Let $N$ be the number of data packets received and buffered in the transmission queue during the time interval $]t_0, t_0 + \Delta]$. The aggregation technique operates by blocking the transmission queue and aggregating all the $N$ packets buffered during the time interval $\Delta$. Note that all the non-data packets arriving during this time interval, e.g. routing messages, signaling messages, and high priority packets are not delayed but transmitted immediately.

At time $t = t_0 + \Delta$, a new packet, called AP (Aggregated Packet), is created. The payload of the new packet contains the payloads of the original data packets; moreover, new fields other than the previous ones in the packet header are added. These new fields carry information about the original source nodes, the number of original packets aggregated and the identity of the aggregator node. Assuming that $t_p$ is the time interval to create the packet, with $t_p \ll \Delta$, at time $t_s = t_0 + \Delta + t_p$ the new packet is transmitted and the $N$ source packets are deleted from the queue.

The AP packet, if correctly received from the sink, will be recognized as an aggregated packet and appropriately processed.

Let us observe that, by choosing to aggregate packets at fixed time intervals of length $\Delta$ seconds, we pursue the target of limiting the end-to-end delay thus avoiding large fluctuations in the delivery time. Instead, if a fixed number of packets was always aggregated we could guarantee a more efficient exploitation of bandwidth resources but could not satisfy any constraint on the delivery delay. Accordingly, we preferred to focus on applications with delay constraints such as common in sensor network applications but different applications could have been chosen.

# 3 Performance Evaluation

## 3.1 Simulation Scenario

We consider a sensor network where $N$=[50, 100, 150] sensor nodes are distributed according to a square grid topology. Each node is located as shown in Fig. 1 where we represent the three network topologies being considered. In all topologies the distance between two adjacent nodes is set equal to 35 m and the sink is located in the down left corner of the grid. Sensors are considered static, as usual in several sensor application scenarios. Two sets of source nodes have been considered. The first set is located along the upper and the right perimeter of the area, while the second set is located in the core of the area. Sources are shown in black in Fig. 1. Each source generates constant bit rate traffic with rate $r = 2$ [packets per second]. The simulation duration is set equal to 200 s. To characterize the signal propagation model, we used the widely adopted two-ray ground model. In our simulations we used the 802.15.4 (ZigBee) MAC protocol [4] such as implemented in ns-2.30 version of the network simulator [5]. To better understand the effects of data aggregation on the system we purposely assumed a static routing protocol. This protocol has been implemented in ns-2 and is called NOAH (NO Ad Hoc routing agent) [6]. This protocol only supports direct communication between wireless nodes. In the scenario depicted, the routing scheme has been designed as follows: if the destination of the packet lies on the same line, the packet is routed along the horizontal line, otherwise the packet is routed towards the neighbor node located along the same vertical line. The main simulation parameters are reported in Table 1.

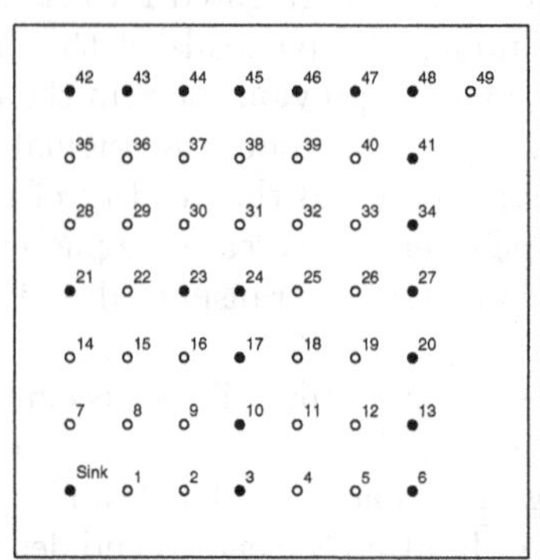


(a) Case I: $N$= 50 nodes

(b) Case II: $N$=100 nodes

(c) Case III: $N$=150 nodes.

**Fig. 1.** Simulation scenario

**Table 1.** Simulation parameters

| Simulation parameter | Value |
|---|---|
| Propagation model | Two-ray ground |
| MAC protocol | 802.15.4 |
| Packet size | 2 byte |
| Transmission range | 40 m |
| Initial energy | 1 J |
| Transmission/Reception power | 54 mW |
| Routing protocol | NOAH |
| Transport protocol | UDP |
| Application protocol | CBR traffic |

### 3.2 Simulation Results

In this section we investigate on the effects of the discussed data aggregation mechanism. As a preliminary step we consider $N$=100 nodes and we show in Fig. 2 the residual energy per node without performing any data aggregation. In this case all nodes act as relays and they simply forward all the data packets originated from the source nodes towards the sink node. We observe that all nodes close to the sink have a very low value of residual energy. In particular this is more evident from Fig. 3 where we observe that the lower value of energy is reached 1 or 2 hops away from the sink located in x=0, y=0. This is because nodes one or 2 hops away from the sink are involved in

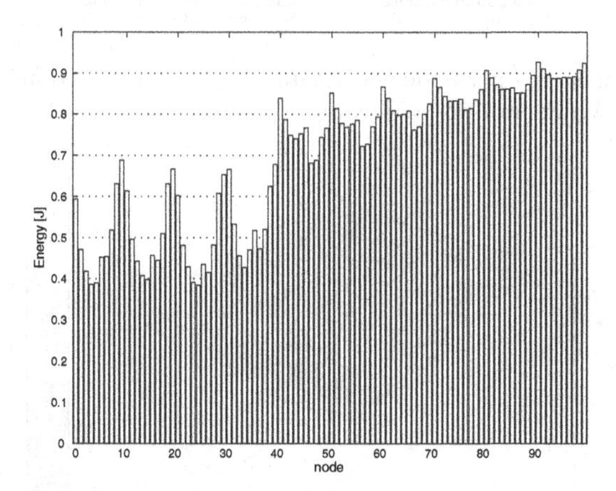


**Fig. 2.** Residual energy per node without data aggregation when $N$=100 nodes are considered

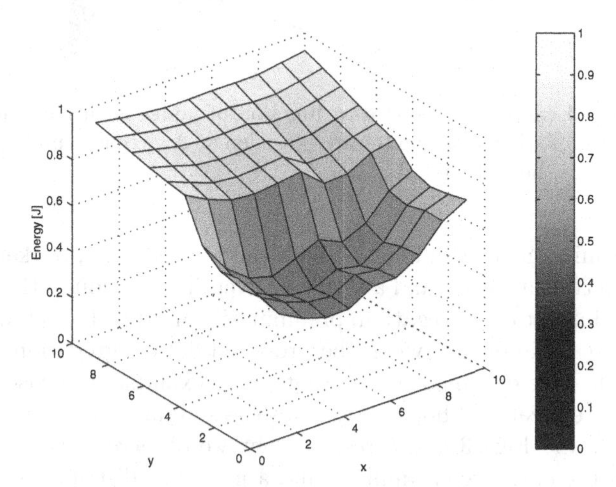


**Fig. 3.** Residual energy per node as a function of nodes' location without performing data aggregation when $N$=100 nodes are considered

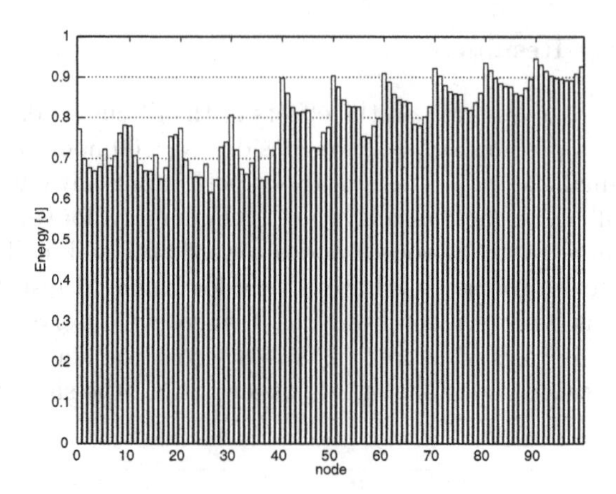


**Fig. 4.** Residual energy per node when data aggregation is used and $N$=100 nodes are considered. Aggregator nodes' location is 8 hops away from the sink and $\Delta$=2 s

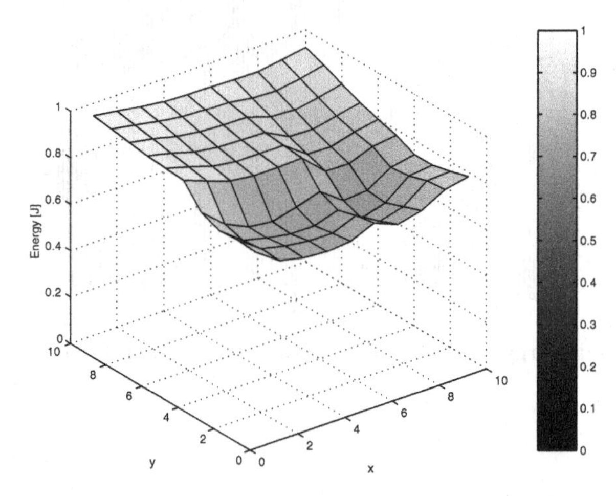


**Fig. 5.** Residual energy per node as a function of nodes' location when data aggregation is used and $N$=100 nodes are considered. Aggregator nodes' location is 8 hops away from the sink and $\Delta$=2 s

several transmissions to neighbors. These cause congestion, packet collisions, subsequent retransmissions and a resulting overall reduction of the energy. To estimate the impact of aggregation, preliminarily, in Figs. 4 and 5 we show the value of the residual energy when the proposed data aggregation mechanism is used. Results are obtained considering, as an example, the case when data aggregators are located 8 hops away from the sink and $\Delta = 2$ s. Comparing Figs. 2 and 4, and Figs. 3 and 5 respectively, we observe in the latter case a reduction in the energy consumption and a more fair distribution of residual energy in the network. In particular the proposed technique, by reducing the

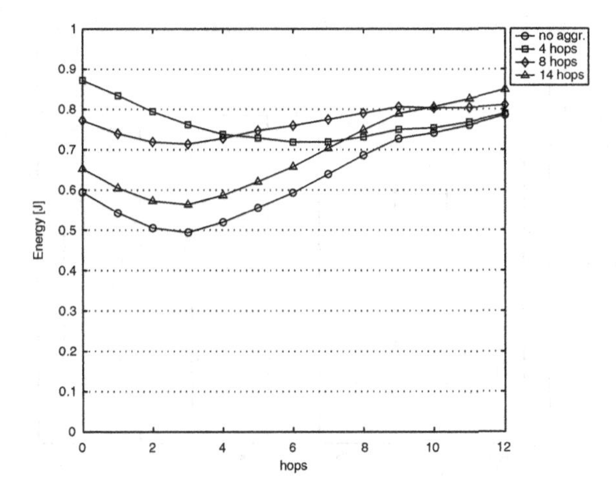


**Fig. 6.** Residual energy vs. nodes' distance from the sink (in hops) for different data aggregator nodes positioning when $N$=100, and $\Delta$=2 s

number of packets in the network, decreases the number of packet lost due to queue overflow and collisions, and consequently the number of retransmissions.

In Fig. 6 we show the residual energy vs. the distance from the sink in hops, when different approaches in the aggregator nodes positioning are considered. In particular, we show the results without aggregation in the following conditions:

- When the aggregator nodes are located 4 hops away from the sink
- When the aggregator nodes are located 8 hops away from the sink
- When the aggregator nodes are located 14 hops away from the sink
- When no aggregation is performed

By comparing these plots, we observe that when considering the residual energy within the first 4 hops from the sink the case when aggregators are located 4 hops away from the sink is the most advantageous; however, both in terms of higher residual energy and fairness in the distribution of the energy among nodes, the case when aggregators are located 8 hops away from the sink performs better. This corresponds to the case when aggregator nodes are located on the main diagonal of the considered grid. This is a significant result since it means that if aggregator nodes are located too far away from the sink their action in reducing traffic and energy consumption is useless; instead, if aggregators are located too close to the sink their action can be too late.

Figure 7 shows the average residual energy in the four cases examined in all the network. The figure confirms the results obtained above, showing that the best location of the aggregator nodes, in terms of reduction in energy consumption, is in the middle of the network (diagonal of the grid).

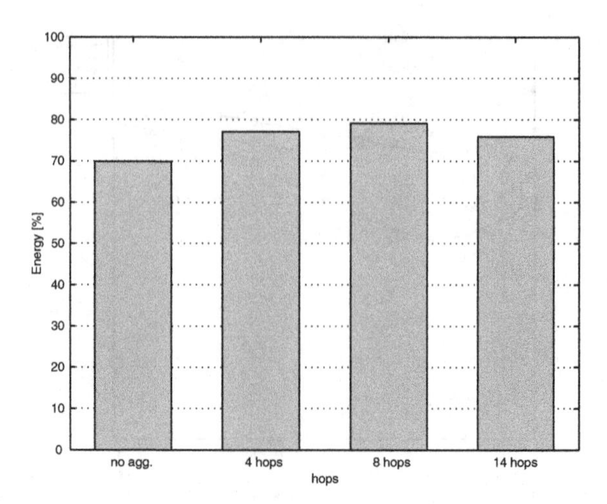


**Fig. 7.** Average residual energy vs. aggregator nodes' positioning (in hops from the sink) when $N$=100 nodes are considered and $\Delta = 2\,\mathrm{s}$

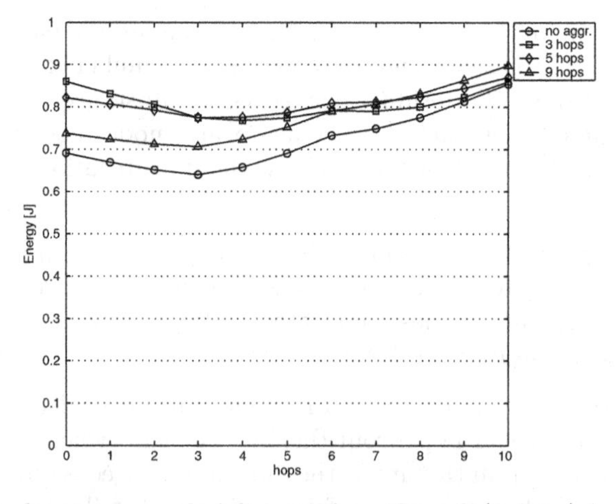


**Fig. 8.** Residual energy vs. nodes' distance from the sink (in hops) for different data aggregator nodes positioning when $N = 50$ nodes and $\Delta = 2\,\mathrm{s}$

In order to show the effects of the number of nodes in the network performance, we have carried out other simulations varying the value of $N$.

In Figs. 8 and 9 we show the residual energy vs. the distance from the sink in hops for different aggregator nodes' position, when $N = 50$ and $N = 150$ respectively. Results confirm that the best aggregators' location is again in the middle of the network (which corresponds to 5 hops away from the sink when $N = 50$ and 10 hops away from the sink when $N = 150$). Moreover, when $N = 50$ we observe a more fair distribution of the energy among nodes and a higher average value of the residual energy if compared to the cases

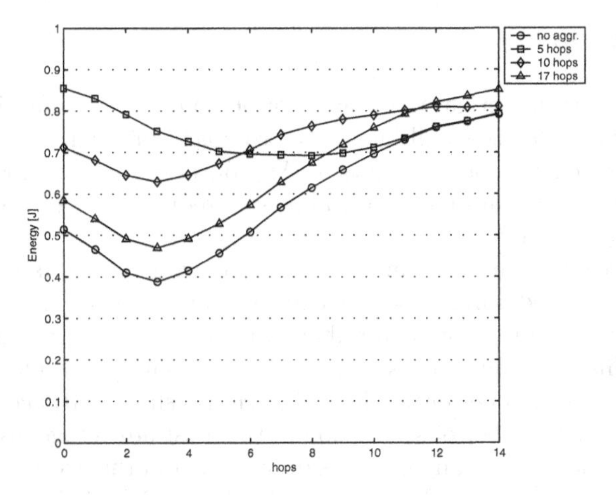


**Fig. 9.** Residual energy vs. nodes' distance from the sink (in hops) for different data aggregator nodes positioning when $N$=150 nodes and $\Delta = 2\,\mathrm{s}$

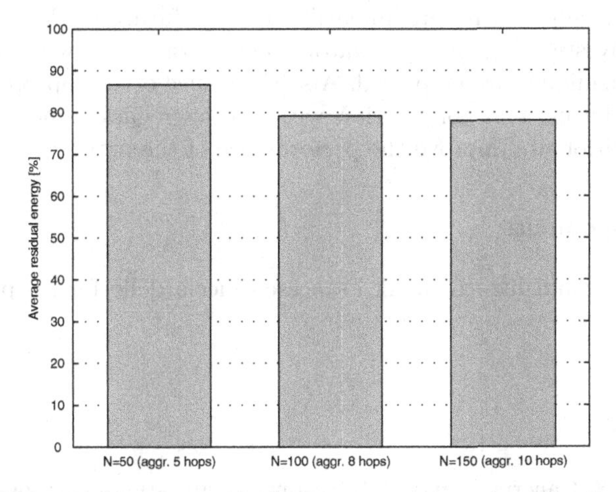


**Fig. 10.** Average residual energy vs. the number of nodes

$N = 100$ and $N = 150$ nodes. This is due to the lower number of nodes which produces a lower number of collisions, consequent retransmissions, and thus a lower waste of energy.

In Fig. 10 we show the average residual energy in percentage vs. the number of nodes, $N$, when aggregators are located in the advantageous positions identified and discussed above. Accordingly, when $N$ increases, the average value of the residual energy inside the network decreases consequently. This is because more source nodes are found in the network, thus leading to more traffic traveling and, consequently, higher energy consumption.

# 4 Conclusions

In this paper we investigated on aggregator nodes positioning in a wireless sensor network. This is a well known NP complete problem and, thus, a simulation approach has been proposed. Appropriately positioning aggregator nodes in a wireless sensor network allows to reduce energy consumption so increasing network lifetime.

The aggregation mechanism considered in this paper is assumed to work in delay constrained wireless sensor network applications. Thus, aggregation is performed by aggregating a variable number of data packets at fixed time intervals. This technique allows to limit the maximum tolerable delay.

Simulation results performed in grid scenarios show that, according to the network size and number of sensor nodes, the most advantageous location for aggregator nodes positioning in terms of reduction in the energy consumption is usually along the diagonal of the square grid topology being considered. In these positions aggregator nodes allow to significantly reduce network traffic and, consequently, energy consumption.

As a follow-up of this preliminarily activity, future work will be mainly related to investigating the aggregator nodes' positioning when different aggregation techniques are employed. Also the choice of an appropriate routing protocol to be used in conjunction with the aggregation mechanism could allow to additionally improve the performance of the network.

## Acknowledgements

The authors would like to thank Francesco Licciardello for his precious help in simulations.

# References

1. A. Boulis, S. Ganeriwal, and M. Srivastava. Aggregation in sensor networks: an energy-accuracy tradeoff. *Proc. of* IEEE SNPA. Anchorage, AL. May 2003
2. M. Lotfinezhad and B. Liang. Effect of partially correlated data on clustering in wireless sensor networks. *Proc. of* IEEE SECON 2004. Santa Clara, CA. October 2004
3. B. Krishnamachari, D. Estrin and S. Wicker. The impact of data aggregation in wireless sensor networks. *Proc. of DEBS*, Wien, Austria. July 2002
4. IEEE Standard 802.15.4: Part 15.5: Wireless Medium Access Control (MAC) and Physical Layer (PHY) Specifications for Low-Rate Wireless Personal Area Networks (LR-WPANs)
5. NS-2 Network Simulator: http://www.isi.edu/nsnam
6. NO Ad-Hoc Routing Agent (NOAH): http://icapeople.epfl.ch/widmer/uwb/ns-2/noah/

27

# SignetLab$^2$: A Modular Management Architecture for Wireless Sensor Networks

Riccardo Crepaldi[1], Albert F. Harris III[2], Andrea Zanella[1], and Michele Zorzi[1]

[1] Department of Information Engineering, University of Padova, Padova, Italy
[2] Department of Computer Science, University of Illinois at Urbana-Champaign, Urbana, IL, USA
riccardo.crepaldi@dei.unipd.it
zorzi@dei.unipd.it,zanella@dei.unipd.it, aharris@cs.uiuc.edu

**Summary.** Large-scale deployed sensor networks have a wide range of possible applications, including a number of types of environmental monitoring. The primary challenge to designing deployable sensor network applications lies in the difficulty of building and managing testbeds. The main contribution of this work is a sensor network management system, called SignetLab$^2$, that allows the control, visualization, and reprogramming of deployed sensor networks. This system facilitates rapid testing and debugging of sensor network applications on deployed networks. To show the ability of our management system, we also present the design of a fire alarm application for sensor networks. This application was designed and tested through the use of the SignetLab$^2$ management system.

## 1 Introduction

The application space for wireless sensor network technology is growing at a rapid rate, driven primarily by the availability of a wide array of devices capable of monitoring a variety of events, from large temperature changes to gas leaks. However, the development of protocols to support specific applications of these sensors requires the ability to perform rapid testing on actual hardware. To this end, a number of testbed tools have been designed with the goal of easing testbed experimentation.

The primary focus of early testbed management systems was facilitating time-sharing on a particular testbed [4,5]. While this is an important component of any sensor network testbed management system, it does not address the fundamental problem of facilitating the experiments themselves. To this end, we developed SignetLab, a testbed with a management system that provides support for the control and programming of nodes [1,2]. While the first version of SignetLab achieved the goal of supporting compilation, node reprogramming, code debugging, and data collection, it relied on a USB backbone.

The main contribution of this work is the design and implementation of SignetLab$^2$, a testbed management system that provides the functionality of the original SignetLab, but does not rely on a USB backbone. Therefore, our new testbed management system is suitable not only for testbed control, but also for controlling and managing deployed sensor networks.

To demonstrate the power of our management system, we present the design of a sensor network built to detect fires within a building. The fire alarm application must be resilient to packet loss and sensor loss, while avoiding data storms at the sink which could prevent alarms from being triggered. Additionally, false alarms should be kept to a minimum, requiring a protocol that takes into account the ability of the sensor nodes to accurately measure temperature changes as well as the failure rate of the nodes themselves. We show that the application is able to accurately trigger fire alarms while transmitting a minimal amount of messages and maintaining a low false alarm rate.

The rest of this paper is as follows. Section 2 presents the challenges faced when designing sensor network management systems and applications. Section 3 presents a description of the SignetLab$^2$ management system. Section 4 presents our fire alarm application and discusses its development and testing using SignetLab$^2$. Section 5 describes the implementation of the fire alarm system on our sensor network and Sect. 6 presents an evaluation of the application. Finally, Sect. 7 presents some conclusions and future directions.

## 2 Challenges

Sensor networks are typically made up of numerous small, inexpensive devices that have low-bandwidth links and limited energy storage. Therefore, there are a number of challenges to developing a management system for deployed sensor networks. Any control and monitoring data must be passed over the wireless links, but the low data rate demands that minimal control data is present on the network. Additionally, since there is no backbone to send the control data across, such data may interfere with the functions of the applications themselves. The transmission and reception of any control and management data will drain the battery supply of the nodes. Since the expected number of nodes in a real deployment is large, and battery replacement is often not feasible, the amount of this control data must be further minimized.

Furthermore, the types of applications run on sensor networks differ dramatically in terms of the demands they have on a management system. We developed the SignetLab$^2$ management system with the goal of supporting many different types of sensor network applications without making it so general that it becomes useless for any application. A desired capability for this kind of management system is also to have fine-grained control over the network, being able to send or receive data from every single device in it. To achieve both these goals, we provide a general-purpose application, and a number of plugins that connect to it, for the control and monitoring of a wide variety of applications.

Finally, developing applications that can tolerate hardware failure and poor network connections is itself challenging. The management system must not make the situation worse by consuming unnecessary resources.

## 3 SignetLab$^2$

SignetLab$^2$ is a modified version of our SignetLab testbed management system [1] to accommodate control of sensor networks deployed with no wired backbone. The goal of the system is to support researchers in the implementation of sensor network applications and protocols, both in testbeds and in actual deployments.

SignetLab$^2$ maintains the modular design of the original management system, using various plugins to provide the interface to the sensor network. An application programming interface (API) is provided allowing researchers and network administrators to easily develop custom interfaces to meet applications needs. This API provides a number of basic functionalities to ensure the rapid design of new plugins.

Examples of plugins are the network programming controller, which automates the dissemination of new applications on the network using an algorithm that has been developed in our research group, and an environment monitoring plugin [2], which collects information about temperature, light intensity and battery level from each node in the network, displaying it on the network map. This last plugin also allows the user to set a threshold for each sensed value. If these values exceed the given threshold the plugin can change the node image on the map and alert the user. Figure 1 is a screenshot of the GUI of SignetLab$^2$, while running the temperature monitoring plugin.

To support multiple data sinks in a single sensor network, SignetLab$^2$ provides a server component which is run on each sink. The client allows a single researcher to maintain access and control of the network through one or more of the sinks. Additionally, multiple researchers can gain access to the same server; however, their actions need to be coordinated to insure no race-conditions between the users occur.

The SignetLab$^2$ architecture is depicted in Fig. 2. The main application interface includes a topology map that allows researchers to easily visualize the sensor network. Through plugins, users can display various data from each of the nodes on the map, either graphically, or in table format. SignetLab$^2$ accepts simple topology files that define the locations of nodes and can be updated via localization algorithms during runtime. Other panes can be rapidly developed by users through use of the API to match the monitoring needs of specific applications. Additionally, nodes can be programmed and controlled

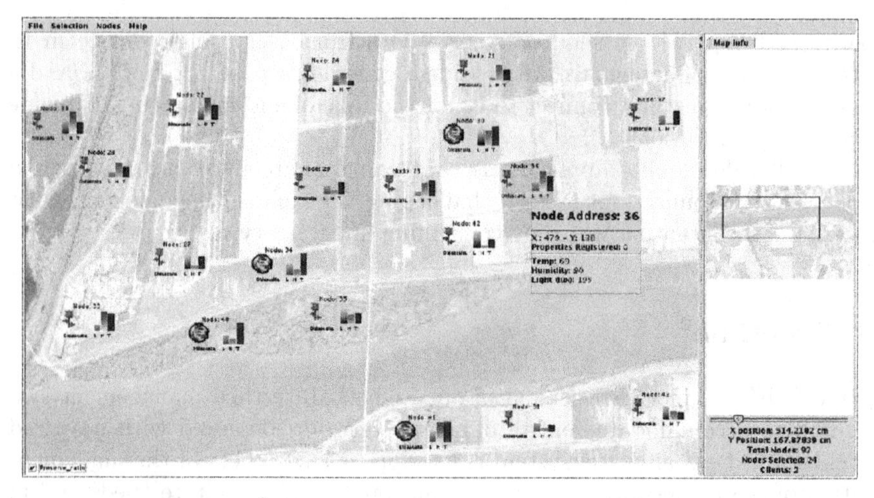


**Fig. 1.** SignetLab$^2$ GUI running a temperature monitoring application

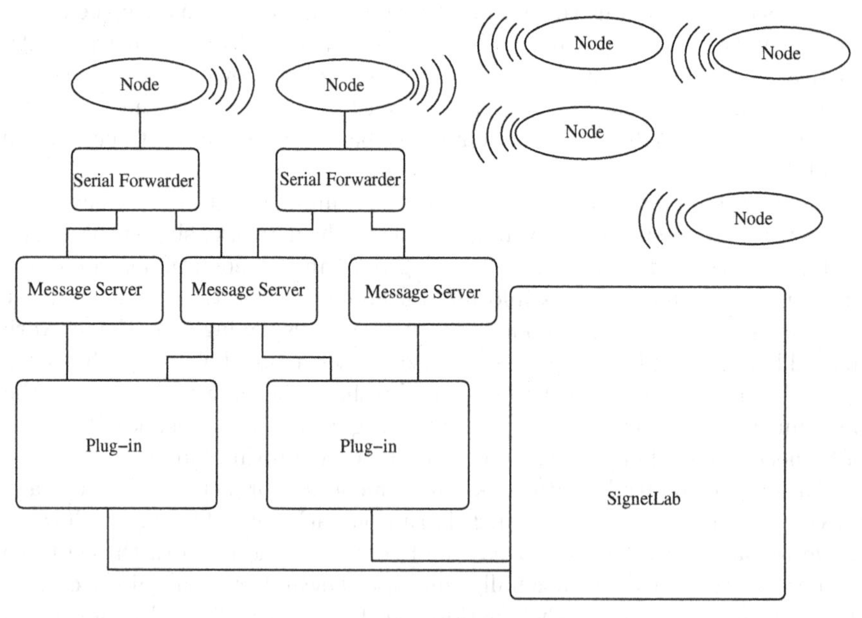


**Fig. 2.** SignetLab$^2$ architecture

via SignetLab$^2$; however, this interaction necessarily interferes with any data being collected at the sinks as there is no separate control plane.

The server architecture allows one or more sensor nodes to be connected to each server for use as interfaces to the sensor network. The server typically connects to its sink nodes through USB cables and uses a TCP connection for the transfer of messages. Clients connect to the server via TCP connections and can register interests for specific data types. The server then encysts each message matching the interests to each of the clients in its interest database. Currently, a server can only forward a single message type at a time; however, we are currently extending this functionality to allow a single server managing multiple data types.

## 4 Fire Detection Application

As a case study to test the management ability of SignetLab$^2$, we implemented a fire alarm application. The fire alarm system is composed of a number of smoke and heat detectors spread throughout a facility. In general, a fire detection system shall be able to promptly detect the start of a fire in the area equipped with sensors, for instance by revealing the event that a heat threshold is crossed or smoke is detected. In this case, the system shall react in some way, such as for instance activating an alarm bell to alert occupants, while sending an alarm notice to the fire department. Additionally, some automated response may be triggered (e.g., fire sprinklers become active).

When designing a fire alarm application, a number of considerations must be taken into account. To define the requirements of our system, we began by studying how a fire starts in terms of its heat signature (see Fig. 3). Experts divide the lifetime of a fire into four phases. The first phase, called the ignition

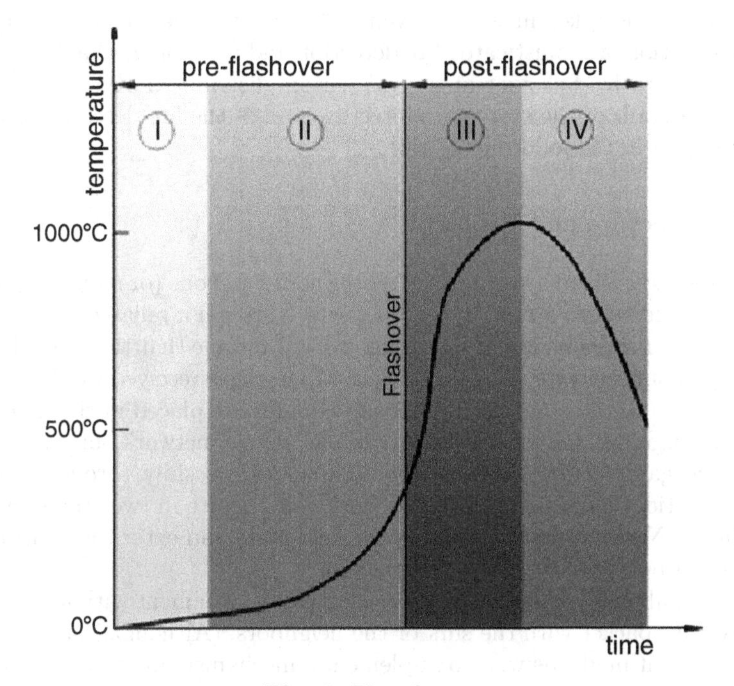


**Fig. 3.** Fire phases

phase, has a low rate of temperature increase. The second phase, called the propagation phase, is characterized by a rapid increase in temperature, as well as a rapid increase in the area covered by the fire. The third phase, called the generalized fire phase, is the time in which all objects in the fire are ignited. The fourth phase occurs after most of the flammable material is burned and the fire slows and is extinguished. The moment between the second and the third phase is called flashover. Ideally, fire sprinklers are activated before the flashover occurs. In most cases not involving arson, the time the fire is in phase two is on the order of several minutes. Therefore, an alarm raised in phase two should have time to reach the sink before flashover occurs. Phase two fires are easier to identify due to the rapid increase in temperature.

Accordingly, the specific design goals that our application should achieve are as follows:

- Lifetime of the network > 6 months.
- Total time between the start of a fire and the signal of an alarm less than 2 min.
- Low false-alarm rate.

The fire detection devices that can be currently found on the market offer rather limited functionality. Most of them simply react when the temperature exceeds a given threshold, and rarely support the use of routing algorithms to monitor a wider area. The fire alarm system we have designed offers much more flexibility in the detection of a hazardous situation and in the reactions that can be undertaken in such an event. The architecture, in fact, permits the implementation of sophisticated fire detection and reaction mechanisms, based on the cooperative behavior of several sensors distributed over the patrolled area. The details of the system architecture and of the fire detection protocol are described in the following section.

### 4.1 Architecture and Protocol

The system consists of a set of static sensor nodes and one (or more) sink nodes directly connected to a remote control center through a reliable network. For the sake of conciseness, in the following, we will assume that the control center is directly connected to the sink, so that all messages received by the sink are also available to the control center. Sensor nodes are placed in the monitored area and organize themselves in a wireless sensor network, in such a way each node gains connectivity to the sink nodes, possibly through a multi-hop connection. Each node in the network can operate in two states: *Normal* and *Alarm*. Nodes usually operate in Normal state and enter the Alarm state whenever a potential fire event is detected.

In Normal state, nodes keep monitoring the environment without exchanging any data packet with the sink or the neighbors. (Although not considered in our demo, it might be wise to implement a mechanism for checking whether nodes are still alive, for instance by using short control packets sporadically sent by each node to the sink.)

Specifically, each sensor node follows a periodic active–sleep pattern: in active (ON) periods, all the sensing and communication capabilities are active, whereas in sleep (OFF) periods the radio transceiver and other hardware modules are powered off to save energy. Hence, a node is active for only the fraction of time $d$ given by:

$$d = \frac{T_{ON}}{T_{ON} + T_{OFF}} = \frac{T_{ON}}{T_c},$$

where $T_{ON}$ and $T_{OFF}$ denote the time duration of an ON and OFF period, respectively, while their sum gives the cycle time $T_c$. Such system parameters, as well as those introduced in the following, are shared by all nodes in the network. Nonetheless, to avoid the cost of keeping time synchronization among nodes in a multi–hop wireless sensor network, nodes are assumed to perform their activity cycles in an asynchronous manner.

In the Alarm state, a node either acts as *master* or as *slave*. Masters are in charge of sending to the sink (and, in turn, to the control center) updated information gathered by the nodes in the area. Ideally, a single master per area should be active, in order to avoid contention for the channel and extra traffic over the network. Slaves, in turn, cooperate with the master by sending, upon request, reports on their current alarm status and on the monitored values. Furthermore, slaves are in charge of forwarding data packets to the sink on the basis of a suitable routing algorithm.

Transition from Normal to Alarm state occurs when a node either triggers an alarm on the basis of the monitored data or it senses a radio transmission on the channel by one of its neighbors. In the latter case, the node takes on the role of slave. In the former case, the node becomes a master if no radio signal is detected within a time interval $T_s$, otherwise it becomes a slave. The sensing delay, $T_s$, trades off the alarm notification delay with the number of simultaneously active masters. The procedure followed by masters and slaves is described in the following.

### Master Procedure

Upon becoming a master, a node sends a `WARNING` packet to the control center (sink), which can react in different ways, for instance switching the control monitors to the interested area, activating alarm signs in the control panels and so on. To increase the confidence of the alarm detection, the master starts querying the surrounding nodes for their alarm status. To this end, the master broadcasts special `QUERY` packets, separated by a time interval $T_Q$ during which the neighbors can send back to the master their status report. (Notice that $T_Q$ shall not exceed $T_{ON}$, in order to avoid the situation where a node misses the `QUERY` messages during its ON period.) Every $N_Q$ queries, the master sends a compound `ALARM` packet to the control center, in which it includes the status reports received by its neighbors. The lack of expected ALARM packets can also be interpreted by the control center as a confirmation of the alarm, thus triggering suitable reactions. The query procedure lasts for a period $T_A = T_c + T_Q$, after which the master sends an ALARM_END packet to the sink, notifying the completion of the alarm procedure, and re-enters the Normal state.

### Slave Procedure

When a node gets the slave role, it keeps listening to the channel until a valid packet is received or a cycle time $T_c$ elapses without sensing any other events on the radio channel. If the slave receives a packet addressed to the sink, as in the case of WARNING, ALARM, or ALARM_END messages, then the routing module is invoked and the slave node acts accordingly. Instead, in case the slave receives a QUERY message from the master, then it returns a REPORT message, which can include different types of information, such as the alarm state of the node, the temperature read by the node in the last period, and so on.

To reduce the risk of collision among slave transmissions, channel access is managed through the following contention scheme. The time interval $T_Q$ is partitioned in $W_Q$ slots, each of which can host the transmission of a REPORT packet. Each slave that wishes to reply to a query, transmits its REPORT message in a random slot uniformly picked in the $W_Q$ window. An acknowledgment field in the following QUERY message notifies the slaves about the successful or unsuccessful reception of their REPORT message. Slaves that get a positive acknowledgment return in Normal state, while the others repeat the procedure in the following contention window.

Finally, if a slave receives and ALARM_END message then it first invokes the routing module, as early described, and then returns to the Normal state.

## 5 Implementation

For our implementation we deployed a testbed of 48 EyesIFX nodes, using SignetLab[2] to monitor the network. The EyesIFXv2 nodes were developed during a three year European research project on self-organizing energy-efficient sensor networks [3]. An image of the board is shown in Fig. 4. The nodes use an ultra-low power MSP430 processor with 10 KB on chip RAM, 48 KB flash/ROM, and an additional 512 KB serial EPROM. The radio chip is a low power FSK/ASK transceiver, providing half-duplex, low data rate communication in the 868 MHz ISM band. It operates using FSK modulation, with a sensitivity of $< -109$ dBm, enabling up to 64 Kbps, half-duplex, wireless connectivity. The platform is also equipped with an on-board stripline antenna and an SMA-connector for an external antenna. The transceiver can accept a supply voltage of up to 5.5 V. The typical current is $I_s = 9$ mA in

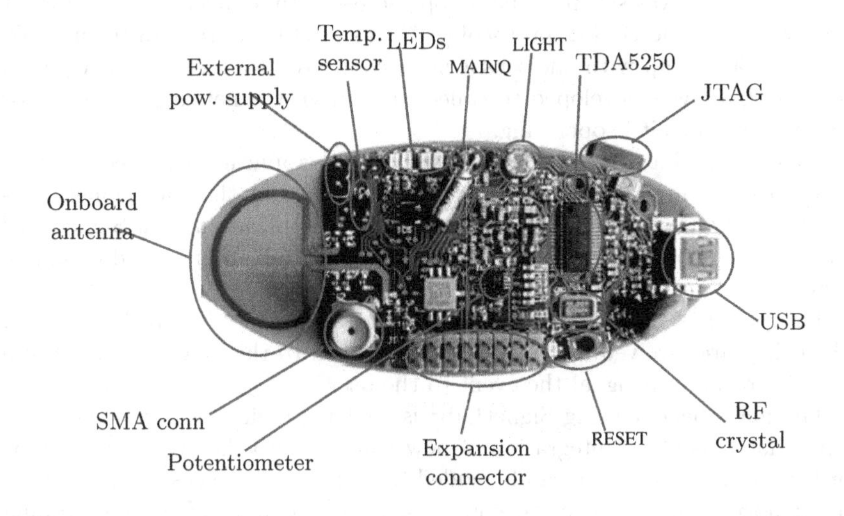


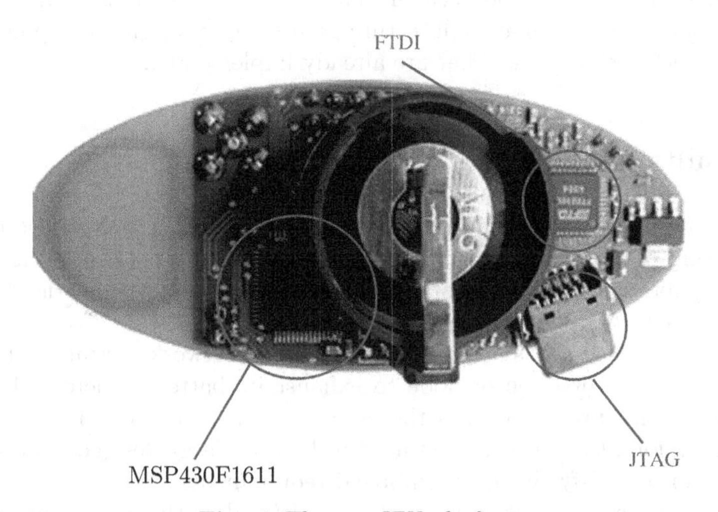


**Fig. 4.** The *eyesIFX* platform

receive mode, and $I_s = 12\,\mathrm{mA}$ in transmit mode. The transmit power can be modulated by means of a digital potentiometer with 255 settings. The nodes come with onboard temperature and light sensors as well as an SPI expansion port that can be used for additional sensing capabilities. The SPI bus is shared between the expansion port, the radio, and the processor. Therefore, there is a hard restriction on the amount of resources that can be used at a time. The nodes can be powered either by batteries with a capacity of 1,000 mAh or through a power supply connected via an external polarized connector or a USB connection.

The network was set up to be 5 hops across with an average neighbor set of eight nodes. The GeRaF protocol [6, 7] was used to route data to the sink. A server was set up with one sensor node connected to it via USB. A plugin for SignetLab$^2$ was developed to collect data from the network and monitor the application as it is operating.

Using SignetLab$^2$ to implement the control application helped because most of the features that were required are provided by the management system. In fact we just installed a standard gateway application, that SignetLab$^2$ provides, on the node connected to the server, and programmed a simple plugin.

The plugin signals to the server its interest in the ALARM messages. When they are received the server forwards them to the plugin, that can use the network map to signal the event to the user.

The main benefit using SignetLab$^2$ is that it provides a flexible environment, that makes the integration of new features into the system very easy. For instance, a new plugin can be added to monitor the status of the network during normal operation or to control a fire extinguishing system. Adding these new features will be cheaper than writing a brand new application. Additionally, the modular architecture guarantees that the new plugin will not interfere with the ones that are already implemented.

# 6 Results

In this section, we report the analysis on the basis of which we have determined the setting of the system parameters, which are: duty cycle $d$, cycle period $T_c$, query interval $T_Q$, contention window size $W_Q$, number of queries between `ALARM` packets $N_Q$, and sensing interval $T_s$.

The duty cycle $d$ has a direct impact on the network lifetime, defined as the time for the first sensor node to exhaust its battery, whereas the cycle duration $T_c$ mainly determines the latency of the event notification to the sink. Therefore, these parameters need to be accurately designed in order for the system to satisfy the aforementioned requirements.

Let $D$ be the normalized network size, defined as the maximum distance between any static sensor node and the sink, normalized to the nominal transmission range $R$ of the sensor nodes.

Furthermore, let us denote by $H$ the average number of hops a message needs to go through to cover the distance $D$, and by $\tau$ the average hop delay. As worst case we assume that all the messages are generated from the farthest end of the network. Therefore, the average message latency can be computed as

$$\delta = H \times \tau \, .$$

For a given the routing algorithm, $H$ and $\tau$ are functions of the cycle period $T_c$ and the duty cycle $d$. Then, a constraint on the message delivery delay $\delta$

reflects on the possible values that can be assigned to the system parameters $d$ and $T_c$.

A second constraint is given by the desired network lifetime, defined as the time in which a node completely discharges its battery, operating in normal conditions. A simple model to evaluate the network lifetime consists in assigning an energy budget to each basic operation performed by the nodes, and then computing the energy spent in an activity cycle. The total number of cycles that can be performed with the nominal battery charge will then give the network lifetime. To be more specific, let $P_{ON}$ and $P_{OFF}$ denote the average power drained by the node in ON and OFF mode, respectively. Furthermore, let $E_{SW}$ be the average energy cost of mode switching. Then, an activity cycle will have an energy cost equal to:

$$E_c = P_{ON}T_{ON} + P_{OFF}T_{OFF} + E_{SW} .$$

Denoting by $C$ the overall energy capacity of a battery, the network lifetime $T_{net}$ can be expressed as:

$$T_{net} = \frac{C}{E_c} \times T_c = \frac{C}{P_{ON}d + P_{OFF}(1-d) + \frac{E_{SW}}{T_c}} .$$

Since $P_{ON}$, $P_{OFF}$, $E_{SW}$, and $C$ are given by the characteristics of the hardware used to realize the system, while $T_{net}$ is a system requirement, then this equation provides another constraint to the choice of the system parameters $d$ and $T_c$.

Figure 5 depicts the estimated network lifetime as a function of the duty cycle. The curve has been obtained by the nominal energy consumption parameters as given in the data sheet of our sensor nodes and a cycle time of $T_c = 30\,\mathrm{s}$, which has been considered as appropriate for the type of application considered here. In any case, results do not change significantly for a rather large set of $T_c$ values.

To explore the effect of the duty cycle on the delay before an alarm is received at the sink, we assumed a nominal coverage range of $R = 5\,\mathrm{m}$ and a network size of $D = 20$. Nodes were assumed to be randomly distributed in the monitored area, with an average density of $N = 10$ nodes for coverage area. Packets were routed using the GeRaF routing algorithms [7], for which there exist analytical expressions for the average hop delay, $\tau$, and the average Number of hops, $H$, to cover a given distance $D$. Figure 6 shows the first alarm delivery latency, i.e., the delay between the sensing of the fire and the reception of an alarm packet by the sink, as a function of the duty cycle.

From these curves, we can notice that a duty cycle of $d = 2\%$, with a cycle period $T_c = 30\,\mathrm{s}$, gives a network lifetime of approximately $T_{net} \simeq 4{,}900\,\mathrm{h}$, corresponding to more than 6 months, while the alarm latency is of $\delta \leq 90\,\mathrm{s}$, which appears adequate for a fire alarm application. Notice that, with this parameters setting, a node remains active for only $T_{ON} = 600\,\mathrm{ms}$ every $T_c = 30\,\mathrm{s}$.

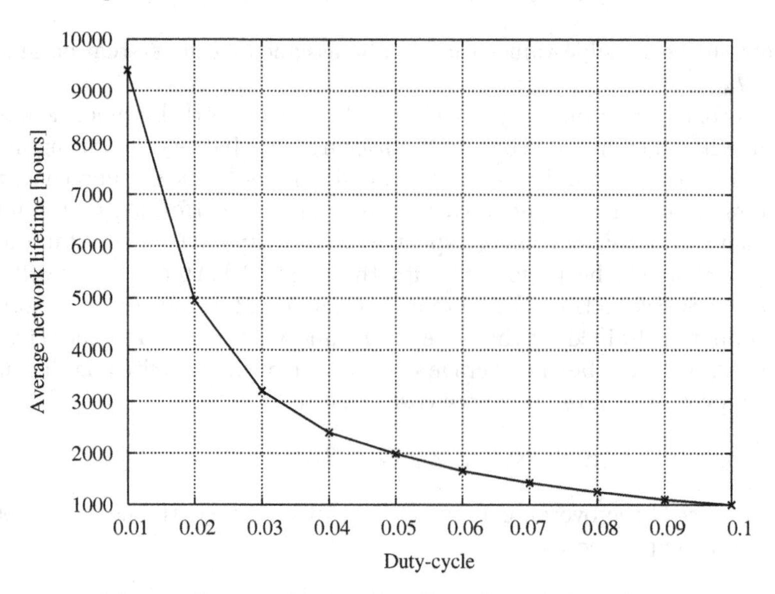


**Fig. 5.** Network lifetime $T_{net}$ [hours] vs. duty cycle $d$

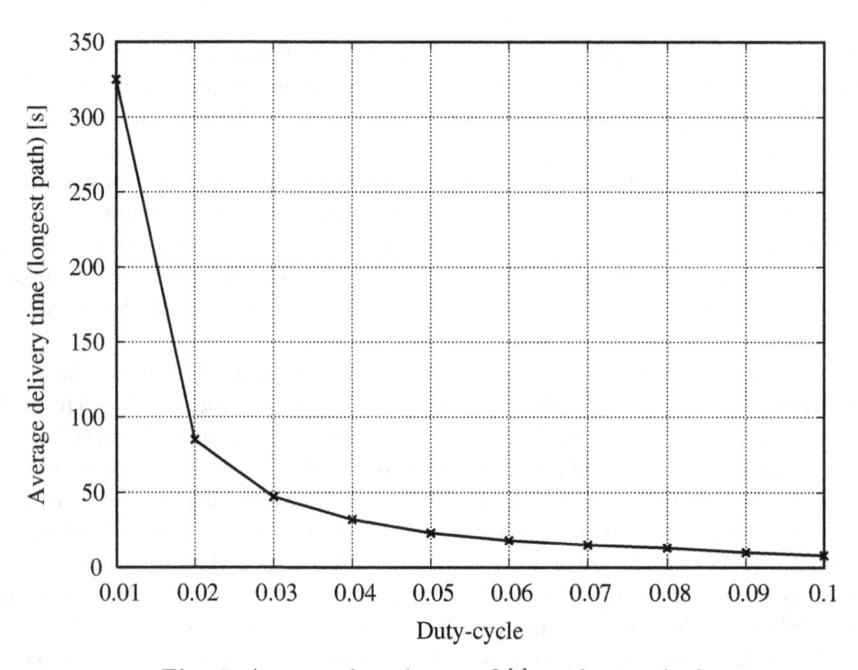


**Fig. 6.** Average alarm latency $\delta$ [s] vs. duty cycle $d$

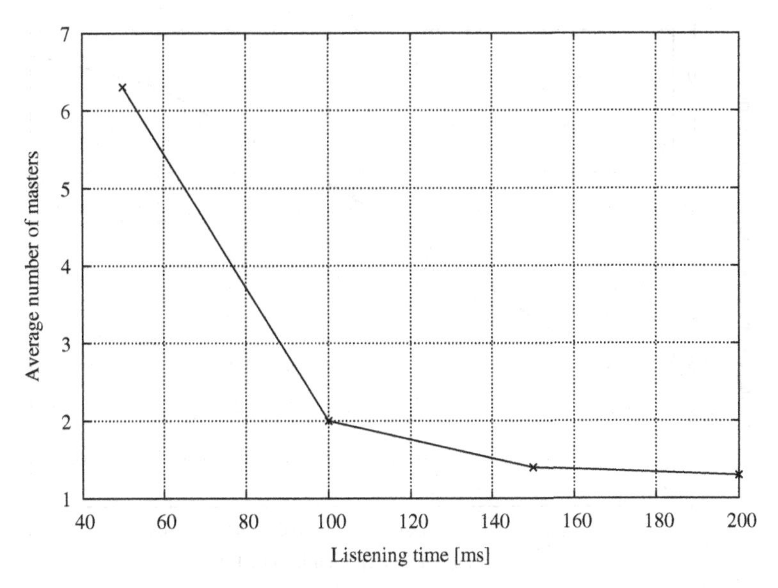


**Fig. 7.** Average number of masters in the zone where the fire is simulated, varying the listening time $T_s$ before the node rise to the master state

To gather data on the accuracy of the fire alarm application, an artificial increase in the temperature on all the nodes in a zone where all devices are in coverage range was caused about 5 min after the experiment has started.

In an ideal situation there should be only one master, to avoid collisions due to multiple ALARM and QUERY packets being sent. These collisions would cause the loss of packets, with a dangerous increase in the reaction time to a critical condition. Therefore, the parameter $T_s$, which trades off the risk of having multiple masters and the delay of first warning message delivery to the sink, has to be accurately set. We tested the average number of simultaneous masters while varying the listening time $T_s$ before rising to the master status in a node that reveals a danger condition. Figure 7 shows the results. A listening time of $T_s = 150$ ms provides a low number of masters while still maintaining the ability to notify the sink within an appropriate time, since in the worst case (no one is in the MASTER state yet), we just add a 150 ms delay to the delivery of the ALARM messages to the sink.

Finally, the query interval $T_Q$ has been set to $T_Q = 120$ ms, which has been divided in $W_Q = 12$ slots, each capable to host the transmission of a data packet of up to 20 bytes. A new `ALARM` message was generated every $N_Q = 25$ queries, that is every $N_Q \times T_Q = 3$ s, for a total of 10 `ALARM` packets before the conclusion of the alarm procedure by that master. Figure 8 depicts the percentage of nodes that respond to a master's query as a function of the elapsed time normalized to the period $T_c$. After about 50% of the period, 60% of the nodes have responded to the query. This number of responders should be enough to determine whether the alarm is reliable or not.

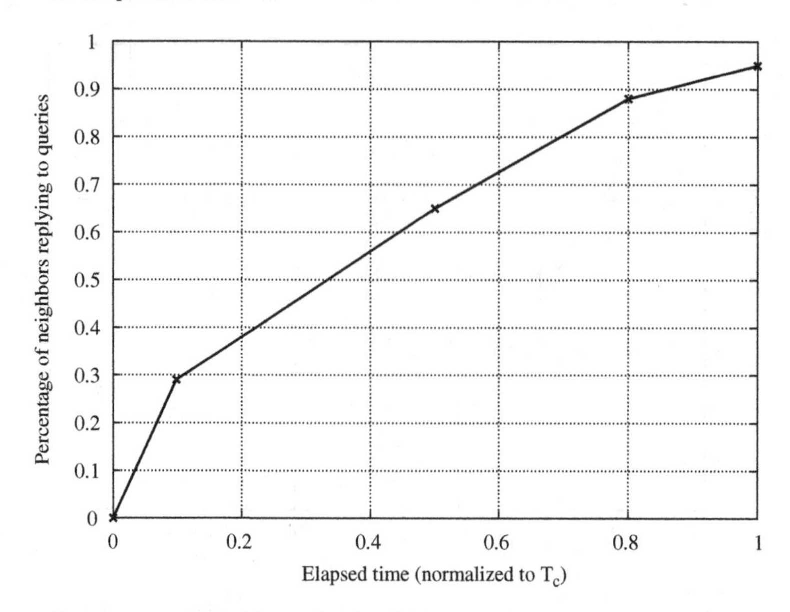


**Fig. 8.** Percentage of neighbors that replied to a QUERY message, after a certain fraction of the period $T_c$

## 7 Conclusion and Future Directions

Sensor networks provide the possibility to support a great variety of applications. However, testing sensor network protocols and monitoring sensor network deployments require a management system capable of providing simple access without interfering with the functionality of the applications run on the sensor network.

In this work we have presented the design of SignetLab$^2$, a sensor network management system that supports the goals of ease of monitoring and minimal interference. SignetLab$^2$ also provides an API to allow for rapid extension to support any sensor network application.

To demonstrate the power of SignetLab$^2$, we presented the design and deployment of a fire alarm system. This application was designed to rapidly detect and react to fires. The application monitoring system and experimentation was designed and performed using our sensor network management system.

Future work includes extending the server functionality to allow it to monitor multiple message types from a single sensor network. Additionally, the system should be extended to handle more sensor network hardware boards. Finally, a comprehensive in-network reprogramming solution would round out the capabilities of the system.

## Acknowledgment

We wish to thank Alessandro Zampieri for his important contribution in designing and testing the Fire Alarm application.


## References


1. R. Crepaldi, S. Friso, A. I. Harris, M. Mastrogiovanni, C. Petrioli, M. Rossi, A. Zanella, and M. Zorzi. The design, deployment, and analysis of signetlab: A sensor network testbed and interactive management tool. In *IEEE Tridentcom*, May 2007
2. R. Crepaldi, A. Harris, A. Scarpa, A. Zanella, and M. Zorzi. Signetlab: deployable sensor network testbed and management tool. In *SenSys '06: Proceedings of the 4th international conference on Embedded networked sensor systems*, pages 375–376, 2006
3. Infineon, Ltd. EyesIFXv2 version 2.0. http://www.infineon.com
4. D. Johnson, T. Stack, R. Fish, D. Flickinger, L. Stoller, R. Ricci, and J. Lepreau. Mobile Emulab: A Robotic Wireless and Sensor Network Testbed. In *IEEE INFOCOM*, April 2006
5. G. Werner-Allen, P. Swieskowski, and M. Welsh. Motelab: A wireless sensor network testbed. In *IEEE/ACM IPSN/SPOTS*, April 2005
6. M. Zorzi and R. Rao. Geographic random forwarding (GeRaF) for ad hoc and sensor networks: energy and latency performance. *IEEE Trans. on Mobile Computing, vol. 2, pp. 349–364*, October–December 2003
7. M. Zorzi and R. Rao. Geographic Random Forwarding (GeRaF) for ad hoc and sensor networks: multihop performance. *IEEE Trans. on Mobile Computing, vol. 2, pp. 337–348*, October–December 2003

# Index

2 Gbps, 195
3G long term evolution, 65
3GPP, 153, 165
4G network, 245
60GHz band, 195

access point, 213
adaptively learning algorithms, 136
adaptive modulation and coding, 179
ad-hoc network, 51, 81, 127
admission control, 258
aggregator nodes positioning, 321
AKA, 174
AMR encoder, 235
application, 331
application debugging, 331
automatic retransmission query, 214, 235
automatic retransmission query proxy, 214
autonomous mobile robots, 310

bandwidth allocation, 191, 283
bandwidth reclaiming, 254
BCJR approach, 22
beamforming, 65, 81
beyond-3G, 153
binary PSK, 20
bit error rate performance, 40
blocking probability, 281
Bluetooth, 98
broadband satellite multimedia, 226

capacity bounds, 127
CBR traffic, 324
CDMA, 135
CDMA 2000, 265
cellular network, 97
cellular system, 65
channel capacity, 35, 282
channel model, 196
channel prediction, 71
channel state information, 65
cheap devices, 295
compression, 201
computational overhead, 153
computer simulation, 195
confidentiality, 168
connectivity, 81
content customization, 201
cooperative wireless network, 97
coverage rate, 286
cross-layer design, 225
cross layer scenario, 140
cross-layer solutions, 136

data aggregation, 321
data rate detection, 19
degree of dissatisfaction, 281
delay end-to-end, 323
detection and estimation, 54
direct sequence CDMA, 19
diversity, 39
downlink, 65

EAP, 172
ECMA 368, 51

EM algorithm, 22
encryption, 201
energy consumption, 97, 321
energy efficiency, 135
energy efficient resource allocation, 142
environment monitoring, 333
equivalent bandwidth, 225
error recovery, 213
experimental measurements, 111
experimentation, 331
EYES project, 312

fading counteraction, 231
false alarm, 55
fault tolerance, 201
file sharing, 101
fire detection, 335
flexible web content, 201
flow deviation method, 128
FMIP, 172
forward error correction, 195
fountain codes, 29
frequency division duplexing, 65
frequency resource, 281

generalized likelihood ratio test, 54
generalized multi-protocol label switching, 180
grid scenario, 330

hash value, 216
HMIP, 172
HOKEY, 174
hop-by-hop routing, 127
hop distance, 187
http, 101
hybrid ARQ, 30

IEEE 802.11, 213, 253
IEEE 802.15.3c, 195
IEEE 802.15.4, 323
IEEE 802.16, 179
IMS, 153
incremental redundancy, 29
indoor, 309
integrated radio transceiver, 295
integrity, 166
interference reduction, 90
internetworking, 268
interworking, 225
IP multimedia subsystem, 265
IPSec, 153
iterative techniques, 20

joint detection techniques, 20

key management, 165
kiosk file downloading, 197

large scale deployment, 331
layered space-time coding, 181
layered space-time receivers, 3
least mean square rule, 136
limited feedback, 65
link capacity matrices, 128
LLE-TCP, 218
localization, 309
LTE, 165

MAC header, 189
mesh network, 111
MIB table, 303
millimeter-wave, 195
MIMO systems, 5
MIP, 172
missed detection, 55
MMSE criterion, 22
mmWave, 195
mobile ad-hoc network, 310
mobile beacon, 309
mobile communications, 281
mobile IP, 265
mobile network, 108
mobility, 166
mobility management, 266
model, 154
modular tool, 333
monitoring, 331
multi band mobile communications, 281
multihop network, 95
multilateration, 314
multilevel coding, 3
multimedia communication network, 245
multimedia streaming, 204
multiple-input multiple-output, 40, 180
multistage decoding, 3
multiuser detection, 20

Nash equilibrium point, 141
network throughput capacity, 127
next generation mobile network, 265
non cooperative game theory, 140
ns-2, 323
nulling, 81

open wireless architecture, 245
optimum threshold selection, 58
orthogonal frequency division multiplexing, 65, 182

packet detection, 51
packet identification, 216
packet suppression, 217
peer-to-peer network, 97, 129
performance evaluation, 101, 233
performance optimization, 213
performance results, 114
power control, 135
proactive schemes, 114

QoS cross-layer framework, 246
QoS management, 255
QoS mapping, 225
quality of service, 225, 245, 253, 265
queue management, 227

rate control, 235
rateless coding, 32
Rayleigh fading channels, 26
real-time, 235
recursive least square rule, 136
reliability, 180
remote buffering, 201
remote caching, 201
resource allocation, 230, 291
resource reservation, 190
Ricean MIMO channel, 39
routing, 111, 295
routing tree, 186

SAE, 165
scheduling, 65, 186, 251
scheduling: contract-based, 245
security, 153
security issues, 165
security requirements, 166
self organizing network, 338
self-positioning, 309
sensor network, 296
serial search, 54
service level agreement, 225
service oriented architecture, 247
session initiation protocol, 266
session management, 266
set partitioning, 5
SIGNET lab, 331
simulation, 111, 321
single carrier, 195
single input single output decoders, 41
SIP, 153
smart antennas, 81
SNMP, 296
soft adaptability, 284
space-time codes, 40
spatial multiplexing, 180
stochastic game, 135
stream customization, 201
successive interference cancellation, 4
symbol-error rate, 184
synchronization, 53

TCP acknowledgement, 213
TCP optimization, 215
technology independent service access point, 225
testbed management, 331
throughput, 185
traffic forwarding, 111
transceiver design, 51
transmission queue, 322
transmit antenna selection, 39
turbo decoding techniques, 19

UDP, 324
ultra wide band communications, 51
universal mobile telecommunications system, 265
universal mobile telecommunication system, 154
upper bounds, 127
usage model, 195
user relocatability, 201
user selection, 65
utility functions, 139
UTRAN LTE, 30

voice multiplexing, 236
VoIP, 235

web middleware system, 201
web stream customizer, 201
WiFi, 213
WiMAX, 179
WiMedia, 51
wireless access, 246
wireless communications, 3, 236
wireless constant bandwidth server, 254
wireless control plane, 334
wireless data network, 135
wireless local area network, 213, 253, 265
wireless mesh network, 111, 235
wireless multihop network, 111
wireless multimedia applications, 245
wireless network, 111
wireless personal area network, 195
wireless sensor network, 309, 323, 331
wireless web access, 201
wireless/wired integration, 180

## Signals and Communication Technology

(continued from page ii)

**Digital Signal Processing with Field Programmable Gate Arrays**
U. Meyer-Baese ISBN 3-540-21119-5

**Neuro-Fuzzy and Fuzzy Neural Applications in Telecommunications**
P. Stavroulakis (Ed.) ISBN 3-540-40759-6

**SDMA for Multipath Wireless Channels**
Limiting Characteristics and Stochastic Models
I.P. Kovalyov ISBN 3-540-40225-X

**Digital Television**
A Practical Guide for Engineers
W. Fischer ISBN 3-540-01155-2

**Speech Enhancement**
J. Benesty (Ed.)
ISBN 3-540-24039-X

**Multimedia Communication Technology**
Representation, Transmission and Identification of Multimedia Signals
J.R. Ohm ISBN 3-540-01249-4

**Information Measures**
Information and its Description in Science and Engineering
C. Arndt ISBN 3-540-40855-X

**Processing of SAR Data**
Fundamentals, Signal Processing, Interferometry
A. Hein ISBN 3-540-05043-4

**Chaos-Based Digital Communication Systems**
Operating Principles, Analysis Methods, and Performance Evaluation
F.C.M. Lau and C.K. Tse
ISBN 3-540-00602-8

**Adaptive Signal Processing**
Application to Real-World Problems
J. Benesty and Y. Huang (Eds.)
ISBN 3-540-00051-8

**Multimedia Information Retrieval and Management Technological**
Fundamentals and Applications D. Feng, W.C. Siu, and H.J. Zhang (Eds.)
ISBN 3-540-00244-8

**Structured Cable Systems**
A.B. Semenov, S.K. Strizhakov, and I.R. Suncheley
ISBN 3-540-43000-8

**UMTS**
The Physical Layer of the Universal Mobile Telecommunications System
A. Springer and R. Weigel
ISBN 3-540-42162-9

**Advanced Theory of Signal Detection**
Weak Signal Detection in Generalized Obeservations
I. Song, J. Bae, and S.Y. Kim
ISBN 3-540-43064-4

**Wireless Internet Access over GSM and UMTS**
M. Taferner and E. Bonek
ISBN 3-540-42551-9

# Signals and Communication Technology